T0201676

Vehicular Networking

Intelligent Transportation Systems

VANET Vehicular Applications and Inter-Networking Technologies

Hannes Hartenstein and **Kenneth P. Laberteaux (Eds.)**

Vehicular Networking: Automotive Applications and Beyond

Marc Emmelmann, Bernd Bochow and **C. Christopher Kellum (Eds.)**

Vehicular Networking

Automotive Applications and Beyond

Editors

Marc Emmelmann

Technical University Berlin, Germany

Bernd Bochow

Fraunhofer Insitute for Open Communication Systems (FOKUS), Germany

C. Christopher Kellum

John Deere, USA

A John Wiley and Sons, Ltd, Publication

This edition first published 2010
© 2010 John Wiley & Sons Ltd

Registered office
John Wiley & Sons Ltd, The Atrium, Southern Gate, Chichester, West Sussex, PO19 8SQ,
United Kingdom

For details of our global editorial offices, for customer services and for information about how to apply
for permission to reuse the copyright material in this book please see our website at www.wiley.com.

Library of Congress Cataloging-in-Publication Data
    Vehicular networking : automotive applications and beyond / Marc Emmelmann, Bernd Bochow,
C. Christopher Kellum, eds.
        p. cm.
    Includes bibliographical reference and index.
    ISBN 978-0-470-74154-2 (cloth)
1. Vehicular ad hoc networks (Computer networks) I. Emmelmann, Marc. II. Bochow, Bernd. III.
Kellum, C. Christopher. IV. Title.
    TE228.37.V398 2010
    388.3'12–dc22
                                                                              2009053148

A catalogue record for this book is available from the British Library.

ISBN 9780470741542 (H/B)

Set in 10/12pt Times by Sunrise Setting Ltd, Torquay, UK.
Printed and Bound in Great Britain by CPI Antony Rowe, Chippenham, Wiltshire.

Contents

3   Communication Systems for Car-2-X Networks   45

Daniel D. Stancil, Fan Bai and Lin Cheng

List of Contributors

Dr. Fan Bai
General Motors

David Bateman
EDF R&D

Bernd Bochow
Fraunhofer Institute for Open Communication
Systems

Benoît Bouchez
Bombardier Transportation France

Lin Cheng
Trinity College, Hartford, CT

Luc de Coen
Bombardier Transportation France

Marc Emmelmann
Technical University Berlin

Rajni Goel
Howard University

Karine Gosse
CEA LIST, Département des Technologies et des
Systèmes Intelligents

Mark Hartong
Office of Safety, Federal Railroad Administration
US Department of Transportation

Dr. Gavin Holland
HRL Laboratories, LLC

Christophe Janneteau
CEA LIST, Département des Technologies et des
Systèmes Intelligents

Mohamed Kamoun
CEA LIST, Département des Technologies et des
Systèmes Intelligents

Mounir Kellil
CEA LIST, Département des Technologies et des
Systèmes Intelligents

Gerdien Klunder
TNO Built Environment and Geosciences

Dr. Hariharan Krishnan
General Motors

Anthony Maida
EF Johnson Technologies

Alexis Olivereau
CEA LIST, Département des Technologies et des
Systèmes Intelligents

Panos Papadimitratos
Ecole Polytechnique Fédérale de Lausanne

Jean-Noël Patillon
CEA LIST, Département des Technologies et des
Systèmes Intelligents

Alexandru Petrescu
CEA LIST, Département des Technologies et des
Systèmes Intelligents

Pierre Roux
CEA LIST, Département des Technologies et des
Systèmes Intelligents

Tom Schaffnit
Schaffnit Consulting, Inc.

Daniel D. Stancil
Carnegie Mellon University

Bart van Arem
Delft University of Technology

Duminda Wijesekera
George Mason University

Sheng Yang
Supelec

Isabel Wilmink
TNO Built Environment and Geosciences

Preface

From a high-level point of view, vehicular networking is simply the communication of information between a vehicle and either another vehicle or the infrastructure. This high-level view of vehicular communication is decades old, and, one could say, includes applications such as voice communication between military vehicles and bases as well as entertainment such as AM/FM radio broadcast to the public in their automobiles. The applications of interest today go beyond voice and extend into digital data, where users simply configure their applications (or take preconfigured versions) and use the basic output, which could be anything from an entertaining sitcom video clip to time-critical train controls. For the purpose of this book, we consider vehicular networking as the new application of wireless communication technologies to the domains of the automobile and train control.

Considering the socio-economic importance of transportation in the world, Intelligent Transportation Systems (ITSs) are envisioned to play a significant role in the future, making transportation safer and more efficient. Better efficiency is expected to result from better utilization of the limited capital assets we have, such as highways and railways. In order to employ these assets better, we have to understand why they are not being fully used, and determine how to improve this. One simple example is a traffic jam. Often traffic jams are the result of an accident on the roadway. At other times they are simply the result of an over-used resource. In other words, there are too many cars on the same road at the same time. Vehicular networking can help us use these resources better than we do today, by collecting and propagating information in new ways. In the case of traffic jams this is done by avoiding the traffic accident altogether, or, in the case of overuse, by routing vehicles around a traffic jam before it even starts.

Vehicular networking is a complicated topic, and it is sometimes easy to lose sight of the end goal and the purpose of the work being done. To make the motivation of the effort clear, Chapters 1 and 2 provide an overview of the applications that researchers and developers are actively pursuing. Here the authors classify and categorize the applications and present reasons why they are important from a market perspective. When reading these two chapters, it becomes clear that scenarios addressed by commercial use, public safety, and military are very similar except for an increased demand on communication system availability and security by the latter two. In addition to motivating the topic, these two chapters provide requirements that must be fulfilled by the communication system in order for the applications to be successful.

From a market point of view, the applications presented in the first two chapters would be great. Every user wants perfect information presented in the perfect way every time. From a wireless communication perspective, the question of whether or not this is possible is not yet answered. The intention of the next two chapters, Chapters 3 and 4, is to discuss some of the detailed technological limitations and work done to date. For instance, one cannot assume that legacy systems do not exist. Any new system has to be backwards compatible. Similarly, communication losses will exist. If an application's developer does not understand the communication channel properties in use case scenarios, its users may be dissatisfied due to communication loss. Undoubtedly, thousands of technical papers have been published discussing the ins and outs and corner conditions of wireless communication. These two chapters do an outstanding job of introducing the reader to some of the most important technological details related to channel propagation, the protocol, and mobile Internet Protocol (IP) solutions.

As in all communication systems, security, privacy, and dependability are important aspects that need to be considered in the context of the application. Chapter 5 elaborates on potential security threats and on suitable methods to detect and counter attacks on the integrity of vehicular communications, specifically in the Vehicular Ad hoc Network (VANET) environment. The chapter elaborates on topics such as issuing and revoking credentials to vehicular network nodes and detecting attacks on sensory input for positioning based on satellite navigational systems. Since the discussion is focused on commercial and public use communications, the chapter closely relates security and privacy threats. Chapter 6 continues the discussion by focusing on security for railway applications such as Automatic Train Stop (ATS) for collision avoidance. The chapter details potential security threats and attack types, and analyzes security requirements with respect to current and upcoming Automatic Train Control (ATC) systems. Taking into account the centralized nature of ATC systems, the chapter also looks into management issues on performance, configuration and security.

At this point in the book it should be clear to the reader that the technology for vehicular communication is complicated, but the entire challenge has not yet been explained. In addition to the points already addressed, deployment of cooperative technology such as this requires some level of concert among multiple parties. In the case of vehicular communication, the method for cooperation is via numerous standards and regulatory bodies governing matters that range from (at one end of the scale) communication frequencies to (at the other end) direct interactions with the user. Chapters 7 and 8 present the topics and progress of regulation and standardization. Admittedly there is overlap between these two chapters, but Chapter 7 focuses mostly on vehicle networking between mobile nodes while Chapter 8 has a specific focus on Internet Engineering Task Force (IETF) activities in vehicular application of mobile IP, and related activities based on ad hoc networking principles such as Mobile Ad hoc Network (MANET).

In addition to the standards activities and efforts, many groups are assembling simulations and concepts to help understand these complex systems better. The earlier chapters concentrate mostly on theoretical ideas built upon pieces of information generated from multiple tests, surveys, and studies. In contrast, Chapters 9 and 10 focus on taking an application from idea into reality, and demonstrate the remaining steps before deployment. Chapter 9 presents information regarding methods used to simulate systems in order to confirm their operation and measure their benefits before major capital investment takes place. Chapter 10 provides insight into the next step of deployment, working with prototype devices in a small-scale system in order to gain more confidence in the theories and assess some of the real-world problems that can result. The steps towards deployment beyond what is described in these two chapters involve making production components for sale in the marketplace. However, once that process begins, many of the researchers and engineers who are currently focused on this generation of vehicular communication will begin looking at next generation systems, and this leads to the final chapter of the book. Chapter 11 takes a look at the technologies and ideas that can shape our world in the years beyond the next decade. It provides some insight into the discussions and fledgling ideas that will continue to evolve over the coming years and form future research projects.

It would be difficult for any small group of people to cover such a broad topic with the appropriate depth and expertise. For this book, we, the editors, looked to our colleagues to find those who are well-respected experts in each area. You can see that the author list for the chapters has grown quite large and contains a good mixture of academics, consultants, and those working in private industry. We think that this phenomenon demonstrates the complexity of vehicular communications and how widespread the topic is. We are very, very fortunate to have received contributions from the authors in this book, and we are extremely grateful. We applaud their efforts and hope to see their work continue so that the magnificent technology presented here can soon be experienced in our everyday lives!

Chris Kellum
Cedar Falls, Iowa, U.S.A.

Marc Emmelmann
Bernd Bochow
Berlin, Germany

1

Commercial and Public Use Applications

Dr. Hariharan Krishnan and Dr. Fan Bai

General Motors

Dr. Gavin Holland

HRL Laboratories, LLC

Together, the Dedicated Short Range Communications (DSRC) and Vehicular Ad hoc Network (VANET) technologies provide a unique opportunity to develop various types of communication-based automotive applications. In this chapter, we focus primarily on four major aspects: (a) description of communication-based automotive applications, (b) investigating the application characteristics and network attributes, (c) classifying the applications into categories, and (d) defining market perspectives and deployment challenges for each class of applications. To date, many applications have been identified by the automotive research community. From a value or customer benefit perspective, these applications can be roughly organized into three major classes: *safety-oriented*, *convenience-oriented*, and *commercial-oriented*, and they vary significantly in terms of application characteristics. We begin by describing communication-based automotive applications that span both the Vehicle-to-Vehicle (V2V) and Vehicle-to-Infrastructure (V2I) communication modes. We follow a systematic classification methodology for such applications that goes through two major steps: characterization and classification. We focus on a rich subset of representative applications and characterize them with respect to plausible application- and networking-related attributes. The characterization process not only strengthens our understanding of the applications but also sets the stage for the classification step, since it reveals numerous

Vehicular Networking   Edited by Marc Emmelmann, Bernd Bochow, C. Christopher Kellum
© 2010 John Wiley & Sons, Ltd

application commonalities. The applications are classified into several generic classes, with the consideration of balancing the trade-off between exploiting as many application similarities as possible and preserving their salient differences. Such classification is of paramount importance in bridging the gap between the automotive and wireless networking communities. We also try and define market perspectives and deployment challenges for each representative class of applications.

1.1   Introduction

The needs for significant reduction in both highway traffic congestion and vehicle crashes are serious challenges throughout the world (Chen and Cai 2005; Reumerman et al. 2005). In order to address these challenges, expensive sensors, radars, cameras, and other state-of-the-art technologies are currently integrated into vehicles to improve vehicle safety and driver comfort during travel. Recently, communication-based applications based on Vehicle-to-Vehicle/Infrastructure (V2X) communications have attracted more attention from industry and governments in the United States, Europe, Japan, and Australia, because of their unique potential to address vehicle safety and traffic congestion challenges at lower operational costs (Sengupta et al. 2007; VSCC 2006). In addition to safety and traffic efficiency applications, wireless communication can also be shared by commercial and vehicular 'infotainment' applications to, for instance, enhance the occupants' driving experience. Thus, wireless communication can be used not only to enhance transportation safety (ElBatt et al. 2006; Torrent-Moreno et al. 2004; Xu et al. 2004; Yin et al. 2004) and traffic efficiency (Anda et al. 2005), but also to create commercial value to vehicle owners and automotive Original Equipment Manufacturers (OEMs) by providing infotainment applications (Das et al. 2004; Nandan et al. 2005).

The United States Department of Transportation (DOT) has recognized the importance of having a dedicated wireless spectrum for improving vehicle safety and traffic efficiency. Hence, in the USA, the Federal Communications Commission (FCC) has allocated 75 MHz of licensed spectrum at 5.9 GHz as the DSRC band for Intelligent Transportation System (ITS) services (FCC 2003). This is used in the rest of North America also. In Europe, the Commission of the European Communities has specified harmonized use of radio spectrum in the 5875–5905 MHz frequency band for safety related applications of ITS. In Japan, the deployment of Electronic Toll Collection (ETC) uses the 5.8 GHz spectrum. The future plan is to expand the 5.8 GHz spectrum use for V2I communication-based applications, and the 700 MHz spectrum for V2V communication-based applications. The Australian government is considering a similar allocation of radio spectrum at 5.9 GHz for ITS. The deployment of ITS, with its V2V and V2I constituents, is supported under major DOT initiatives (CICAS 2009; VII 2009). The Physical (PHY) and Medium Access Control (MAC) portions of the DSRC standard are currently being addressed by the IEEE 802.11p Task Group (IEEE 802.11 1999; IEEE P802.11p/D6.0 2009), which is widely considered as the leading technology for communication-based automotive applications. Major automotive OEMs, wireless device manufacturers, research institutions, public agencies, and private enterprises are conducting research on various topics pertaining to V2X communications, such as wireless channel modeling (Taliwal et al. 2004; Yin et al. 2006), mobility modeling (Bai et al. 2003; Lin et al. 2004), routing protocols (Chennikara-Varghese et al. 2006; Korkmaz et al. 2004; LeBrun et al. 2005; Lochert et al. 2005), security (Picconi et al. 2006;

Raya et al. 2006), and market penetration mitigation strategies (Kosch 2005; Shladover and Tan 2006). There has also been some limited attention dedicated to better understanding, modeling and analyzing communication-based automotive applications as the major driving force for VANET-focused technologies.

1.1.1 Motivation

This work is motivated by the need for a systematic and thorough analysis of communication-based automotive applications from a networking point of view. As a preliminary study, we attempt not only to raise awareness about the performance requirements of the automotive community, but also to attract sufficient attention from the networking research community.

The Vehicle Safety Communication Project (VSCC 2006) has identified a number of applications for potential deployment along with projected user benefits (VSCC 2005). The applications of interest vary significantly in terms of their characteristics, requirements, and constraints, ranging from safety/warning applications to content download/streaming applications (for entertainment) and free-flow payment applications (for improving highway traffic efficiency and driver convenience). Analyzing and developing wireless networking solutions tailored to such a large number of diverse applications, in an exhaustive manner, is a cumbersome and inefficient task. Obviously there is a gap between developing communication-based automotive applications (the focus of the automotive community) and developing VANET protocols (the focus of the wireless networking community). To bridge this void, we aim at categorizing communication-based automotive applications, not only from the application characteristics perspective, but more importantly from a wireless networking perspective.

1.1.2 Contributions and benefits

To the best of our knowledge, this chapter is the first study of classifying communication-based automotive applications from the perspective of wireless networking design. Towards this objective, we are interested in answering the following questions:

(a) What are the key application characteristics and networking attributes in the design space of automotive application development?

(b) How should these applications be categorized into generic classes, from the viewpoint of network designers?

(c) What are the market perspectives and deployment challenges associated with each class of applications?

Part of the challenge in this study is to create a rich set of application characteristics and network attributes, which capture the major dimensions of the design space of V2X applications in a systematic and thorough manner. With deep insight into the application design space, we have categorized a set of applications into several *generic* classes based on their identified commonalities. We focus primarily on three major aspects: (a) investigating the application characteristics and network attributes, (b) classifying the applications into generic classes, and (c) understanding market perspectives and deployment challenges for each class of applications.

Our aim for this study is not only to simplify large-scale simulation efforts, which play an important role in understanding the performance limits of VANETs in realistic scenarios, but also to shed light on designing network protocol stack(s) and system integration for different applications. For instance, using the analysis, network designers may focus on just a few abstract classes of V2X applications, rather than designing for individual applications in an exhaustive manner. Also, evaluating the performance trends of generic classes of applications with the same mechanisms or tools simplifies the task and reveals valuable insights at a reasonable cost. If necessary, individual applications can be further studied and analyzed as simple extensions of the proposed generic classes. Finally, it should be noted that the proposed classes are not meant to be comprehensive, but they constitute an essential first step that could be refined and extended in the future as automotive applications emerge, dominate or subside.

This classification serves as a potential road-map for developing the VANET technology needed to support different applications. A generic class of applications is more likely to have a similar set of protocols and mechanisms in the network stack because similar application characteristics and performance requirements tend to implicitly mandate the same technical solution. Thus, network designers should be able to maximize the reusability of common mechanistic 'building blocks' (or modules) for a specific class of applications with similar application characteristics and performance requirements.

1.1.3   Chapter organization

This chapter is organized as follows. In Section 1.2, we introduce a set of V2X applications as a representative of the *connected vehicle* vision. After that, we introduce the attributes used for characterizing those applications in Section 1.3. In Section 1.4, we characterize each application according to the introduced attributes; this constitutes a fundamental step towards identifying a few *generic* application classes. Next, we introduce the market perspectives and challenges for deployment for each representative application class in Section 1.5. Finally, we conclude the chapter and lay out potential directions for future research in Section 1.6.

1.2   V2X Applications from the User Benefits Perspective

Research on VANET technology has been driven mainly by the demand for the provision of network support for application development. So far, the DSRC research community has developed a large number of potential V2X applications for future deployment, ranging from safety/warning applications and highway traffic management to commercial applications. Since it is difficult to analyze a large number of applications, we chose 16 representative ones based on criteria such as customer value, near-term feasibility of deployment, technical novelty, and diversity of enabling technologies. The chosen applications (shown in Tables 1.1, 1.2, and 1.3) constitute the basis for our study.

From a value or customer benefit perspective, these applications can be roughly organized into three major categories: *safety-oriented*, *convenience-oriented*, and *commercial-oriented*. These categories are derived from the characteristics and customer benefits of the applications. Note that, among those listed, safety-oriented applications are of special interest because they are expected to reduce the fatalities and economic losses caused by traffic crashes.

- *Safety-oriented applications* (shown in Table 1.1) actively monitor the nearby environment (the state of other vehicles or road conditions) via message exchanges between vehicles, so that applications are able to assist drivers in handling the upcoming events or potential danger. Some applications may automatically take appropriate actions (such as automatic braking) to avoid potential crashes, while others provide only advisory or warning information as configured by the driver. The latter category of applications is very similar to the former, even though the system requirements (such as reliability, latency, etc.) are less stringent. However, both types of applications aim to improve the level of vehicle safety.

Table 1.1   V2X *safety-oriented* applications of interest

Acronym	Name	Description
SVA	Stopped or Slow Vehicle Advisor	A slow or stopped vehicle broadcasts warning messages to approaching vehicles while it is slow/stopped. Approaching vehicles notify their drivers of the slow/stopped vehicle.
EEBL	Emergency Electronic Brake Light	A vehicle braking hard broadcasts warning messages to approaching vehicles during the hard braking. Approaching vehicles notify their drivers of the hard braking event.
PCN	V2V Post Crash Notification	A vehicle involved in a crash broadcasts warning messages to approaching vehicles until the crash site is cleared. Approaching vehicles notify their drivers of the crash.
RHCN	Road Hazard Condition Notification	A vehicle detecting a road hazard (e.g. a pothole or ice) broadcasts warning messages to vehicles within the affected region. Approaching vehicles notify their drivers of the hazard.
RFN	Road Feature Notification	A vehicle detecting an advisory road feature (e.g. sharp curve, steep grade) broadcasts warning messages to approaching vehicles, which notify their drivers of the advisory road feature.
CCW	Cooperative Collision Warning	A vehicle actively monitors kinematics status messages that are broadcast from other vehicles in its neighborhood to warn the driver of potential collisions.
CVW	Cooperative Violation Warning	A roadside unit actively broadcasts signal phase, timing and related information to approaching vehicles. The vehicles use this information to warn drivers of potential signal violations.

- *Convenience-oriented (traffic management) applications* (shown in Table 1.2) share traffic information among roadway infrastructure, vehicles on the road, and centralized traffic control systems, to enable more efficient traffic flow control and maximize vehicle throughput on the road. Ultimately, these applications not only enhance traffic efficiency, but also boost the degree of convenience for drivers.

- *Commercial-oriented applications* (shown in Table 1.3) provide drivers with various types of communication services to improve driver productivity, entertainment, and satisfaction, such as web access and streaming audio and video.

1.2.1   Application value

Driver assistance and safety features are capable of providing comprehensive information about the surrounding environment to drivers. For example, such features can inform drivers

Table 1.2   V2X *convenience-oriented* applications of interest

Acronym	Name	Description
CRN	Congested Road Notification	A vehicle detects road congestion and broadcasts the information to other vehicles in the region so that other vehicles can use the information for alternate route and trip planning.
TP	Traffic Probe	A probe vehicle aggregates traffic probe information by actively monitoring kinematics status messages that are broadcast from other vehicles and transmits this information through roadside units to a traffic management center.
TOLL	Free Flow Tolling	A vehicle entering a highway toll gate receives a beacon from the toll gate's roadside unit. The vehicle establishes unicast communication with the toll gate roadside unit and an E-payment transaction is completed for toll payment. This enables roadway and congestion pricing via non-stop tolling.
PAN	Parking Availability Notification	A driver looking for an available parking facility drives the vehicle within communication range of a Parking Availability Notification roadside unit. The vehicle sends a request to the roadside unit for a list of nearby parking locations, and the roadside unit responds with a list of parking locations and availability. The vehicle may sort responses according to distance from the current location, and presents a list showing available parking lots within a certain geographical region to the driver.
PSL	Parking Spot Locator	A vehicle entering a parking area or structure sends a request to the Parking Space Locator roadside unit for a list of open parking spaces. The roadside unit sends a list of open spaces and an optional map to the vehicle, and the vehicle notifies the driver of their locations.

of construction zones, speed limits, curves, and broken down vehicles well in advance. In terms of performance when compared to traditional sensors, communication-based safety features take advantage of the relatively long communication range provided by DSRC in order to sense distant events that occur outside the sensing range of traditional sensors. This advance notice can provide drivers with ample time to plan maneuvers around obstacles, or to plan changes in their route. Some aspects of communication-based features have already been described in terms of their safety value by VSCC (2005). The value of a system that relies entirely on safety communications can be bounded by the features implemented, the crash statistics pertaining to that feature, and the market penetration. For instance, a forward collision alert application can help to reduce rear-end crashes, which account for roughly 30% of all crashes in the USA each year. Vehicular communication can also enable numerous types of new applications. For example, since a radar sensor cannot see through objects, there is no way for it to provide an indication that a vehicle a few cars ahead has braked hard. An Emergency Electronic Brake Light (EEBL) feature has the potential to assist drivers of both equipped and unequipped vehicles by providing useful advance information and potentially reducing traffic shockwaves and pile-up accidents.

The value of convenience and traffic efficiency features has currently not been studied in much detail. Transportation efficiency features improve the customer's experience very indirectly by making the roadways and the organizations that maintain them more efficient. There are many individual services that can be provided by this set of features, and each may require a different value model. Traditional ETC systems have been shown to dramatically improve throughput on a per-lane basis when compared with manual toll collection techniques. In Japan it was found that 30% of roadway congestion was due to

Table 1.3   V2X *commercial-oriented* applications of interest

Acronym	Name	Description
RVP/D	Remote Vehicle Personalization/ Diagnostics	When a vehicle is within communication range of the driver's home, the driver may initiate a wireless connection of the vehicle to the home network and download (or upload) the latest personalized vehicle settings. This allows drivers to personalize their vehicle settings remotely. Also, when a vehicle is within the service bay at a dealership, the driver may initiate a wireless connection between the vehicle and the dealer service network in order to upload the latest vehicle diagnostics information and download any new updates.
SA	Service Announcement	Businesses (e.g. a fast food restaurant) may use roadside infrastructure to announce services to vehicles wirelessly as they pass within communication range. A vehicle may inform its driver of the services based on the driver's subscription or request for such information.
CMDD	Content, Map or Database Download	When a vehicle is within proximity of a home or hot-spot, the driver may initiate and connect wirelessly to the home network or hot-spot so that the vehicle can download content (e.g. maps, multimedia or web pages) from the network to the hard drive radio/navigation system.
RTVR	Real-time Video Relay	A vehicle may initiate transmission of real-time video that may be useful to other drivers in the area (e.g. a traffic jam scene). Other vehicles may display this information to their drivers and also relay the real-time video via multi-hop broadcasts in order to extend the range of the communicated information to other vehicles or roadside units.

toll gates. A reduction in roadway congestion has a direct benefit to society in terms of fuel consumption and CO_2 emissions. Additional benefits such as increased productivity and personal time may also be realized, though perhaps at a level that is not as noticeable to the customer.

Commercial features may be able to create substantial demand from consumers since they span many aspects of a customer's daily life. However, commercial features will in general require a strong non-OEM provider with necessary infrastructure support to enable such services. Some of the services and activities enabled by commercial features include news, traffic and weather updates, Internet and email access, music downloads, and drive-through payments. Currently there has been little work performed that shows the value of these types of applications. Again, there are many individual services that can be provided through this set of features, and each may require a different value model. Commercial applications certainly have an opportunity to increase many aspects of customer convenience. In Japan, ETC systems based on DSRC have been used to make payments at parking garages. Electronic drive-through payment systems may be able to substantially increase vehicle throughput just as ETC systems have. By reducing the wait time at fast food restaurants, car washes or parking garages, applications should be able to benefit customers directly in terms of time saved. Likewise, there are many types of information-access services that could be provided to consumers in a personalized manner, something traditional radio services can currently not provide. For example, instead of the traditional broadcast news, a custom version of the news, based on customer preferences, could be downloaded from the customer's home to the car before a morning drive.

1.3 Application Characteristics and Network Attributes

In this section, we define the application and networking criteria used in our classification. Careful selection of these criteria is critical to adequately capture the subtle, yet important, differences between various applications, and their diverse networking requirements. Thus, our approach was first to enumerate the characteristics of the applications in Tables 1.1, 1.2, and 1.3 in a systematic and thorough manner so that we could gain important insight into the applications, and then to use this insight to explore the demands these applications place on network design and enumerate their common network-related attributes. We group these criteria into two major divisions, application-related characteristics and network-related attributes, which are discussed in the next sections.

1.3.1 Application characteristics

In this section, we introduce the application-related characteristics that we identified and used as the basis for our proposed classification. These characteristics, summarized in Table 1.4, describe properties directly related to the applications themselves, such as user benefit and affected geographical region. As mentioned previously, the goal is to develop key characteristics that cover the various design aspects of the set of applications that we have explored. While we have attempted to be as general and as thorough as possible, we acknowledge that future analysis of a broader set may uncover other important characteristics. Indeed, it is our hope that the work presented here will inspire others to research and expand the list as future applications are explored and developed. However, as we will show, this list covers a sufficiently broad range to be a useful reference tool for application and network designers. In the remainder of this section we discuss these characteristics in more detail.

Table 1.4   Candidate criteria to characterize and classify applications (application characteristics)

Application Characteristics	Description	Choices
User benefit of application	What benefit does the application bring to users?	Safety, Convenience, Commercial
Participants of application	What entities participate in the application?	V2V, V2I
Application Region-of-Interest	What is the size of the affected geographical region of the application?	Long, Medium, Short
Application trigger condition	When and how is the application triggered?	Periodic, Event-driven, User-initiated
Recipient pattern of application message	What is the pattern of recipients for the application messages?	One-to-one, One-to-many, One-to-a-zone, Many-to-one
Event lifetime	How long does the event last?	Long, Short
Event correlation	What is the degree of event correlation in the Region-of-Interest?	Strong, Weak, None
Event detector	How many hosts can detect/generate the event?	Single host, Multiple hosts

- *User benefit* describes the type of benefit or value the application provides to the end customer, as defined in a number of studies (VSCC 2005) (and discussed in Section 1.2). Overall, there are three widely accepted types: *safety-oriented* applications, *convenience-oriented* applications, and *commercial-oriented* applications.

- *Application participants* specifies the entities that may potentially be involved in the application. Some applications only require communication among vehicles, while others require coordination between vehicles and roadside infrastructure. Hence, communication-based automotive applications can be categorized as either V2V or V2I applications.

- *Application Region-of-Interest (ROI)* is the size of the geographical region covered by those entities participating in an application. Different kinds of applications have different ROI sizes. For example, in some safety applications, vehicles need to be aware of the kinematics status of other vehicles in their direct neighborhood (i.e. a few hundred meters), whereas in other safety applications vehicles need to know the hazard situation of a stretch of road that lies ahead (i.e. up to 1 kilometer). Likewise, for some convenience applications, vehicle occupants may want to know the status of road congestion far ahead (i.e. several kilometers) for trip planning. Qualitatively, the application ROI can be classified into three major types: *short-*, *medium-*, and *long-range*. Quantitative characterization of the shape and dimensions of the ROI, for various applications, is an important topic that requires interdisciplinary research in system reliability, driver behavior, and traffic/road dynamics to name a few.

- *Application trigger condition* specifies how applications are triggered. This is generally either *periodic*, *event-driven*, or *user-initiated*. Implicitly, it also specifies the kind of communication methods used by the application. For example, the vehicular kinematics status messages used for collision detection are normally broadcast periodically, whereas warning messages for events such as panic braking are usually event-driven, and request messages for on-demand convenience services from vehicle occupants are generally user-initiated.

- *Recipient pattern of application message* specifies the pattern of potential message recipients for an event, which varies across applications. For instance, for safety applications like Cooperative Collision Warning (CCW) and Cooperative Violation Warning (CVW), it is critical for all neighboring vehicles to hear broadcast safety alert messages to avoid potential collisions (a *one-to-many* pattern), whereas for safety applications such as EEBL, Stopped or Slow Vehicle Advisor (SVA), and Post Crash Notification (PCN), only vehicles in the region being affected (vehicles behind the event originator) need to hear the safety alert message (a *one-to-a-zone* pattern). Likewise, a *point-to-point* communication pattern is often used in many convenience and commercial applications, and a *many-to-one* pattern is also sometimes used. Thus, the pattern of event message recipients can be grouped into four categories: *one-to-many*, *one-to-a-zone*, *one-to-one*, and *many-to-one*.

- *Event lifetime* illustrates how long an application event (e.g. traffic crash or road congestion) persists over time. Among the criteria discussed so far, event duration is one application characteristic that may directly affect network system design. Among all applications, event lifetime may differ significantly. For example, some events have relatively short durations (e.g. EEBL events may last only a few seconds on average), while others may have relatively long durations (e.g. a PCN event may take hours before the crashed vehicles are cleared from the roadway). Among the applications we studied, most fell into one of two general categories: either a *short* event lifetime on the order of seconds or a *long* event lifetime on the order of minutes to hours.

- *Event correlation* specifies the degree to which events generated by entities within a geographical region of interest are correlated with each other. For example, the occurrence of an EEBL event in a vehicle may be highly correlated to EEBL events generated by other vehicles in front of it. Another example is Road Hazard Condition Notification (RHCN), where RHCN events in nearby vehicles may be highly correlated since they are caused by the same road hazard condition. Qualitatively, applications can be grouped into three categories: those with *strong* event correlation, *weak* event correlation, and *no* event correlation.

- *Event detector* specifies how many vehicles generate event messages in response to the same event. For instance, for applications such as SVA or PCN, where a vehicle reports its kinematics status, the vehicle is the sole event detector (i.e. of its kinematics state) and event message host (originator), whereas for applications such as RHCN and Road Feature Notification (RFN), where a vehicle reports road hazards, many vehicles may detect the same event (i.e. the same road hazard) and serve as event message hosts. Therefore, we classify application event detection as either *single host* or *multiple hosts*.

As mentioned previously, we believe these are the key defining characteristics, among the 16 applications that we studied, that are of most relevance to network design. However, we acknowledge that further application analysis may reveal other characteristics to add to the list, and we hope that it inspires others to do so. For the purposes of this study, however, these are the basis for the application characteristics portion of our classification effort. In the next section we present our group of key network-related attributes and their relation to the application characteristics above.

1.3.2   Network attributes

In this section we introduce the key network-related attributes that we used in our classification to characterize the fundamental aspects of network design for communication-based automotive applications. These attributes, summarized in Table 1.5, are somewhat related to the application characteristics discussed in the previous section, as we will show. In the remaining part of this section we discuss these network attributes, and their relationship with the corresponding application characteristics, in detail.

- *Channel frequency* specifies the type of physical-layer channels that may be used to support communication-based automotive applications. Following FCC regulations in the USA, safety-oriented applications are normally assumed to use a single DSRC-Control Channel (CCH), whereas convenience-oriented applications use one of six DSRC-Service Channels (SCHs). On the other hand, commercial-oriented applications can either occupy DSRC-SCH channels, or any other channel frequency in the unlicensed Industrial, Scientific, and Medical (ISM) bands (e.g. Wi-Fi 2.4 and 5.8 GHz). In other words, the choice of channel is largely determined by the value of

Table 1.5   Candidate criteria to characterize and classify applications (network attributes)

Application Attributes	Description	Choices
Channel frequency	What channel does the application use?	DSRC-CCH, DSRC-SCH, Wi-Fi
Infrastructure	Is infrastructure required?	Yes, No
Message time-to-live	How far do messages propagate?	Single-hop, Multi-hop
Packet format	What type of packet is used?	WSMP, IP
Routing protocol	How are messages distributed?	Unicast, Broadcast, Geocast, Aggregation
Network protocol initiation mode	How is a network protocol initiated?	Beacon, On-demand, Event-triggered
Transport protocol	What form of end-to-end communication is needed?	Connectionless, Connection-oriented
Security	What kind of security is needed?	V2V security, V2I security, Internet security

the *user benefit* characteristic of the application. While there are many other channels that can be used (such as cellular telephony or WiMAX), in practice the choice of channel is generally one of either *DSRC-CCH, DSRC-SCH*, or *Wi-Fi*. In other regions of the world, a similar choice needs to be made for the applications to be effective.

- *Infrastructure* specifies whether the application needs infrastructure (i.e. a roadside unit) for its operation. Obviously, this is needed if the *participants of the application* characteristic involves a roadside unit. Otherwise, it may not be required.

- *Message Time-to-Live (TTL)* specifies how far a message is propagated by the network, and what type of packet forwarding/routing functionality (i.e. single-hop or multi-hop) is needed by the network layer. This attribute is partly determined by the *application region of interest* characteristic. Single-hop communication is sufficient for short-range applications, while medium- or long-range applications require multi-hop packet forwarding/routing functionality for extended reachability. Thus, design choices include either *single-hop* or *multi-hop* routing.

- *Message packet format* describes the format of the network packets that are used to encapsulate the application messages. This attribute is partly influenced by the *user benefit* characteristic of the application. In general, the automotive industry (VSCC 2005) and the IEEE standard community (IEEE 802.11 1999) have promoted the idea that safety and convenience applications are more likely to use relatively constant and small-sized packets, whereas commercial applications are more likely to use variable and large-sized packets. In the DSRC standard developed in the USA, the WAVE Short Message Protocol (WSMP) is proposed for safety and convenience use. It is essentially a simplified version of the IP protocol, with a smaller packet header to reduce per-packet overhead for improved network efficiency. For commercial applications, it is assumed that the traditional IP packet format will be used. Thus, we classify packet formats into two types: either *WSMP* format or *IP* format.

- *Routing protocol* is a design choice that illustrates what kind of network routing protocols are used for the various applications. Obviously, this network attribute is closely related to the *recipient pattern of application message* characteristic. For instance, most safety applications use *broadcast* routing (one-to-many) or *geocast* routing (one-to-a-zone), while convenience and commercial applications normally use *unicast* routing (one-to-one) or *aggregation* routing (many-to-one).

- *Network protocol initiation mode* describes how the network protocol is triggered. Some safety applications mandate periodic broadcast 'beaconing' of status messages, like CCW and CVW (i.e. *beacon* mode), whereas other safety applications, like EEBL and PCN, send messages only when a critical event is detected (i.e. *event-triggered* mode). For a portion of convenience and commercial applications, it is the vehicle occupants that initiate message announcements and service request (i.e. user-initiated *on-demand* mode).

- *Transport protocol* is a design choice that indicates whether or not a reliable end-to-end connection is needed to support the application. As we discovered, safety and convenience applications generally follow the *connectionless* paradigm (e.g. WSMP, UDP), while commercial applications often use the traditional *connection-oriented* paradigm (e.g. TCP).

- *Security* considers what kind of security solution is needed for the application. The choices include *V2V security*, *V2I security* and *Internet security*. Safety applications require high-level V2V security preventing vehicles from malicious attacks, convenience applications also mandate the stringent V2I security solution because financial transactions could be involved at roadside infrastructure, and most commercial applications require efficient collaboration between V2X security solutions and existing security solutions for the Internet.

As indicated earlier, many of these network attributes are closely related to specific application characteristics. Intuitively, a given application characteristic or performance requirement normally requires a given networking mechanism or capability. In the next section we show how sets of applications with similar characteristics and requirements lead to the same network solutions, resulting in a very useful and intuitive general classification.

1.4   Application Classification and Categorization

In this section we present the results of characterizing and classifying the set of 16 applications introduced in Section 1.2. We then compare and contrast these applications, first with respect to the application characteristics presented in Section 1.3.1, and then with respect to the network attributes presented in Section 1.3.2. Afterwards, we show how, by combining the applications with similar characteristics and network functionalities, we can group these applications into seven generic classes (from the perspective of network design).

1.4.1   Characterization based on application characteristics

The process of application characterization is divided into two steps: *characterization of application attributes* and *characterization of network attributes* (i.e. network design), as shown in Tables 1.6 and 1.7 respectively. By first exploring all the relevant application characteristics for each one, we gain a more complete understanding of the fundamental properties and functionality requirements of these applications. Later, we show how this effort gives rise to application characterization from the network design point of view.

Table 1.6 lays out the main characteristics of each application based on the selected application-related attributes summarized in Section 1.3.1. Given the limited space, we are unable to discuss the characteristics of all 16 applications. Instead, we only highlight a few important application characteristics, illustrating how these criteria help to differentiate the often subtle differences between various applications, as follows.

- Notice that most of the safety applications have a medium-sized effective application range (i.e. a few hundred meters to 1 kilometer), since safety messages, such as vehicular kinematics status or road conditions, are only relevant to other vehicles within a moderate geographical region. Exceptions are the CCW and CVW applications, which have a small application effective range because they require closer monitoring of vehicles in their direct neighborhood (i.e. within 200 meters). Conversely, convenience applications generally require a medium to large effective range (i.e. up to a few kilometers), because it is vital for drivers to know the congestion situation or traffic conditions at this range for effective detour or trip planning decision making. Similarly, commercial applications tend to have a large effective range in order to access remote commercial service providers. For example, a fast food restaurant is willing to announce its service to vehicles a long distance (i.e. several kilometers) from its location.

- Most safety applications (e.g. EEBL, RHCN and SVA) and few convenience applications (e.g. Congested Road Notification, or CRN, Traffic Probe, or TP, and Free Flow Tolling, or TOLL) are initiated by the events happening on the road, such as vehicle collisions, detection of road hazards (e.g. ice or oil), sudden braking, or detection of traffic congestion. If no such events happen, these applications will not be called upon. Among safety applications, CCW and CVW are unusual because they rely on the periodic message updates to monitor the neighboring vehicles' driving status, regardless of safety events. On the other hand, most convenience and commercial applications are triggered on demand by vehicle occupants, rather than by any safety event on the road or the vehicle itself.

- The potential recipients of application messages, in most safety applications (e.g. SVA and EEBL), are vehicles within a specific zone (i.e. behind the vehicle that detects the event and originates the safety message). Thus, safety applications can be summarized as having *one-to-a-zone* recipient patterns. Again, CCW and CVW do not follow this general trend. In these two applications, all the vehicles in the neighborhood are supposed to receive the periodic update in order to avoid the potential crash from any direction. So these two applications have one-to-many recipient patterns. At the same time, convenience and commercial applications vary from application to application: some convenience applications (e.g. TOLL) and commercial applications (e.g. Remote Vehicle Personalization/Diagnostics, or RVP/D, Content, Map or Database Download, or CMDD, and Real-time Video Relay, or RTVR) belong to the point-to-point (*one-to-one*) communication paradigm, while other convenience (e.g. CRN) and commercial applications (e.g. Service Announcements, or SAs) are fundamentally *one-to-a-zone* in nature.

Table 1.6 Application characterization based on application characteristics

Acronym	User Benefit	Application Participants	Application Region-of-Interest	Application Trigger Condition	Recipient Pattern	Event Lifetime	Event Correlation	Event Detector
SVA	Safety	V2V	Medium	Event	One-to-a-zone	Long	None	One
EEBL	Safety	V2V	Medium	Event	One-to-a-zone	Short	Weak	Many
PCN	Safety	V2V	Medium	Event	One-to-a-zone	Long	None	One
RHCN	Safety	V2V	Medium	Event	One-to-a-zone	Long	Strong	Many
RFN	Safety	V2V	Medium	Event	One-to-a-zone	Long	Strong	Many
CCW	Safety	V2V	Short	Periodic	One-to-many	N.A.	N.A.	N.A.
CVW	Safety	V2I	Short	Periodic	One-to-many	N.A.	N.A.	N.A.
CRN	Convenience	V2V	Long	Event	One-to-a-zone	Long	Strong	Many
TP	Convenience	V2I	Long	Event	One-to-one	Short	None	Many
TOLL	Convenience	V2I	Short	Event	One-to-one	Short	None	One
PAN	Convenience	V2I	Long	User-initiated	One-to-one	N.A.	N.A.	N.A.
PSL	Convenience	V2I	Short	User-initiated	One-to-one	N.A.	N.A.	N.A.
RVP/D	Commercial	V2I	Short	User-initiated	One-to-one	N.A.	N.A.	N.A.
SA	Commercial	V2I	Long	User-initiated	One-to-a-zone	N.A.	N.A.	N.A.
CMDD	Commercial	V2I	Long	User-initiated	One-to-one	N.A.	N.A.	N.A.
RTVR	Commercial	V2I	Long	User-initiated	One-to-one	N.A.	N.A.	N.A.

- 'Event' is an important concept in safety applications, and inxs a few convenience applications, because it is an event that initiates the application operations. In our study we also characterize safety events via several properties, including event duration, event correlation, and event detectors. Consistent with our conjecture, we find that safety events drastically vary from application to application. For example, sudden braking (EEBL) is a one-shot event, while road hazard/feature events (RHCN or RFN) are persistent ones. Also, different instances of RHCN or RFN events caused by the same road hazard/feature are more likely to be correlated with each other, in contrast with the totally independent PCN events. Even though the study of event characteristics is not directly used in the network design conducted in Section 1.4.2, we believe that such an analysis can assist future network designers better to capture the data traffic patterns induced by event-driven safety applications.

From an application benefits point of view, different applications have different functionalities, providing different usages for customers. Interestingly enough, we realize that many applications exhibit highly similar application characteristics, with the exception of a few minor differences. To validate whether such an observation is valid from a network design perspective, we conduct an application characterization based on the relevant network attributes in Section 1.4.2.

1.4.2   Characterization based on network attributes

As mentioned in Section 1.3.2, for each application we discovered that its characteristics tend to mandate a certain design in the network protocol stack. For example, applications with one-to-many recipient patterns are more likely to use broadcast routing protocols, while unicast routing protocols are suitable for applications with one-to-one recipient patterns. Similarly, a single-hop packet dissemination mechanism is adequate to support applications with small application ROIs (i.e. a few hundred meters). In contrast, multi-hop routing protocols are needed for applications with medium or large application ROI. Accordingly, we are capable of determining the potential design choices for various components in the network stack by referring to their corresponding application characteristics and requirements. At the same time, we also notice that some of the network attributes are purely the choices of network designers, since different technical approaches are able to achieve the same objective.

Table 1.7 lays out the main network attributes of each application based on the selected network attributes summarized in Section 1.3.2, from the lower physical layer to the upper transport layer. These network attributes cover design issues such as the physical layer channel frequency, the usage of infrastructure, message TTL, routing protocol and network protocol triggers at the network layer, transport layer design, and security solutions. Again, we emphasize only a few important network attributes, discussing the potential impact of application characteristics on these network design issues.

- The message packet format is determined by the type of application (from the perspective of user benefit). Normally, safety and convenience applications use lightweight short messages in the WSMP format, to improve network resource efficiency. Commercial applications, on the other hand, generally prefer the traditional heavyweight IP format to be compatible with existing Internet commercial services.

- The network-layer routing protocol is one essential component in a network stack, differentiating the reachability and recipient patterns of various applications.

Table 1.7   Application characterization based on network attributes

Acronym	Channel Frequency	Infra-structure	Message TTL	Packet Format	Routing Protocol	Network Trigger	Transport Protocol	Security Solution
SVA	DSRC-CCH	No	Multi-hop	WSMP	Geocast	Event-triggered	Connectionless	V2V security
EEBL	DSRC-CCH	No	Multi-hop	WSMP	Geocast	Event-triggered	Connectionless	V2V security
PCN	DSRC-CCH	No	Multi-hop	WSMP	Geocast	Event-triggered	Connectionless	V2V security
RHCN	DSRC-CCH	No	Multi-hop	WSMP	Geocast	Event-triggered	Connectionless	V2V security
RFN	DSRC-CCH	No	Multi-hop	WSMP	Geocast	Event-triggered	Connectionless	V2V security
CCW	DSRC-CCH	No	Single-hop	WSMP	Broadcast	Beacon	Connectionless	V2V security
CVW	DSRC-CCH	Yes	Single-hop	WSMP	Broadcast	Beacon	Connectionless	V2I security
CRN	DSRC-SCH	No	Multi-hop	WSMP	Geocast	Event-triggered	Connectionless	V2V security
TP	DSRC-SCH	Yes	Multi-hop	WSMP	Unicast	Event-triggered	Connection-oriented	V2I security
TOLL	DSRC-SCH	Yes	Single-hop	WSMP	Unicast	Event-triggered	Connection-oriented	Internet security
PAN	DSRC-SCH	Yes	Multi-hop	WSMP	Unicast	On-demand	Connection-oriented	V2I security
PSL	DSRC-SCH	Yes	Single-hop	WSMP	Unicast	On-demand	Connection-oriented	V2I security
RVP/D	DSRC-SCH, Wi-Fi	Yes	Single-hop	IP	Unicast	On-demand	Connection-oriented	V2I security
SA	DSRC-SCH, Wi-Fi	Yes	Multi-hop	IP	Geocast	On-demand	Connectionless	Internet security
CMDD	DSRC-SCH, Wi-Fi	Yes	Single-hop	IP	Unicast	On-demand	Connection-oriented	Internet security
RTVR	DSRC-SCH, Wi-Fi	Yes	Multi-hop	IP	Unicast	On-demand	Connection-oriented	Internet security

Most safety applications utilize multi-hop geocast routing protocols, because of the one-to-many communication nature of safety applications. Geocast routing distributes packets within a given zone or region. Thus, a geocast routing protocol can be viewed as a special case of broadcast routing. CCW and CVW applications, instead, use the single-hop broadcast scheme to announce the periodic update in their direct neighborhood. Convenience and commercial applications either use geocast/broadcast protocols to announce messages in a region (for advertisement services such as SAs, or traffic congestion notifications such as CRN), or exploit unicast protocols to forward packets to a given destination (for financial transactions like TOLL, or data downloads from infrastructure as with CMDD).

- How the network routing protocol is triggered is another interesting design choice to be examined in our study. Event-driven safety applications (e.g. SVA, EEBL, and CRN) require the event-triggered mechanism in network protocols, periodic-based safety applications (e.g. CCW and CVW) mandate the periodic beacon (or 'hello message') mechanism, and user-initiated convenience and commercial applications (e.g. SAs, RVP/D, and Parking Spot Locator, or PSL) are triggered in an on-demand fashion.

- The involvement of infrastructure in network design and application development is another key issue for consideration. Please note that the term 'infrastructure' in this chapter only refers to roadside units along the roads, rather than base stations in cellular systems. Two minor differences exist between infrastructure and vehicles: (a) infrastructure is stationary while vehicles are mobile; (b) infrastructure may have a direct connection to the Internet, but vehicles do not. Both infrastructure-oriented approaches and non-infrastructure approaches (or even a combination of both approaches) are used to achieve the objective of supporting the applications or services discussed above. Deployment of infrastructure-oriented services depends on considerations such as availability of infrastructure, costs, and technology. The availability of infrastructure would facilitate the design and deployment of convenience applications. Having infrastructure that provides a gateway to the Internet would enable the design and deployment of commercial applications. As a side note, the involvement of infrastructure also complicates the design of security solutions. We believe that security solutions for V2V applications are different from those for V2I applications. Also, the gateway to the Internet requires compatibility of V2X security solutions with the existing Internet security solutions.

Throughout our study we found that Tables 1.6 and 1.7 reveal a number of interesting observations. Generally speaking, many applications exhibit highly similar application characteristics, resulting in similar protocol design across various layers in the network stack. For instance: (a) RHCN and RFN are nearly the same, except that the type of safety warning messages are different – RHCN is about road hazards, while RFN is about road features; (b) PCN and RHCN are also similar except for the number of event originators – PCN has a sole message host, while RHCN has multiple message hosts – and even though this difference gives rise to different levels of data traffic burstiness from event generation, the network protocol stacks for these two applications are still similar to each other; (c) CCW and CVW applications can be categorized into the same type, although the former is a V2V application whereas the latter is a V2I one.

In summary, the first seven safety applications (SVA, EEBL, PCN, RHCN, RFN, CCW, and CVW) all utilize broadcast/geocast routing protocols to distribute safety/warning messages in the WSMP format. On the other hand, some convenience applications mostly rely on user-initiated unicast routing protocols to deliver non-safety messages in the WSMP format, while commercial applications may exploit IP protocols to enable enhanced functionality such as Quality of Service (QoS) routing. This, in turn, suggests that the studied applications naturally lend themselves to smaller numbers of generic/abstract classes, and this is the subject matter of the next section.

1.4.3   Application classification

With a deep understanding of application characteristics and network attributes applied to all studied applications, we are able to classify them into a number of generic classes. Notice that application classification can be conducted at different levels, depending on the design granularities. For example, simple classification and few abstract classes are adequate for high-level concept design of automotive communication applications. On the other hand, empirical design of prototype systems normally mandates an exhaustive effort, resulting in sophisticated multi-level application categories.

At the initial stage of this emergent research field, we believe that a high-level classification is sufficient to serve the purpose of distilling the major concepts and identifying the synergy among various applications, without unnecessarily complicating the problem formulation. Later on, the empirical prototype system can be designed and implemented based on the refined and enriched version of this study. Here we present such a way of classifying the aforementioned applications from the perspective of network design (as shown in Figure 1.1), among other alternatives. Generally speaking, V2X applications can be classified into two broad generic classes, namely *short message communications* and *large-volume content download/streaming*. Most safety and convenience applications belong to the first class, since the messages in these applications are lightweight WSMP messages. Considering that the IP message format is appropriate for large-volume data (such as Internet web access or video/audio streaming), most commercial applications fall into the second category.

Short message communication

First we discuss the class of short message communication, which uses lightweight WSMP packets. This class can be divided, depending on the recipient pattern and routing protocol, into either broadcast/geocast or unicast applications. Clearly, most safety applications require message announcements be sent to a large number of nodes (one-to-many or one-to-a-zone); hence they would fall into the broadcast/geocast-oriented type. On the other hand, many convenience applications (including payment-type applications) would fall into the unicast-oriented type.

According to the network protocol triggering condition, *broadcast/geocast*-oriented applications can be further classified as event-driven, scheduled (periodic) and on-demand approaches. The event-driven class is used to model safety applications focusing on life-threatening events, and the scheduled class is suitable for safety applications requiring periodic message updates, whereas the on-demand class is appropriate for convenience applications such as a parking spot locator. As a side note, high-level V2V security solutions are required to protect safety applications from malicious hackers. These three sub-classes of broadcast/geocast-oriented applications are:

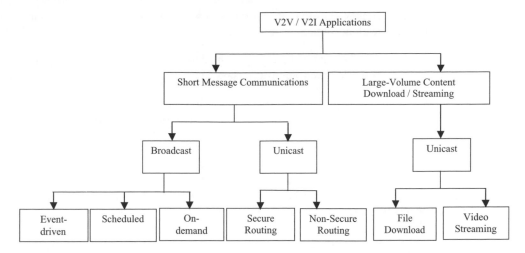

Figure 1.1 Classification from the perspective of network design

- *Event-driven broadcast/geocast class:* SVA, EEBL, PCN, RHCN and RFN applications, as well as CRN application, belong to this category (class 1).

- *Scheduled (periodic) broadcast/geocast class:* CCW and CVW applications fall into this category (class 2).

- *On-demand broadcast/geocast class:* Some convenience or commercial applications, such as SAs, belong to this category (class 3).

The secure routing of financial transactions in convenience applications also plays an important role in *unicast*-oriented applications. Thus, these unicast-oriented applications can be divided into those that involve stringent secure routing for financial transactions, and those that do not involve secure routing. Thus we list these two sub-classes of unicast-oriented applications:

- *Secure unicast class:* One example of this approach is a TOLL application (e.g. drive-through payment or free-flow tolling). RVP/D also falls into this category since it is potentially related to the control components of vehicles (class 4).

- *Unsecured unicast class:* TP, Parking Availability Notification (PAN) and PSL applications fall into this category. Some commercial applications (e.g. V2V online chatting or social networking applications) also belong to it (class 5).

Large-volume content download/streaming

Next we focus on the second major class of applications, namely large-volume content download/streaming, which is normally implemented in the IP format for compatibility. These applications often utilize unicast protocols because of their one-to-one communication nature. This class is further divided depending on the content type: either file download or media streaming. The former type allows short-term disruption in network service, so it is inherently latency-tolerant. The latter type requires a relatively smooth streaming transfer, so it is fundamentally latency-sensitive. It is straightforward to notice the types of some large-volume content download/streaming applications:

- *File download:* a CMDD application (e.g. map database download or web access/ browsing) is one example of this approach (class 6).

- *Video continuous streaming:* RTVR (e.g. video/MP3 streaming among vehicles, or from roadside infrastructure for entertainment) falls into this category (class 7).

These seven types of V2X applications and their key considerations in network design are summarized in Table 1.8. From the above discussion, we conclude that the given set of applications can be grouped into seven generic classes. Since these applications are carefully chosen to represent many others, we believe that our classification methodology and its results can also apply to a large number of V2X applications.

Table 1.8   Network design considerations for seven types of applications

Application Type	Channel Frequency	Packet Format	Routing Protocol	Internet	Transport Protocol	Security
Event-driven broadcast/ geocast	DSRC-CCH	WSMP	*Event-driven* multi-hop broadcast/geocast	No	Connection-less	V2V security
Scheduled broadcast/ geocast	DSRC-CCH	WSMP	*Scheduled* multi-hop broadcast/geocast	No	Connection-less	V2V/V2I security
On-demand broadcast/ geocast	DSRC-SCH, or Wi-Fi	WSMP, or IP	User-initiated *on-demand* multi-hop broadcast/geocast	No	Connection-less	V2V/V2I security
Secure unicast	DSRC-SCH	WSMP	Multi-hop unicast with *secure routing*	No	Connection-oriented	Stringent V2V/V2I security
Normal unicast	DSRC-SCH	WSMP	Multi-hop unicast	No	Connection-oriented	V2V/V2I security
File download	DSRC-SCH, or Wi-Fi	*IP*	Multi-hop unicast	Yes	Connection-oriented	V2V/V2I/ Internet security
Media streaming	DSRC-SCH, or Wi-Fi	*IP*	Single-hop unicast with *QoS routing*	Yes	Connection-oriented	V2V/V2I/ Internet security

The potential benefits of application classification are:

- The classification effort not only contributes to capturing the common features and technical requirements of applications, but also helps to develop common networking stacks for the identified generic classes. In the near future, with the deeper understanding of these generic and abstract classes, we shall be able to increase the module reusability of wireless networking solutions for the given set of applications with similar characteristics.

- The classification effort helps to identify common requirements and performance metrics relevant to each application class. It also eases application modeling in simulation studies targeted at the performance evaluation of a large number of applications. By appropriately isolating generic network design from different application instantiations, we argue that it is much more efficient to model these seven generic classes than it is to model all 16 applications in an exhaustive manner without exploiting their noticeable commonality. Thus, a generic model should suffice for gathering statistics for the performance metrics defined for a specific class. Gathering performance results for a particular application, for the purposes of detailed analysis, could be achieved by deriving the application of interest as a simple extension from its generic model.

1.5 Market Perspectives and Challenges for Deployment

1.5.1 Fleet penetration

According to the DOT, Federal Highway Administration, there were a total of 237 million registered vehicles in the USA in 2004. Since the effectiveness of communication-based applications will require a certain level of market penetration, it will be a challenge to deploy a large number of equipped vehicles in a short time frame. Using production information from vehicle manufacturers, it is possible to project the length of time it will take for a sufficient level of market penetration to be achieved. A simple projection was performed using the figure of 237 million registered vehicles, together with the total number of vehicles produced by selected automobile manufacturers for 2005. It was assumed that the number of total vehicle registrations as well as the market share of the selected manufacturers will remain constant over time. The projection also assumes that every vehicle produced by the selected maker will be equipped with the V2X system.

As Figure 1.2 indicates, partnering strategies between automobile makers can result in an increased market penetration over a shorter period of time. At the very beginning of deployment, manufacturers will be adding technology to vehicles even though it has no immediate value to consumers. For this reason, it is likely that some kind of partnering strategy will emerge so that the features enabled by V2X become effective as soon as possible.

1.5.2 System rollout options

There are various options when it comes to the deployment of V2X systems.

- *Standalone system solution:* In a standalone implementation of an in-vehicle V2X system, we envision an additional embedded module, which executes the V2X functionality, added to the vehicle. This requires a processor with sufficient computation resources, a GPS receiver and wireless transceiver, vehicle network access, and, optionally, a human–machine interface. This is the most expensive approach, since the complete in-vehicle system infrastructure has to be added to execute the V2X functionality. The average cost would depend on the vehicle manufacturer; however, it should be clear that this option would be costly.

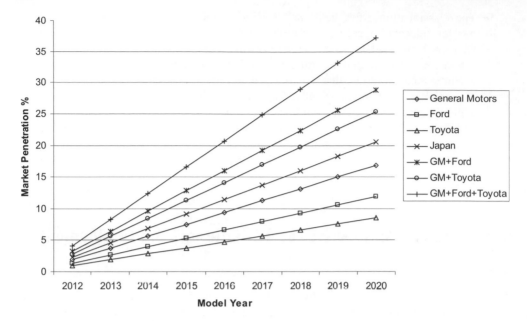

Figure 1.2 Projected US market penetration level of equipped vehicles (best case)

- *Navigation system solution:* Navigation systems will provide sufficient computing capabilities, since they are equipped with large amounts of memory, vehicle network access, and GPS. In terms of hardware, only the wireless radio unit and the antenna have to be added. In terms of software, the applications have to be added and the communication protocol has to be enabled (including security). However, growth of V2X would be directly proportional to and limited by the growth of navigation systems. In 2006 it was expected that there were between 1.5 million and 2 million in-vehicle navigation systems shipped in North America (ABI Research 2009). By 2011 this number is expected to climb substantially. Conservative estimates place the total at around 6.5 million, but it is possible that the volume of in-vehicle navigation systems could reach nearly 9 million by 2011 (ABI Research 2009). This corresponds to an overall fleet penetration of less than 4%.

- *Telematics system solution:* This option is similar to the navigation system solution in that the system architecture required for V2X already exists for the most part. In terms of hardware, only the radio unit and the antenna have to be added. In terms of software, the applications have to be added and the communication protocol including security has to be enabled. The advantage of the telematics system solution is that the number of installed telematics units is much larger than the number of navigation systems. Telematics service provider OnStar currently has a total of 4 million subscribers, and for the 2008 model year OnStar systems will be standard on 50 out of 54 vehicles produced by General Motors. The OnStar service is also available for Acura and Lexus vehicles. Independent telematics service provider ATX currently has a total of 700,000 subscribers. ATX customers currently include Mercedes (Tele Aid) and BMW (Assist).

- *Aftermarket transceiver:* This option is an aftermarket solution that enables older vehicles to function as communicating ones. The aftermarket transceiver solution has two potential types of embodiment: a passive version, and an integrated one that is capable of interfacing with the vehicle system in order to provide customers with a complete add-on solution. In the passive solution, everything required to support V2X communication is supplied on a device that reports GPS information to listening vehicles. However, consumers of a standalone transceiver do not receive any direct benefit from the device, other than making their vehicle visible to other equipped vehicles, since no information is displayed back to the customer. In the integrated solution, the device is capable of providing customers with a fully featured V2X system. In this type of implementation, a device that connects to the vehicle on-board diagnostic connector could interface with the GPS system of the vehicle and activate vehicle displays or audible chimes. An example of an integrated aftermarket transceiver would be a small device that plugs into the On-Board Diagnostic systems (OBDII) connector of the vehicle and provides advisory information to drivers through a Driver Information Center (DIC).

1.5.3 Market penetration analysis

Currently there is no common view about how to deploy communication technology economically in vehicles along with a suitable market introduction strategy to overcome the problems of the initial market penetration that many of the applications require. Communication-based systems are directly subjected to network effects (i.e. the value of this technology to the customer will increase as market penetration increases). In particular, for a V2V system, the customer will only perceive value once there is sufficient QoS provided by a minimum market penetration. Different application features will require different thresholds in terms of the minimum market penetration in order for this technology to become useful to consumers. Deployment decisions in all cases will be based on market penetration to ensure that customers receive a quantifiable benefit from the deployed applications. In all situations, the perceivable benefit to the consumers will increase as the market penetration level increases. With this in mind, the initial features to be deployed will be simple applications with very low market penetration level requirements. This deployment strategy will also allow us to gain valuable learning experience and help build customer confidence in the system and applications. The estimated minimum market penetration levels for the different broad application families defined in Section 1.2 will now be discussed. These estimates are based on preliminary research into this area and on engineering judgments about the penetration levels at which features become useful to consumers.

Convenience (traffic management) and commercial applications

The minimum required market penetration required for such applications is estimated to be very low. Communications infrastructure would serve as a catalyst for these applications and therefore the benefit to the vehicle customer would be seen even with low market penetration. While this may suggest that these types of applications might appear first, there must also be a roadway infrastructure available that provides services to the vehicles. Assuming such an infrastructure exists, some features will be able to provide immediate benefits to consumers, since the vehicle and the infrastructure will be able to communicate. Certain convenience (traffic efficiency) features may require a minimum penetration before there is sufficient collected data from equipped vehicles to make traffic management features viable. The latency of the information received from and sent to the vehicles could be relatively large without compromising the reliability of the feature.

Safety applications

For safety applications providing advisory information, it is expected that the driver would perceive value from the technology beginning at low market penetration levels. While these numbers seem promising, benefits of these types of features need to be studied to determine their actual and perceived value to consumers in terms of safety and convenience. It is suggested that simulation studies be used to measure the benefits of these types of applications in terms of crash reductions and increased traffic efficiency due to the reduction in crashes. The low minimum market penetration for these types of advisory features is primarily due to the fact that these features are uniquely enabled by communication technology and offer added value to the driving experience at a very low cost. However, while drivers would receive information they would not otherwise have, they may question the reliability or usefulness of a system that appears to function so infrequently. For these features to have value, a minimum system penetration would be required so that advisory information generated by an equipped vehicle is relayed to other vehicles often enough that consumers appreciate the benefit of it. If application information needs to be distributed over a large distance, an intelligent store and multi-hop message-forwarding scheme may be required.

For safety applications providing warning information, a higher estimated minimum market penetration level is required. Communication technology is not an enabler for these features since autonomous object detection sensors would achieve this objective and provide significant reliability to the driver. However, a collection of autonomous object detection sensors offering these warning features would be very expensive (since multiple sensors would be needed) and therefore autonomous sensors cannot be widely deployed in every vehicle segment. The trend shows that the market penetration of object detection sensors is extremely low and they can only be targeted for luxury vehicle segments. Thus it could be argued that since V2X technology is capable of providing these features at significantly low cost based on a single sensor, a majority of the drivers would benefit from these features at the market penetration levels indicated, and the system would provide more reliability as the market penetration increases. The societal benefits of transportation safety would be very significant even at modest penetration levels. Because current trends show very low market penetration levels of warning systems based on autonomous sensing, V2X may be able to provide a low cost technology alternative in a society that demands a significant reduction in annual vehicle crashes.

For safety features, it will be challenging to identify the market penetration levels at which drivers begin to perceive that the system is providing them with useful warnings. In terms of societal benefits, however, for an equipped vehicle pair that is involved in a potential collision there should be measurable reductions in vehicle crashes, property damage and personal injury. Collision avoidance control features would naturally be introduced only when the system is highly reliable, which would imply a very large market penetration. Since the requirement for fleet penetration is extremely high, it is likely these types of applications will only become available in the near term if the government mandates the technology and system, and all unequipped vehicles are also retrofitted.

1.5.4   System rollout

Based on the maturity of wireless technology, the advancements in positioning technology and optimistic market penetration forecasts, it is anticipated that the first V2X applications could reach the market by 2012. Further development of the V2X technologies as well as even further increases in fleet penetration levels will enable V2X crash warning features. The evolution of V2X feature deployment with respect to technology advancement and fleet penetration is shown in Figure 1.3.

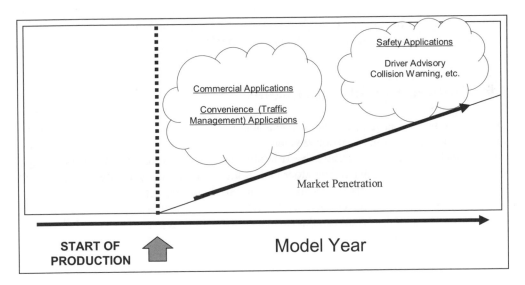

Figure 1.3 Evolution of V2V features with respect to fleet penetration

1.5.5   Role of infrastructure

Applications that can interface with infrastructure elements such as instrumented intersections or roadside units will be deployed as these structures become available. Infrastructure-based features are appealing to automakers since customers can potentially derive immediate benefits, given the availability of infrastructure hardware. Thus, in the deployment of V2I features, the only market penetration problem that needs to be solved is the deployment of the infrastructure hardware itself. Given that the number of hardware units required to support V2I features is much less than the number of communicating vehicles required to support V2V features, this proposition seems quite favorable. However, the deployment and maintenance of infrastructure hardware requires initiatives by government authorities or the establishment of agreements between automakers and private entities. The nature of these interactions can make it difficult to anticipate the availability of these services. This is in contrast with the deployment of V2V features, where automakers have control over the availability and timing of deployed systems.

The Vehicle Infrastructure Integration (VII) is a US government-sponsored activity that is oriented toward the nationwide deployment of DSRC-based communication infrastructure. This effort is leading toward a proof of concept demonstration and will ultimately lead to a field operational test. While VII is actively supported by all automakers, some have expressed concern about the ability of the government to deploy and maintain such an extensive infrastructure. Thus far, this uncertainty has made it difficult for automakers to build a business case around features that utilize communicating infrastructure.

A less ambitious (but more readily achievable) alternative that targets high risk locations (e.g. high-incident intersection locations or curve locations that have a large number of road departures) may be more feasible. Deployment of infrastructure to high risk locations will provide immediate safety benefits to early adopters. It would also be highly desirable that any national infrastructure that is deployed be able to support a nationwide security solution and provide positioning assistance in challenging environments.

1.6 Summary and Conclusions

In this chapter we have analyzed the characteristics of various communication-based automotive applications in a systematic manner and classified them into several major generic classes. Such an application characterization and classification effort facilitates the design and implementation of a network protocol stack for these applications. We first proposed a rich set of attributes of the applications, including both application characteristics and networking attributes, to better capture the properties of various applications. We then carefully investigated and analyzed the attributes of 16 V2X applications. We showed that these applications can be categorized into three major classes: short message broadcast type (for safety applications), on-demand short message unicast type (for convenience applications) and large-volume content download/streaming (for commercial applications). Finally, it should be clear that lists of performance metrics and QoS requirements for each type of application have to be developed; these are used to evaluate the performance trend of applications and network protocols.

The analysis of application characteristics and networking attributes, the classification of various vehicular communication applications, and the need for key performance metrics for each category of applications presented in this chapter, shed some light on our future task of developing a network protocol stack for various communication-based automotive applications. As the next step, the current effort of investigating the potential network solutions for these seven generic types of vehicle-related communication applications should continue, with the consideration of reusability of network protocol modules (or building blocks). To be specific, one should look into designs that decompose the network protocol stack into a set of mechanistic building blocks for different types of applications, so as to maximize the reusability of common building blocks for various applications.

In the development and deployment of communication-based V2X systems, there will be many other challenges to overcome. Some of the challenges that are ahead relate to the evolution of communication technology, positioning, and application specification and development. Deployment of a small number of driver assistance applications early in the next decade would facilitate development of additional applications afterwards that could take advantage of increased market penetration and the increasing availability of communication infrastructure.

Acknowledgements

The authors would like to thank our colleagues, C. Christopher Kellum, Donald K. Grimm, Priyantha Mudalige, Tamer Elbatt, and Varsha Sadekar, for their insightful discussion in the early phase of this work.

References

ABI Research (2009) Consumer Navigation Systems and Devices, The Changing Dynamics of Portable, Converged, OEM and Aftermarket Navigation. http://abiresearch.com.

Anda, J., LeBrun, J., Ghosal, D., Chuah, C.N. and Zhang, M. (2005) VGrid: Vehicular Ad Hoc Networking and Computing Grid for Intelligent Traffic Control. *Proceedings of the 61st IEEE Vehicular Technology Conference (VTC-2005-Spring)*, Stockholm, Sweden.

Bai, F., Sadagopan. N. and Helmy. A. (2003) The IMPORTANT Framework for Analyzing the Impact of Mobility on Performance of Routing for Ad Hoc Networks. *Ad hoc Networks* 1(4), 383–403.

Chen, W. and Cai, S. (2005) Ad Hoc Peer-to-Peer Network Architecture for Vehicle Safety Communications. *IEEE Communications Magazine* 43(4), 100–107.

Chennikara-Varghese, J., Chen, W., Altintas, O. and Cai, S. (2006) Survey of routing protocols for inter-vehicle communications. *Proceedings of the 3rd Annual International Conference on Mobile and Ubiquitous Systems – Workshops (mobiquitous)*.

CICAS (2009) Cooperative Intersection Collision Avoidance Systems (CICAS), USDOT Major Initiative. http://www.its.dot.gov/cicas/index.htm.

Das, S., Nandan, A. and Pau, G. (2004) SPAWN: A Swarming Protocol for Vehicular Ad Hoc Networks. *Proceedings of the 1st ACM International Workshop on Vehicular ad hoc Networks (VANET 2004)*, Philadelphia, PA, USA.

ElBatt, T., Goel, S., Holland, G., Krishnan, H. and Parikh, J. (2006) Cooperative Collision Warning Using Dedicated Short Range Wireless Communications. *Proceedings of the 3rd ACM International Workshop on Vehicular ad hoc Networks (VANET 2006)*, Los Angeles, CA, USA.

FCC (2003) Federal Communications Commission Notice of Proposed Rulemaking and Order FCC 03-324 NPRM.

IEEE 802.11 (1999) IEEE 802.11 WG, Part 11: Wireless LAN Medium Access Control (MAC) and Physical Layer (PHY) Specifications IEEE 802.11 standards.

IEEE P802.11p/D6.0 (2009) Draft Standard for Information Technology – Telecommunications and information exchange between systems – Local and metropolitan area networks – Specific requirements, Part 11: Wireless LAN Medium Access Control (MAC) and Physical Layer (PHY) specifications. Amendment 7: Wireless Access in Vehicular Environments IEEE 802.11 WG.

Korkmaz, G., Ekici, E., Ozguner, F. and Ozguner, U. (2004) Urban multi-hop broadcast protocol for inter-vehicle communication systems. *Proceedings of the 1st ACM International Workshop on Vehicular ad hoc Networks (VANET 2004)*, Philadelphia, PA, USA.

Kosch, T. (2005) Phase-Transition Phenomena with Respect to the Penetration Rate of DSRC Enabled Vehicles. *Proceedings of the 12th World Congress on Intelligent Transport Systems (ITS 2005)*, San Francisco, CA, USA.

LeBrun, J., Chuah, C.N., Ghosal, D. and Zhang, H.M. (2005) Knowledge-based opportunistic forwarding in vehicular wireless ad hoc networks. *Proceedings of the 61st IEEE Vehicular Technology Conference (VTC-2005-Spring)*, Stockholm, Sweden.

Lin, G., Noubir, G. and Rajaraman, R. (2004) Mobility Models for Ad hoc Network Simulation. *Proceedings of the INFOCOM 2004.*

Lochert, C., Mauve, M., Füssler, H. and Hartenstein, H. (2005) Geographic routing in city scenarios. *Mobile Computing and Communications Review* **9**(1), 69–72.

Nandan, A., Das, S., Pau, G., Gerla, M. and Sanadidi, M. (2005) Cooperative downloading in Vehicular Ad Hoc Wireless Networks. *Proceedings of the 2nd Wireless On-Demand Network Systems and Services (WONS'05)*, St. Moritz, Switzerland.

Picconi, F., Ravi, N., Gruteser, M. and Iftode, L. (2006) Probabilistic Validation of Aggregated Data for V2V Traffic Information Systems. *Proceedings of the 3rd ACM International Workshop on Vehicular ad hoc Networks (VANET 2006)*, Los Angeles, CA, USA.

Raya, M., Papadimitratos, P. and Hubaux, J.P. (2006) Securing vehicular communications. *IEEE Wireless Communications* **13**, 8–15.

Reumerman, H.J., Roggero, M. and Ruffini, M. (2005) The Application-based Clustering Concept and Requirements for Intervehicle Networks. *IEEE Communications Magazine* **43**(4), 108–113.

Sengupta, R., Rezaei, S., Shladover, S., Cody, D., Dickey, S. and Krishnan, H. (2007) Cooperative Collision Warning Systems: Concept Definition and Experimental Implementation. *Journal of Intelligent Transportation Systems* **11**(3), 143–155.

Shladover, S.E. and Tan, S. (2006) Analysis of Vehicle Positioning Accuracy Requirements for Communication-Based Cooperative Collision Warning. *Journal of Intelligent Transportation Systems* **10**(3), 131–140.

Taliwal, V., Jiang, D., Mangold, H., Chen, C. and Sengupta, R. (2004) Empirical determination of channel characteristics for DSRC vehicle-to-vehicle communication. *Proceedings of the 1st ACM International Workshop on Vehicular ad hoc Networks (VANET 2004)*, Philadelphia, PA, USA.

Torrent-Moreno, M., Jiang, D. and Hartenstein, H. (2004) Broadcast reception rates and effects of priority access in 802.11-based vehicular ad-hoc network. *Proceedings of the 1st ACM International Workshop on Vehicular ad hoc Networks (VANET 2004)*, Philadelphia, PA, USA.

VII (2009) Vehicle Infrastructure Integration (VII), USDOT Major Initiative. http://www.its.dot.gov/vii.

VSCC (2005) Vehicle Safety Communications Project, Task 3 Report, Identify Intelligent Vehicle Safety Applications Enabled by DSRC. http://www-nrd.nhtsa.dot.gov/pdf/nrd-12/1665CAMP3web/images/CAMP3scr.pdf.

VSCC (2006) Vehicle Safety Communications Project Final Report CAMP IVI Light Vehicle Enabling Research Program, DOT HS 810 591. http://www-nrd.nhtsa.dot.gov/pdf/nrd-12/060419-0843/.

Xu, Q., Mak, T., Ko, J. and Sengupta, R. (2004) Vehicle-to-vehicle safety messaging in DSRC. *Proceedings of the 1st ACM International Workshop on Vehicular ad hoc Networks (VANET 2004)*, Philadelphia, PA, USA.

Yin, J., ElBatt, T.A., Yeung, G., Ryu, B., Habermas, S., Krishnan, H. and Talty, T. (2004) Performance evaluation of safety applications over DSRC vehicular ad hoc networks. *Proceedings of the 1st ACM International Workshop on Vehicular ad hoc Networks (VANET 2004)*, Philadelphia, PA, USA.

Yin, J., Holland, G., Elbatt, T., Bai, F. and Krishnan, H. (2006) DSRC channel fading analysis from empirical measurement. *Proceedings of the 1st International Conference on Communications and Networking in China (ChinaCom '06)*, Beijing, China.

2

Governmental and Military Applications

Anthony Maida

EF Johnson Technologies

Wireless technology has been a fast growing consumer-based industry since its introduction in the 1990s. It was not until recently that the public safety and military sectors began catching up with this technology. Now that the eyes of a whole new breed of consumer have been opened, the public safety and military sectors have taken the wheel and begun to steer the current direction of its evolution. This chapter provides a background history of the uses and applications of Wi-Fi within these sectors as well as some insight into where it is going.

2.1 Introduction

The market trend in the Wi-Fi industry takes on a horizontal aspect as the same products marketed for commercial and home use are utilized in integrated applications such as the vehicular environment. The vehicular environment is a vertical market that has not been fully developed and may never reach its full potential, since companies do not tend to focus upon a single vertical market. Regardless of the market the end result that needs to be achieved is improved efficiency by way of wireless connectivity to data resources.

This chapter provides insight into the growth of vehicular networking systems for both public safety and military applications. It is structured as a chronological storyline to explain the historical significance of events and the response by the industry to supply the new needs of this market. Different network architectures currently in use are explored, as well as the future applications currently in design or prototype development stages. If not otherwise referenced, the description of all of the following application scenarios is based on the work experience of the author while working at 3eTI/EF Johnson.

Vehicular Networking   Edited by Marc Emmelmann, Bernd Bochow, C. Christopher Kellum
© 2010 John Wiley & Sons, Ltd

2.2   Vehicular Networks for First Responders

2.2.1   Public safety communications

When we think of public safety the first item that comes to mind typically is a police car or an ambulance. These vehicles are the mobile offices of our first responders, and for any office there are basic requirements. The first key requirement of any office is for two-way communications. This requirement can be met in a multitude of different ways. Until 1995, the only way such a communication link was established was through the automobile's radio. This radio provided the first responder with a link back to the central command center as well as to other officers in the vicinity. The shortfall of this communication system is that first responders could never be guaranteed 100% radio coverage within their vehicles.

In many other offices at that time, the personal computer was rapidly becoming a staple. In this vehicular office the cost of a laptop had been prohibitive until this point, but in 1995 the first laptop was installed into a squad car, utilizing what we would now consider primitive systems such as cellular digital packets transmitted at speeds up to 19.2 kbit/s, but that really transmitted the data at rates of 10 kbit/s due to protocol overhead and channel congestion. There were also public two-way wide area packet data networks, which transmitted at rates from 4.8 to 19.2 kbit/s (Miller 2009). These networks were better suited for short bursts of packets to transmit quick messages. There were other technologies present, but the costs were prohibitive for local and state government agencies; they were used by national governments. These systems included various satellite networks such as the IRIDIUM and Teledesic networks.

Many offices enjoyed an always-on Internet communications channel, but the mobile office of the first responders still needed to wait for a dial-up connection via the cellular network to be established. This was time consuming, and the need was there for an always-on Internet connection. Telecommunications companies were already hard at work on creating a vehicular modem for corporate customers, but began to target first responders as well. This modem would act as an intermediary between current vehicular equipment and the rapidly improving cellular networks. With serial interface conversion units, the RS-232 feeds on the in-car radio were able to be routed into and through these modems. Although the radios were not Voice over IP (VoIP) capable yet, it allowed for the radio firmware to be upgraded over the air by the central command office. The in-car laptop could also be routed through this modem, which negated the need for a separate cellular card for each laptop. While these in-car modems provided an always-on Internet connection, the first generation only provided a 14.4 kbit/s connection speed. While it was slower than other Internet connections of the time, it was better than the alternative of not having such a robust network in the vehicle.

Regardless of the network bandwidth, the public safety sector now had an avenue for accessing text-based reports of queries by vehicular patrols in real time (Miller 2009). This technology made first responders much more efficient at crime scenes in their communications, made routine traffic stops go faster and led to the capture of more fugitives wanted with outstanding arrest warrants, and to crimes being solved much more quickly. Furthermore, some basic reports no longer required officers to fill out manual paperwork in the precinct, since they were able to transmit them in text for electronic input into reports. The communications gap seemed to have been filled; however, there were still many more capabilities that could be added in order to go beyond just the requirement to communicate.

This was simply a stepping stone, which would pave the way to many different scenarios where vehicular wireless networks would be the solution.

This system went unchanged in capabilities for nearly seven years. The events that transpired on September 11, 2001 sparked a call to action by governments to the world's technology sector to improve the current infrastructure's ability to respond, save lives, and prevent terrorist attacks. Corporations became more sensitive and aware of the first responders' plight on the front lines as they protected public safety. Many responded to the call to action by sending sales engineers to various law enforcement offices to determine scenarios and requirements so that new products could be developed. One of the scenarios that most of the agencies in the USA had in common was the apparent inability to communicate across state lines or even to other agencies, due to changes in frequency and vehicles not being outfitted with the correct antennas. One product addressing those needs, which was developed in response to that situation, was the AirGuard^{TM} 3E-528Q Wireless Video Server by 3eTI (3eTI 2007). Even before this, the need for common radio communications had been identified: it was in the same year that a new cellular phone concept was introduced with a push-to-talk system (Nortel 2006). This technology allowed the same type of communication that was enjoyed by walkie-talkie users to be experienced by those with cellular phones, on a nationwide basis. Communication was as instant as a phone conversation, and yet the phone minutes were preserved. The vehicular communication system therefore needed somehow to be integrated with the cellular network.

2.2.2 Vehicular communications

Since the majority of American first responder vehicles had in-car modem systems that were aging and needed upgrading, the solution was fully to integrate the vehicular modem communication system with the in-car radio, and to relay the voice as VoIP packets to a central trunking system. This would process the data and route it to another trunking system, which in turn relayed the information back over the air as an analog signal. This analog signal was received by the vehicular communication system without the use of the Code Division Multiple Access (CDMA) network to receive it. When the first responder was out of the vehicle, the vehicle acted as the intermediary, as it was equipped with the analog to CDMA conversion hardware. So if the scenario called for the officers to be out of the vehicle, the radio would first relay through the vehicle and then follow the same path as if they were within it. At the same time, there was a need to be mobile and to maintain the network connection. The newly integrated in-vehicle modem also created a mobile hot-spot utilizing IEEE 802.11b and g networks (Miller 2009). Now the first responders were able to connect to their vehicle, which was the access point, just as they would at home or at the office. The vehicle had evolved from being a simple office to become a mobile routing hub for radio communications.

While the technology was emerging, new additions to current technologies within the vehicles were made, to improve the efficiency of the response times. Until 2003, when a call was placed to the emergency services by a person in distress precious minutes were wasted in attempts to locate the nearest responders and deploy them to the scene. The call center had no insight into the real-time location of their first responders without first putting out an all-call. The advent of in-car GPS and Wi-Fi locator services produced great strides forward in solving this problem. With an integrated approach, each vehicle outfitted with a modem was upgraded to make use of both the Wi-Fi-based triangulation locator algorithms and GPS-based location services. The position of each vehicle was relayed, in real time, back to the call center and displayed prominently upon a screen, complete with a map of the area. When a distress call came through, the address was input automatically by the enhanced call response system, and the real-time location of the closest vehicles was displayed on the screen. The closest responder was notified by radio, information was sent to their in-vehicle computers, and the responder was deployed to the scene in under one minute. Without the routing hub and locator services this step would have never been possible.

This routing hub idea, however, needed another step in order to go beyond just radio communications. Each car was a segregated office; communications needed to be routed to the central hub and then back over the same distance – to a vehicle that might have been only two feet away. The delay in this scheme was not in the best interest of these first responders, and there was a new problem to be solved: interconnection of nearby vehicles.

The first scheme that was tried was to have the first officer at the scene configure the in-car communications system as an access point; all subsequent responders would connect to a single router so that inter-client communication was possible (Miller 2009). While this worked in theory, it was impractical for a first responder to remember to reconfigure the car's network if a life-threatening situation was at hand and every second counted. This led to more investigation into the probable and feasible solutions to the problem.

Around the same time, in 2004, while the solutions were being investigated for inter-connectivity, the IEEE introduced a new standard, which held the key. IEEE Std 802.1D (2004) created a spanning tree protocol for nodes within a wired mesh network. There was no need for reconfiguration each time a new member was connected to the mesh via a wired interface; the mesh would self-reconfigure each time this was done. There would be no need to manually create, or have to reconfigure, a central hub (otherwise known as the root node).

Since these networks were already deployed in offices, the task now was to bring them to the vehicular environment. The spanning tree protocol did not discriminate between types of connection made between network nodes, since the wired and wireless connections looked alike (IEEE Std 802.1D 2004; IEEE Std 802.1Q 2006). There was one caveat: the router in the car needed to be both a wireless client and a wireless bridge simultaneously. The client side connected to the main network and allowed communications back to the central command hub. The bridge side was needed to create an ad hoc local network when other vehicles were in range, to allow for Vehicle-to-Vehicle (V2V) communications. Once first-responder vehicles were equipped with these new modems the capability of forming mesh networks became a reality.

The mesh network allowed first responders the speed they craved to share information in real-time without jitter or lag. Surveillance cameras outfitted within these vehicles were able to stream live video to colleagues, and radio silence could be maintained while communications were held through VoIP meetings in the vehicles. In addition to this, city cameras were being upgraded to include wireless access for first responders. This allowed for real-time situational awareness through the use of strategically placed cameras around the metropolitan area, by streaming the video into the comfort and safety of the vehicles. Life-saving decisions could be made in such a way that no additional risk needed to be taken. The vehicle was now a rolling fortress of an office for first responders; everything was at their fingertips, and situations could be defused from the vehicle.

2.3 The Need for Public Safety Vehicular Networks

Wireless communications in the public safety heavily depends on the robustness, reliability, and availability of the communication system. In the past decades this was achieved at the price of extremely high system cost, and was often based on specialized solutions that lacked interoperability. Faced by severe cost constraints, the need to ensure interoperation of various agencies, and the desire to involve existing infrastructures where available, the public safety community is increasingly attracted by the opportunity to utilize off-the-shelf technology in conjunction with both specialized and commercial communication systems.

The most basic communication need of the public safety community is radio-based voice communications. This type of communication allows dispatchers to direct personnel to areas where incidents have occurred. The trend in this marketplace has been geared towards allowing for inter-agency communication in the case of large scale disasters. The most notable large scale response effort occurred on September 11, 2001, when multiple agencies responded to the terror attacks in New York. The state of the most basic radio technology could not meet the increased demand for radio communications that arose on that day. The crush of radio communications flooded the spectrum, and caused massive failures across the board with regard to the relaying of crucial information, which could have saved more lives. This lack of communications led to more deaths of first responders, since there was no way to relay information about the collapse of the South Tower of the World Trade Center to those still embedded within the North Tower. The most gripping issue regarding the state of the technology at this time was the fact that the same failures had occurred in 1993 and nothing had been done to address the issue. More focus had been put on developing faster and more reliable communications for a more lucrative consumer market, and this area had been forgotten.

The reasons behind this failure cannot be attributed to a single root cause. Many agencies were still using unreliable analog radios within their vehicles, with no backup system. Cellular towers were damaged in the attacks, resulting in loss of alternative communications. The smoke and concrete dust increased the attenuation of the air, creating a communications blackout for those who were at ground zero. The New York Fire Department continued to utilize a single communication channel, which resulted in bandwidth overload. Emergency Medical Service (EMS) workers also relied heavily upon the city-wide channel, and caused similar overloads and communication failures due to spectrum congestion. The radio spectrum issues are noted in Table 2.1.

Each channel in the VHF spectrum uses roughly 12 MHz, and from the table ranges indicated it can be determined that there were two channels in the 25–50 MHz range and two more in the 150–174 MHz range. In the UHF range the channel usage is a little more than doubled, since it requires roughly 26 MHz for each channel. In the 450 to 512 MHz range this amounted to another two available channels for public safety use. The use of the 700 and 800 MHz spectrum was not available in urban areas due to reception problems (Homeland Security, US Dept. of 2009). Realistically, in a catastrophic event for which hundreds of first responders are on the scene, far more than six communication channels would be needed to keep all of them informed of life or death situations.

Radio was the primary medium for the transmission of voice communications. Later developments allowed for the transmission of voice and data over the same radio spectrum. The problem was that the only people capable of receiving these transmissions were other first responders in the same department. There was an inability to communicate across different

Table 2.1   UHF and VHF frequency allocation and issues in the USA

Spectrum	Frequencies	Issues
VHF	25–50 MHz	Used by many commercial applications, and experiences overcrowding
		No public safety quality radios being produced today
	150–174 MHz	Inadequate capacity
		Inefficient allocation
UHF	450–512 MHz	Extremely crowded in metropolitan areas
		Heavily occupied in other areas
	700 MHz	Blocked by TV stations
		Canadian/Mexican border issues
		Commercial equipment causes interference
		Cost prohibitive
	800 MHz	Very limited capacity
		Harmful interferences from commercial users
		Cost prohibitive

Data compiled from FCC Website (2009) and FCC: Office of Engineering and Technology, Policy and Rules Division (2009)

departments or agencies for coordination during a disaster. The conventional radio system typically had three segregated channels: car to station, station to car, and car to car. There was also a shortfall due to the fact that personnel must wait for a transmission to complete prior to being able to send their own transmissions, since the channel only allowed for one speaker at a time. A vehicular mesh network would have allowed for additional channel resources for voice communication. Further, a video channel could have been set up with real-time situational awareness, with a tie in to vehicle cameras. Short message services through the use of private messaging networks would also have been available in the event that a voice channel was unavailable, thus allowing for vital information to be relayed immediately rather than waiting for a chance to transmit.

Trunking systems were developed to be a viable solution to ease the frequency shortages and increase the efficiency of the radio spectrum. This was the first step in allowing multiple agencies to use common radio communications by way of a computer-controlled system, which kept all the available frequencies in a pool and allocated an open frequency when a person tried to talk. Talk groups were created to achieve a one-to-many type of transmission, in which a conference call over the radio could be conducted. While being an improvement to efficiency overall, these trunking systems came with a steep price tag. This was due to their need for an infrastructure change, which completely replaced the relay systems already in place in cities, and to the need to upgrade radios to be compatible with a new trunking system. Some departments were fortunate enough to have compatible equipment, and just needed to pay for the replacement of the radio relay systems; most were not so fortunate, and could not justify the cost. Many agencies were forced to use conventional radio because no other cost-effective alternative was readily available. This was the case until wireless networking entered the marketplace, and the ability to utilize these devices to create vehicular networks was created. Vehicular networks paved the way towards bridging the communication gaps, increasing efficiency, and finally creating an office feel in a vehicle that previously had had limited access to data.

2.4 State of Vehicular Network Technology

Vehicular networks are becoming an increasingly popular area of exploration, since the ability of vehicles to establish V2V communication links will contribute to safer and more efficient emergency response in times of need. The current technologies in use in first responder vehicles address the following challenges faced by public safety agencies:

- Lack of inter-agency voice communication interoperability

- Bandwidth constraints, which limit traffic on proprietary narrowband networks

- Inability to access criminal or medical records during a response

- Necessity for office resources to complete field reports

- Lack of full situational awareness of an incident

- Need for rugged and interoperable standards-compliant networks

Many of the devices designed by companies create an applications-oriented environment with these shortfalls in mind. The current technology that is emerging is driven by the Statement of Requirements (SoR) set forth in 2004 by SAFECOM (Homeland Security, US Dept. of 2009). The main vision of these requirements is to create networks that have the ability to interoperate seamlessly for the transmission and reception of data.

Public safety vehicular environments mainly make use of the following three types of network architecture: Incident Area Networks (IANs), Jurisdictional Area Networks (JANs), and Extended Area Networks (EANs) (Miller 2009).

2.4.1 Incident Area Networks

Incident Area Networks allow for the creation of temporary network infrastructures while responding at a scene. These networks make use of existing IEEE 802.11 wireless local area networks and wireless ad hoc networking technologies that were originally designed with the consumer in mind. IANs are designed to be capable of adapting to the physical circumstances of any given incident. This network's primary intention is to create an architecture capable of providing connectivity among public safety communication devices that enter and leave the scene of an incident. These networks are easily achieved by utilizing wireless local area network devices based upon the IEEE 802.11 set of evolving standards. It is the ad hoc mode of networking that has sparked much interest in the applications of IANs, as more and more public safety officials seek a 'mesh' networking capability with these devices to create IANs effectively (Miller 2009). Typical scenarios faced by first responders are active shooter situations, hostage situations, barricade situations, drug sting operations, Hazardous Materials (HAZMAT) spills, and special events.

As IANs can be set up as secluded from or included in a backbone network, the versatility of such a network proves its worth in emergency response situations. IANs have the flexibility to utilize multiple standards, such as IEEE 802.11a for bridging and IEEE 802.11g for communication. When a backbone network inclusion is desired, a single IAN-capable device in the network would connect via Ethernet to another infrastructure device offering an Internet connection (Miller 2009), which is the premise of forming a JAN – explained later in this chapter (see Section 2.4.2). The ideal IAN is self forming and requires little or no setup in order to be operational. Some vehicular networking devices have this self-forming mesh capability while others do not. Those vehicles that do have the self-forming mesh capabilities are able to create an IAN with zero setup and provide the responders with video, geolocation information, telephony, video conferencing, instant messaging, database translations, paging, file transfers, and web browsing. The IANs combine the benefits of traditional home gateways with the architecture of an ad hoc bridge to create a shared system that allows for seamless integration with larger voice and data systems across different states. Other scenarios have deployable quick kits, which are pre-configured with operating modes, channels, and security keys. These kits will include mobile access points, a battery pack for power, and one or two cameras on a tripod. All that is required for the setup is to position the kits, point the camera, and flip the power switch. Within minutes these kits provide a video and audio stream directly into the vehicular networks of the first responders.

The state of the technology that is used in these IANs is changing slowly. The current state makes use of dual radio access points for the kits and dual radio routers for the vehicles. In the dual radio architecture one radio provides the link to the backbone network and maintains connectivity to outside networks and information. The second radio card provides the ad hoc bridging capability. In order for a mesh to form, each vehicle must be outfitted with same-vendor equipment, since the formation of the mesh protocol is still not standardized and each vendor uses a different methodology for formation and maintenance of these mesh networks. This technology is in the process of moving towards a standard that will allow for the use of a multi-vendor network for the IAN. This step hinges upon the ratification and adopting of the IEEE 802.11s (IEEE 802.11s-D0.3 2006) standard, which addresses the need for standardizing the mesh networks. Other vendors have been making strides in utilizing a single radio platform to achieve both the network connectivity and bridging capability. This is achieved through simultaneous virtual connections using a single radio. The shortfalls of this type of network that are currently being addressed are the reduction in bandwidth capability and the loss of data stemming under heavy traffic loads (Camp and Knightly 2009).

2.4.2 Jurisdictional Area Networks

Jurisdictional Area Networks provide a permanent network infrastructure for the transmission of voice data from in-car radio systems. This network infrastructure type utilizes IEEE 802.16e (incorporated in IEEE Std 802.16 2009) mobile broadband wireless networking and mesh networking technologies as a standards methodology for interoperability across multiple vendors (IEEE 802.11s-D0.3 2006). This network architecture complements existing EANs, which will be discussed later in this chapter (see Section 2.4.3). As the IAN architecture is intended to be a temporary network infrastructure, its facilities are designed to utilize existing permanent EANs or JANs in order to maintain network connectivity to Internet resources. Therefore, by definition an IAN must be capable of setting up and maintaining connectivity to the JAN. The JAN infrastructure has historically consisted of analog voice capabilities set up to cover a specific geographic area using Land Mobile Radio (LMR). In recent years the need has arisen to integrate broadband services by way of cellular communications to complement the range and capabilities in a JAN (Miller 2009).

Systems currently exist that utilize 802.11 technologies in the unlicensed 2.4 GHz and 5.8 GHz bands to provide this networking capability within the JAN. For example, EF Johnson set up such systems for the municipalities of Toronto, Canada, and Baton Rouge, LA, USA, as well as the US Secret Service in Washington, DC. This approach has proved to be limited by connectivity issues in remote areas where a wireless infrastructure access point cannot be contacted to provide Internet connectivity in a JAN. Efforts have been made to provide public-safety-specific frequencies in the 700 MHz and 4.9 GHz spectrum, since these frequencies are not shared with commercial or consumer markets. The behavior in the 4.9 GHz spectrum is not at all different from that of the current 2.4 GHz wireless devices, except that special equipment is needed to access these devices. It is the moves towards a more unified system in the 700 MHz part of the spectrum that have been hindered in the United States by the delays in the transition from analog to digital television (FCC Website 2009). Each delay has reduced the opportunity to create test-beds and to achieve more advancement through the observation and use of a live system.

The JAN has a primary purpose in mind: to provide reliable connectivity in a mobile environment, and to provide network access, when needed, to temporary IANs. One of the most surprising results was a move into the 700 MHz UHF band as an alternative to the microwave GHz frequencies upon which current wireless devices operate. It is the theoretical range that can be achieved by 700 MHz systems that appeals the most to public safety officials tasked with the creation and integration of a JAN that will effectively tie into a nationwide system. When a cell tower is placed in a region with open space and minimal obstruction, a range of approximately 160 km can be achieved in all directions. Even in a metropolitan area or a mountainous region, an impressive 55 to 65 km of range can be achieved through the use of a single cell tower. This reduces the need for multiple cell towers to create stationary infrastructure to which the vehicular networks can connect.

The current state of this technology is still in the development stages (as of the Spring of 2009, when this chapter was written). The most notable development project is being headed by Motorola in Pinellas County, Florida and is known as the Greenhouse Project (Motorola Corporation of Schaumburg, IL 2009). This project makes use of a prototype system that supports transmission speeds of 460 kbit/s to provide wireless voice and data communications in the 700 MHz band. This prototype utilizes what will soon be known as public safety channels 63, 64, 68, and 69 and makes use of an improvement to Time Division Multiple Access (TDMA), known as the Scalable Adaptive Modulation (SAM) air protocol.

The SAM air protocol will be the standard method of wideband transmission as adopted by the Telecommunications Industry Association (TIA). Existing IP networking standards and technologies are utilized as well, including VoIP with full duplexing and streaming video. Potential use cases of the system have been under test since its installation in 2001, and include transmission of a child's picture during an amber alert, surveillance videos of robberies immediately after they occur, blueprints to firefighters, fingerprints of arrested suspects, and live video feed of pursuits; video conferencing between officers on the field and the command center; and the ability to conduct remote situation analysis. The prototype system is installed in selected Pinella's County Sherriff's Office vehicles, an EMS ambulance, a Fire Department Rescue truck, a deputy chief vehicle and a mobile unit. Each of these installations provides the ability to operate and access the system through a touch-screen user interface to communicate and share information with other agencies (Motorola Corporation of Schaumburg, IL 2009).

The continued evolution of JANs is hindered with regard to two distinct aspects. First, the public safety organizations crave a fully interoperable system by which a user can seamlessly transition their vehicle's network from a 700 MHz system to a 4.9 GHz system and finally to a 2.4 GHz system without the need for different devices to achieve each step. The Federal Communications Commission (FCC)'s emission mask standards make this requirement impossible to achieve, since they are incompatible with the emission standards of the 802.11 community. A mask comparison of 4.9 GHz and DSRC-A through D, to illustrate how the signals are currently incompatible, is shown in Figure 2.1.

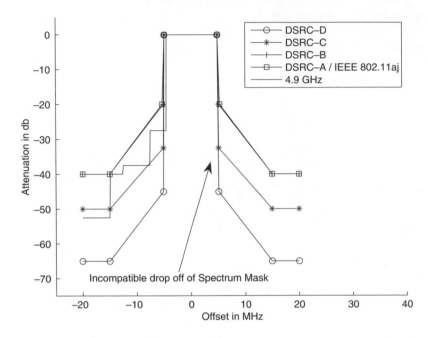

Figure 2.1 Spectrum mask comparison for different DSRC Channels

The other shortfall is that these bands are licensed and the licensing costs imposed by the FCC could be prohibitive to agencies that do not make use of technological advances, such as smaller precincts in country settings in the United States. A further complication is the tight economy that is being faced by leading nations at the time of writing. The leaner budgets of technology companies have hindered the advancement of such technologies as WiMAX and Mesh Networking standardization, two key components to the creation of a cost-effective JAN. The dilemma that public safety officials are facing at the present time is whether to spend their resources on currently available 2.4 GHz systems that do not meet all their needs or to wait for the better solution to arrive in the 4.9 GHz and 700 MHz systems. Even if they waited, the licensing issue might still be prohibitive to the creation of a JAN, since the licenses are exclusive and each end user must have one. If the FCC relaxed the regulations that were imposed and created non-exclusive transmission licenses with no requirement for users to have a license, then the JAN in the 700 MHz and 4.9 GHz spectrum would become a reality. The JAN architecture is a key component to creating an EAN to tie in all resources from coast to coast (Miller 2009).

2.4.3 Extended Area Networks

An Extended Area Network is the integration of regional, state, and national network resources as applicable to public safety. These networks are used instead of, or as a part of, a current LMR system and an IEEE 802.16e protocol device. The EAN encompasses both the JAN and the IAN and provides coverage into the neighboring jurisdictions. Figure 2.2 shows the incorporation of all three types of network infrastructure.

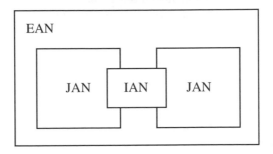

Figure 2.2 Conceptual structure of extended, jurisdictional, and incident area networks

As depicted in Figure 2.2, the JAN and IAN are simply subsets of the EAN. The EAN is predicted to utilize similar infrastructure equipment to the JAN, but the wireless type of EAN is not something that is in existence today. The current role is played by the Internet and regional public safety nets. Until the implementation of a full scale WiMAX system, the EAN will continue to be utilized as a wired venue to which the JAN connects.

Since the network architecture currently in use still relics heavily upon commercially available Wireless Local Area Network (WLAN) products, the need for a vehicular standard has arisen. The IEEE has been working on an amendment to the 802.11-2007 standard by way of 802.11p (IEEE P802.11p/D6.0 2009). The p standard sets in motion a way to add wireless access in vehicular environments. It defines the enhancements to the standard that are required to support the hand-off of vehicles traveling at high speeds between roadside infrastructure operating in the 5.9 GHz spectrum. Once this standard is in place, and commercially available 802.11 products incorporate the enhancements into the baseline, then the creation of JANs utilizing non-public-safety channels for vehicles can be achieved, while the public-safety-specific equipment is created in the years to come. The creation of a fully linked vehicular network capable of handling transmission of high speed data at vehicle speeds will enhance the ability of first responders to share information in real time with other agencies and organizations (Jiang and Delgrossi 2008).

One routine situation that these networks will make a reality is the response of EMS units to a medical situation. The ambulance would be equipped with network connectivity, allowing the paramedic team to create a link to the hospital staff en route, to transmit patient vital signs in real time. The patient information and vital signs would be recorded through wireless equipment, and voice recordings would be cataloged in real time to the hospital system, thus eliminating the need for paper notes and reports to be handed over once the patient arrived. As the monitors and vital equipment would be wireless, they would provide for easy transportation along with the patient into the hospital, where the paramedic team could replace the equipment currently connected to the patient with a monitor and equipment that might have been brought in with a previous patient, thus eliminating the need to disconnect and reconnect a critically ill patient. As time is an important commodity in this type of situation, a saving of precious minutes can be achieved through this scenario, which could thus lead to more lives being saved.

Another routine scenario is the response of firefighters to a residential fire. Once a call reporting the fire was placed to the call center, and the responders were dispatched in a network-equipped emergency vehicle, the GPS location of the responders would be tracked in real time by the call center. EMS support personnel in similarly equipped vehicles would also be tracked as they responded to fire victims, and their location and status would be relayed to the firefighters, hospitals, and incident commander. While the firefighters were en route, the floor plans, unit occupants, fire hydrant locations and pressures, and any pertinent medical information would be relayed automatically to the in-vehicle computers by the incident commander to give a real-time situational awareness. The first responders at the scene would create an IAN with either vehicle-mounted cameras or ready-kits that would have the ability to be set up within minutes to provide a live video feed to other responders and the incident commander. All the items tied in together would create a more efficient response and increased awareness, and lead to the potential saving of more lives in an otherwise unfortunate situation.

The most common scenario is the routine traffic stop by law enforcement officials. A network-connected police vehicle would have the ability to access nationwide information during a traffic stop, to perform a thorough check for outstanding warrants for the suspect without the need to call the command center to process the same information. Should the search yield a positive confirmation of an outstanding warrant, full details including pictures, fingerprints, and crime information would be relayed in real time to the vehicle's computer. The suspect would then be arrested and outfitted with an Radio Frequency Identification (RFID) tag, which would track the suspect should they have the ability to escape custody while in transit. During the arrest, the questioning, Miranda warnings, collection of evidence, and other routine arrest activities would be digitally recorded and transmitted in real time to the command center. Fingerprints would be read into the system at the scene to reduce processing times of arrest suspects. Once en route to the county facility, the officer's vehicle would be tracked in real time, along with a recording of any in-vehicle conversation the suspect might have with the arresting officer. Once the suspect was processed, all recordings would be handed over to the prosecutors for use as evidence during a trial. This would make the need for paper reports during arrests unnecessary, and allow an audit trail to be created with ease.

2.5   Vehicular Networks for Military Use

The public safety use cases discussed above sparked some interest within the military community for other applications utilizing the same type of network infrastructure. The applications in which military users are most interested are integrated detection systems, rapid deployment video systems, in-vehicle information sharing, and vehicular tracking.

In the integrated detection systems, roadside infrastructure would be deployed in a mesh network architecture. Detection equipment such as chemical detectors or motion sensors would be connected to the roadside units through serial or Universal Serial Bus (USB) ports housed within the access points. The roadside units would relay real-time information to base camps as well as to equipped military vehicles. Within an equipped military vehicle a network connection would be established to the mesh network of roadside units. Real-time information would be updated on an in-dash screen, which would be overlaid upon a moving map. Each roadside unit would create a green, yellow, or red area upon the map, based on current and historical information. A green zone is one in which no incidents have occurred and no danger has been detected currently. A yellow zone is one in which a previous incident has occurred or where danger has been detected within the past 24 hours. A red zone is one in which imminent danger exists and sensor alarms are currently activated. The soldier would be able to access red zones through the touch screen user interface to query the sensor reporting the alarm. If the motion sensor had been activated, the soldier would have access to a mounted camera with pan–tilt–zoom capabilities, to perform a remote visual sweep of the area from within the vehicle. If the chemical detector's alarm had been activated, the soldier would have information about the type of chemicals detected. The in-vehicle system would also access the Internet through the same network to find out which weapons could have been created using the detected chemical, in order to create more of a heightened awareness on the part of the soldier. The Internet connectivity would be achieved by ensuring that at least two root units have backbone network connectivity, which would allow access through the use of a multi-hop path within the mesh network.

An extension of this setup would be the ability of convoys of military vehicles to communicate with one other over a secured wireless link utilizing the in-vehicle network and related hardware. Video conferencing between commanding officers and members of the convoy would be conducted to give real-time instructions while en route to hot zones. Tie-in of these conferences with the national defense headquarters would be achieved as well, to allow for direction and information from military intelligence. Real-time downloads of satellite imagery as well as real-time video surveillance imagery from an IAN or even Unmanned Aerial Vehicles (UAVs) could be streamed to the vehicles. The benefits of such an application in a war zone would be tremendous, and it would give the soldiers a better chance of survival when the full picture of a situation was readily available.

It may seem that the military applications differ from the public safety applications, but in reality the same capabilities that are available and used for public safety are also used by the military. There are no added capabilities or features other than a more robust security encryption methodology. Apart from this, the same products that are used by first responders are in turn sold to the military, with an added security encryption module. The security required for military use is what makes the choice of vendors for military applications sparse. Most companies do not want to invest the added cost of going through a certification process in order to gain a military customer: the state and local governments are sufficient to meet the goals of the company. This is the primary reason why military vehicular networks closely mirror those of first responders with respect to capabilities and applications.

However, since security is a major concern for the military, the devices have had to be compliant to a special form of IEEE 802.11i (incorporated in IEEE 802.11 2007) encryption defined by the United States National Institute of Standards and Technology (NIST) as FIPS-140, which stands for Federal Information Processing Standards. This standard has four levels of certification, which are defined as Levels 1 through 4.

- Level 1 is the lowest level of security, and is only for prototype equipment – normally equipment that is being evaluated by the government while security mechanisms are defined.

- Level 2 has physical security by way of tamper-evident security and/or pick-resistant locks, and is ideal for role-based authentication.

- Level 3 has the same means of physical security as Level 2 does, and also provides for identity-based authentication, which is typically achieved through the use of biometrics or a smart card.

- Level 4 is the highest level of security. It provides for a physical envelope of protection around the cryptographic module, and has security requirements during the production stage of the equipment. It utilizes multi-level authentication, which is a combination of role and identity through the use of biometrics and certificates.

The most commonly used level is the second, and this is most commonly referred to as FIPS-140-2 (2001). The FIPS-140-2 cryptographic module can be achieved in one of two ways: a separate FPGA module or a software-based module. In the separate FPGA module setup the throughput speeds can be maintained at between 29 and 33 Mbit/s, while on the software-based module setup the typical throughput speed is between 11 and 17 Mbit/s. Most vendors sell systems that make use of the software-based module, since it does not require a redesign of existing commercial products and the baseline software can be maintained in two branches to simplify operations. As the military applications do not require high bandwidth use in vehicular environments, these throughput speeds are acceptable. The reason behind this lower bandwidth requirement is twofold. First, the military transmits primarily text and binary messages, which require low throughput speed of less than 1 Mbps. Secondly, when voice and video are used in conjunction with specific compression standards the requirement is still less than 1 Mbps. The low bandwidth requirement is also aided by the fact that most battle situations occur in remote areas not congested by Wi-Fi signals, creating less likelihood of packet collisions and loss of data.

2.6 Conclusions

Technological progress in wireless networking of the vehicular environment for the military and for public safety has come a long way. Now that the infrastructure is in place for mobile networking, and that the near future will incorporate the much-needed interoperability features with the introduction and implementation of the IEEE-802.11s standard for mesh networking, companies are looking at the next opportunity to extend the vehicular network: the driver and passengers. For both public safety and military uses the need exists to create wearable wireless devices that will relay vital statistics to the vehicular network, which in turn will relay this crucial information to command centers. Wearable wireless devices will lead to wireless cameras, also integrated into the same wearable system, which will relay real-time video in situations where radio communications cannot be achieved due to risk. As these future advancements are outside the scope of vehicular network applications, it is this type of forward thinking that is alive in the industry today. Vehicular networks have created the pathway necessary to open the door to future technology that further saves lives of the people on the front lines of life.

References

3eTI (2007) AirGuard™ 3E-528Q Wireless Video Server, Product Sheet.
    http://www.efjohnsontechnologies.com/resources/dyn/files/79140/_fn/3e-528Q.pdf.

Camp, J. and Knightly, E. (2009) The IEEE 802.11s Extended Service Set Mesh Networking Standard. http://networks.rice.edu/papers/mesh80211s.pdf.

FCC: Office of Engineering and Technology, Policy and Rules Division (2009) FCC online table of frequency allocations. Technical report, FCC.

FCC Website (2009) 700 MHz Public Safety Spectrum. http://www.fcc.gov/pshs/public-safety-spectrum/700-MHz/.

FIPS-140-2 (2001) Security Requirements for cryptographic modules.

Homeland Security, US Dept. of (2009) SAFECOM Homepage. http://www.safecomprogram.gov/SAFECOM.

IEEE 802.11 (2007) Wireless LAN Medium Access Control (MAC) and Physical Layer (PHY) Specifications.

IEEE 802.11s-D0.3 (2006) ESS Mesh Networking, Draft Amendment to Standard for Information Technology – Telecommunications and Information Exchange Between Systems – LAN/MAN Specific Requirements – Part 11: Wireless Medium Access Control (MAC) and physical layer (PHY) specifications.

IEEE P802.11p/D6.0 (2009) Draft Standard for Information Technology – Telecommunications and information exchange between systems – Local and metropolitan area networks – Specific requirements, Part 11: Wireless LAN Medium Access Control (MAC) and Physical Layer (PHY) specifications. Amendment 7: Wireless Access in Vehicular Environments IEEE 802.11 WG.

IEEE Std 802.16 (2009) IEEE Standard for Local and metropolitan area networks Part 16: Air Interface for Broadband Wireless Access Systems. *IEEE Std 802.16-2009 (Revision of IEEE Std 802.16-2004)*.

IEEE Std 802.1D (2004) IEEE Standard for Local and metropolitan area networks Media Access Control (MAC) Bridges. *IEEE Std 802.1D-2004 (Revision of IEEE Std 802.1D-1998)*.

IEEE Std 802.1Q (2006) IEEE standard for local and metropolitan area networks virtual bridged local area networks. *IEEE Std 802.1Q-2005*. (Incorporates IEEE Std 802.1Q1998, IEEE Std 802.1u-2001, IEEE Std 802.1v-2001, and IEEE Std 802.1s-2002.)

Jiang, D. and Delgrossi, L. (2008) IEEE 802.11p: Towards an International Standard for Wireless Access in Vehicular Environments. *Proc. IEEE Vehicular Technology Conference VTC Spring 2008*, pp. 2036–2040.

Miller, L.E. (2009) Wireless Technologies and the SAFECOM SoR for Public Safety Communications. http://www.antd.nist.gov/wctg/manet/docs/WirelessAndSoR060206.pdf.

Motorola Corporation of Schaumburg, IL (2009) Greenhouse Project Wideband Data Technology http://www.safecomprogram.gov/NR/rdonlyres/A9AA7144-4311-8D7E-7DFD9613A0E3/0/greenhouse_project.pdf.

Nortel (2006) Deliver Converged Multimedia Services across any Access or Device Product Flyer, http://www.nortel.com/solutions/cms/collateral/nn116741.pdf.

3

Communication Systems for Car-2-X Networks

Daniel D. Stancil

Carnegie Mellon University

Fan Bai

General Motors

Lin Cheng

Trinity College, Hartford, CT

This chapter provides an overview of the architectures and environments for Vehicle-to-Infrastructure (V2I) and Vehicle-to-Vehicle (V2V) communication. Antenna requirements and channel properties are discussed with an emphasis on the car-to-car environment. The impact of channel properties on Orthogonal Frequency Division Multiplexing (OFDM) is discussed together with its implications and recommendations for the IEEE 802.11p standard.

The protocol requirements for Dedicated Short Range Communications (DSRC) applications differ substantially from those of conventional wireless networks. The use of multiple channels for medium access is discussed, as well as broadcast and unicast protocols. Single-hop and multi-hop protocols are discussed, including techniques for enhancing the reliability for safety-critical applications. Next, the technical portion ends with a discussion of network

Vehicular Networking Edited by Marc Emmelmann, Bernd Bochow, C. Christopher Kellum
© 2010 John Wiley & Sons, Ltd

layer challenges related to mobility, such as mobile IP and dynamic address allocation schemes. The chapter concludes with a discussion of future directions and challenges.

3.1 Overview of the Vehicle-to-Vehicle/Infrastructure Environment

An aspect of Intelligent Transportation Systems (ITSs) that shows significant promise is wireless communications with cars. This communication could be between cars (V2V) or between cars and a fixed infrastructure (V2I) such as Road-Side Units (RSUs). We refer to these possibilities generically as Vehicle-to-Vehicle/Infrastructure (V2X) environments. These types of applications are also referred to as DSRC applications.

The future growth of DSRC applications is anticipated to take place at frequencies near 5.9 GHz, where various agencies around the world have allocated bands for this purpose. Table 3.1 lists the allocated bands in various parts of the world. The physical layer specification proposed in the draft standard IEEE 802.11p is OFDM patterned from the popular 802.11a/g standards. However, to increase the tolerance to multipath in the outdoor environment, the waveforms are scaled from the 20 MHz used in the a/g standard to either 10 MHz or 5 MHz. Other parameters such as the number of carriers and modulation formats remain unchanged.

Table 3.1   DSRC bands

Region	Standard	Frequency (GHz)
Europe	EN 12253	5.795–5.815
ITU-R	ITU-R M.1453-2	5.725–5.875
Japan	ARIB T55	5.770–5.850
North America	ASTM E 2213-02	5.850–5.925

3.1.1   Vehicle-to-Infrastructure

V2I applications such as toll-booth payments have been deployed for a number of years. Most of these deployments have used the 915 MHz unlicensed band. However, a number of additional applications related to safety and service/infotainment could also be supported. For example, intersections equipped with RSUs could alert a driver as to other cars approaching the intersection, although they may not be visible around a corner. Other RSUs could provide information about local attractions, or provide portals to the Internet.

3.1.2   Vehicle-to-Vehicle

V2V communications have yet to be deployed in other than small-scale experimental configurations, but they also offer a number of attractive potential applications. For example, sudden braking in response to an accident or road hazard could automatically initiate a wireless alert message that is relayed down the highway to warn approaching drivers of the need to slow and approach with caution. Information about major congestion or delays caused by accidents or construction could also be relayed kilometers down the highway, allowing drivers to leave and find alternative routes. Alternatively, cars traveling together could share an Internet portal, or passengers could play multimedia games between cars.

However, V2V communications faces a challenging 'chicken and egg' problem: to offer useful functions, a significant fraction of the cars on the road must be equipped with the wireless capability. There is little motivation for early adopters to purchase wireless capability if there are few other equipped cars on the roadway. In contrast, if a government agency or service provider installed RSU infrastructure, wireless capability in the car would provide immediate benefits. Consequently, deployment of the technologies may require government agencies or service providers to take the first step. V2V communication would then be an emerging capability as the installed base of wireless-enabled cars grows.

3.1.3   Antenna requirements

Traditionally, two-ray radio and mobile telephone antennas on cars have been vertical monopoles or collinear antennas with phasing coils. These provide approximately omnidirectional radiation patterns. Further, the finite size of the ground plane formed by the metallic car surfaces causes the direction of maximum gain to be tilted above the horizontal. These attributes are appropriate for these applications, since the base station could be in any direction, and is usually at an elevation considerably higher than the car. These traditional mobile antenna patterns are illustrated in Figure 3.1, where (a) shows the omnidirectional pattern used by a traditional mobile antenna (dashed) and an elliptical pattern (solid) that would provide better down-road range while still providing adequate cross-road coverage, and (b) shows the vertical pattern for a quarter-wave monopole above a 1 m diameter ground plane at 5.9 GHz. Note that the gain in the horizontal direction is about 4.6 dB lower than the peak gain.

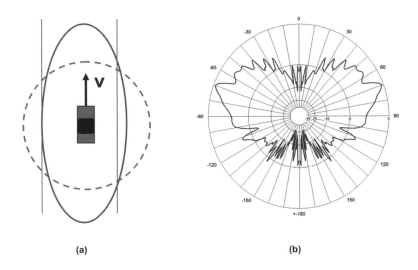

(a)   (b)

Figure 3.1 Antenna patterns for car-2-X communications

V2X applications have significantly different requirements. First, the difference in height of the antennas on each end of the link is typically much smaller. In the case of V2V communication, the upward tilt of the antenna beam could increase the link loss by 6 to 12 dB (Figure 3.1b). If the signal power falls off as $1/r^2$, this reduces the range by a factor of 1/2 to 1/4. Second, the expected distance to a desired point of radio contact is much longer in the direction of the road than it is transverse to the direction of the road. Consequently, some type of elliptical or cigar-shaped beam would be preferred, with the major beam direction aligned to the front and rear of the car (Figure 3.1a). Assuming an additional 6 dB in link loss from the use of omnidirectional antennas, we conclude that antennas designed with horizontal beams favoring the forward and reverse directions could increase the range by as much as a factor of eight. Alternatively, for a given communication distance, the required transmission power could be reduced by a factor of 1/64, or almost two orders of magnitude. At long distances where the path loss exponent may be closer to 4, the required power could be reduced by almost four orders of magnitude. This reduced power requirement will be particularly important for battery-powered vehicles.

In the case of V2I communication, close-range communication may benefit from an upward tilt to the antenna pattern, for example if the RSU is located on a light pole near an intersection. For short-range links, signal strength is not generally a problem. At longer distances, the RSU is likely to be in front of or behind the car at elevation angles approaching the horizontal. In both scenarios the requirements for V2I communication are similar to those for V2V communication.

We conclude that the area of antenna design is one that could provide significant benefits to V2X communications.

3.2   Vehicle-to-Vehicle/Infrastructure Channel Models

3.2.1   Deterministic models

If enough is known about the environment, including building locations, geographical features, trees and vegetation, etc., it is possible in principle to calculate the received signal, including the multipath components, exactly. Such a problem is too complex to solve at present using full-wave electromagnetic techniques. However, significant success has been obtained using ray tracing techniques. Using these, Wiesbeck and coworkers have obtained agreement between simulation and measurements with a standard deviation of 3 dB (Maurer et al. 2004; Wiesbeck and Knorzer 2007).

Although deterministic techniques such as ray tracing offer the most precise channel prediction and modeling, they are also the most computationally intensive. Consequently, a hybrid approach is to use gross geometrical information such as distance, distribution of scattering objects, and ground reflections combined with statistical models of small-scale fading. Models of this type are discussed in the next section.

3.2.2   Geometry-based statistical models

The simplest geometrical model that considers more than one propagation path is the two-ray flat-earth model. In this model, the received signal is composed of a direct path and a specular reflection from the ground. At short distances this model gives rapid signal fluctuations as the ground reflection goes in and out of phase with the direct path. Averaging the fluctuations results in an average received power proportional to $1/r^2$, where r is the distance between the transmitter and the receiver. At long distances the direct and ground-reflected paths become almost equal. Since the reflected path at grazing angles acquires an extra phase shift of π radians, the two paths make contributions that are comparable in amplitude and nearly out of phase. As a result, the received power falls off as $1/r^4$ at long distances. A common estimate of the boundary between these two regions is the distance at which the path length difference corresponds to π radians. This breakpoint, or critical distance, is given by

$$d_c = \frac{4h_t h_r}{\lambda},$$

(3.1)

where h_t and h_r are the heights of the transmit and receive antennas, respectively. A simple approximation suggested by the two-ray model is one in which the received power diminishes as $1/r^2$ up to the distance d_c given by (3.1), and diminishes as $1/r^4$ thereafter. A further generalization of this dual-slope model allows the slopes in the two regions to be parameters, and adds a random fluctuation. The statistics of the fluctuation depend on the physics of the scattering in the environment, but a zero-mean Gaussian distribution when the signal is expressed in dB is a simple assumption that often gives reasonable agreement with measurements. This is referred to as a *log-normal* model. The form of the general dual slope log-normal model is given by

$$P(r) = \begin{cases} P(d_0) - 10n_1 \log_{10}\left(\frac{r}{d_0}\right) + X_{\sigma_1} & \text{if } d_0 \leq r \leq d_c \\ P(d_0) - 10n_1 \log_{10}\left(\frac{d_c}{d_0}\right) - 10n_2 \log_{10}\left(\frac{r}{d_c}\right) + X_{\sigma_2} & \text{if } r > d_c, \end{cases}$$

(3.2)

where d_0 is a reference distance at which the power $P(d_0)$ is known, n_1 and n_2 are the path loss exponents in the two regions, and σ_1 and σ_2 are the standard deviations of the random variables X_{σ_1} and X_{σ_2}, respectively.

The log-normal distribution can be mathematically justified in the case of cellular systems where the signal is assumed to pass through a large number of attenuating materials before reaching the mobile. Passing through these attenuating materials causes variations in amplitude over length scales that are large compared to a wavelength, but sufficiently small that variations in $1/r^2$ or $1/r^4$ are negligible. This phenomenon is referred to as log-normal shadowing. Shadowing of this type usually does not occur for V2X links, since a Line-of-Sight (LoS) path is usually present, and few obstructions are encountered owing to the relatively short range. Nevertheless, we find that the log-normal distribution often gives a reasonable qualitative description of the measured signal variations.

Owing to the short range of typical V2X links, a more precise description of the fluctuations is often given by small-scale fading models where the fluctuations are caused by constructive and destructive interference between multipath components. If the in-phase and quadrature signal components fluctuate as independent Gaussian random variables, Rayleigh or Rician statistics result, depending on whether or not there is a LoS. A distribution that is capable of modeling both types of statistics is the Nakagami distribution given by

$$f(x; \mu, \omega) = \frac{2\mu^\mu x^{2\mu-1}}{\omega^\mu \Gamma(\mu)} e^{\frac{-\mu x^2}{\omega}},$$

(3.3)

where μ is a shape parameter, and $\omega = E[x^2]$ is an estimate of the average power in the fading envelope. When $\mu = 1$ the Nakagami distribution describes a Rayleigh distribution, while for $\mu > 1$ the distribution is Rician.

For moving vehicles, small-scale fading also results in rapid variations in the signal amplitude with time. Multipath components arriving from different directions will be shifted in frequency by differing amounts owing to the Doppler effect. These components with slightly different frequencies will go in and out of phase with each other, causing the amplitude variations. The specific way in which the amplitude varies is determined by the relative amplitudes and phases of the multipath components. The complex amplitude of the received signal with frequency is referred to as the Doppler spectrum. For the V2I case, a Gans spectrum (Gans 1972) using a single speed parameter may be appropriate if the car is surrounded by dense scatterers. For the V2V case, the shape of the spectrum is influenced by the speeds of both cars. Geometry-based models have been proposed in which each car is surrounded uniformly by dense scatterers (Akki and Haber 1986; Patel et al. 2005; Wang and Cheng 2005); and where the cars travel between rows of scatterers (Cheng 2008). A multipath model in which two cars are surrounded by an elliptical ring of scatterers has also been proposed (Liberti and Rappaport 1996), but the Doppler spectrum was not discussed.

3.2.3   Multi-tap models

The multiple propagation paths in a typical wireless environment arrive with different delays, so that the received signal consists of a possible direct path followed by multiple echoes. If the time it takes for the echoes to die out is comparable with a symbol duration, then the detection of one symbol is impaired by echoes from the previous symbol. Consequently, it is important to be able to model this time-dispersive nature of the channel as well – particularly as the transmitted data rates increase.

A common way to model this type of behavior is with a delay line having taps corresponding to typical multipath delays. The signals from all of the taps are then summed to provide a composite signal containing the echoes. Multi-tap models for the V2X environment have been described by Acosta et al. (2004), Acosta-Marum and Ingram (2007), and Sen and Matolak (2008).

In a complete multi-tap model, the amplitude of each tap will fluctuate independently in time in a manner determined by the Doppler spectrum for each of them. The total average power of the multi-tap response would then be determined by a large-scale model such as that given in (3.2). In general, log-normal shadowing described by the variable X_σ would be present in addition to the small-scale fading. However, in our models of the V2X environment we do not distinguish between these two effects. Instead, the log-normal model would be used if the short-term correlations in amplitude were not important, and a small-scale fading model depending on the Doppler spectrum would be used when short-term correlations were important. The correlation in the amplitude variations will be important, for example, when two packets are received during an interval for which the channel is approximately constant.

3.3   Vehicle-to-Vehicle/Infrastructure Channel Properties

The properties of the base-to-mobile channel at 800 MHz and 1.9 GHz have been extensively explored in the context of cellular communications. However, the V2X channel is different in several significant ways. First, the frequency band at 5.9 GHz has very different behavior owing to differences in diffraction, absorption by foliage, and scattering from fixed-size objects. Second, the communication range is much shorter – typically hundreds of meters rather than kilometers. Third, a LoS path is present the majority of the time. And finally, the scattering objects are not randomly distributed in an omnidirectional fashion, but instead tend to be aligned along the roadway.

For these reasons, it is important that new models and insights be based upon and validated against measurements of channel properties in representative environments.

Some of the earliest measurements of the V2V propagation channel were at frequencies below 1 GHz. Davis and Linnartz (1994) performed measurements at 900 MHz in a roadway environment with parked cars, while Punnoose et al. (1999) reported measurements between multiple moving vehicles in an ad hoc network at 915 MHz. Acosta et al. (2004) reported joint Doppler-delay power profiles between vehicles at 2.4 GHz, and broadband V2V measurements at 2.4 GHz were performed by Zajic et al. (2009). Maurer et al. (2002) conducted narrowband measurements of inter-vehicle transmissions at 5.2 GHz, and measurements of the mobile V2V channel at 5.12 GHz in several cities in Ohio were reported by Sen and Matolak (2008). Finally, V2V channel measurements at 5.9 GHz for selected highway sites with high delay spreads were reported by Acosta-Marum and Ingram (2007). A recent overview of the properties of the V2V channel is given by Molisch et al. (2009).

3.3.1   Empirical measurement platform

We have developed a measurement platform for characterizing the V2X environment over instantaneous bandwidths of up to 40 MHz at 5.9 GHz. The system architecture is illustrated in Figure 3.2. At the center of both the transmitting and receiving systems are instruments with capabilities equivalent to those of software-defined radios. An arbitrary transmitting waveform band limited to 40 MHz can be generated with the Agilent E4433B, while the I and Q components of an arbitrary waveform can be recorded and analyzed by the Agilent 89610A vector signal analyzer. Rubidium frequency standards are used to ensure phase stability between the systems, and each vehicle is equipped with differential GPS with one to three meter accuracy. Laptop computer controllers on each vehicle control the instruments and log GPS data. The controller on the receiving system also logs I and Q waveforms received from the channel. The recorded waveforms are post-processed to extract channel properties.

The channel sounding system has been used with Continuous Wave (CW) signals to examine narrowband Doppler spectra, Direct Sequence Spread Spectrum (DSSS) waveforms to obtain power-delay profiles, and OFDM packet waveforms to examine bit error and packet error performance data directly.[1] Measurements were collected in three typical environments: suburban neighborhoods near Carnegie Mellon University, interstate highways, and rural roadways north of Pittsburgh (Cheng et al. 2008b, 2007b).

3.3.2   Large-scale path loss

Narrowband measurements at 5.9 GHz were used to determine the parameters for a dual slope log-normal model as described in Section 3.2.2. The parameter fits were based on a

---

[1]The narrowband measurements were made prior to installing the rubidium clocks, however, so drift corrections were done in post-processing (Cheng et al. 2007b).

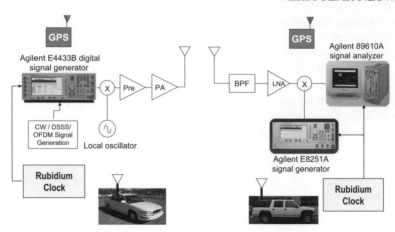

Figure 3.2 Measurement system for car-2-X channel sounding: adapted from Cheng et al. (2008c)

total of 6456 measurements for the suburban environment, and 6247 and 3296 measurements for the highway and rural environments, respectively. The parameters extracted from fits to the data are given in Table 3.2 for a dual slope log-normal model. The ranges of values given for the suburban case result from two separate data sets (Cheng et al. 2007b).

Table 3.2 Dual-slope path loss parameters from Pitts-burgh measurements

Parameter	Suburban	Highway	Rural
n_1	2–2.1	1.9	2.3
σ_1 (dB)	2.6–5.6	5.9	5.5
n_2	3.8–4	4	4
σ_2 (dB)	4.4–8.4	6.6	4.7
d_c (m)	100	220	226

Suburban data from Cheng et al. (2007b); highway and rural data from Cheng et al. (2008b). (Note that the values for σ_1, σ_2 are in error in Cheng et al. (2008b) © 2008 IEEE.)

The path loss exponents shown in Table 3.2 are similar to those of the two-ray flat earth model discussed in Section 3.2.2, suggesting the use of the values $n_1 = 2$ and $n_2 = 4$. The breakpoint distance given by (3.1) with $h_t = 1.51$ m, $h_r = 1.93$ m, and $\lambda = 0.0508$ m is 229 m, in reasonable agreement with the values for the highway and rural cases in Table 3.2. However, the breakpoint distance for the suburban case is about half what is expected. A smaller-than-expected breakpoint distance has also been reported by Masui et al. (2002), and may be due to the presence of vehicles, pedestrians, and other objects giving rise to reflections from points above the ground. With this hypothesis, we can extract an effective ground offset from the suburban data. Assuming the ground offset is a, the breakpoint distance would be

$$d_c = \frac{4(h_t - a)(h_r - a)}{\lambda}. \tag{3.4}$$

Setting $d_c = 100$ m and solving for a gives $a = 0.57$ m. These observations lead us to the proposed large-scale model summarized in Table 3.3, with the breakpoint distance given by (3.4) and ground offsets a as given in the table.

Table 3.3 Proposed dual-slope path loss parameters

Parameter	Suburban	Highway, Rural
n_1	2	2
σ_1 (dB)	5	5
n_2	4	4
σ_2 (dB)	6	6
a (m)	0.57	0

3.3.3 Fading statistics

The fading statistics for two suburban data sets were analyzed by dividing the received signal strength data into distance bins, and fitting a Nakagami distribution to the data in each bin (Cheng et al. 2007b). It was observed that the fading was Rician ($\mu > 1$) at short distances, and became more severe as distance increased. The statistics became Rayleigh ($\mu = 1$) at $r \leq 100$ m, and became worse than Rayleigh ($\mu < 1$) beyond about 100 m. The severe fading beyond 100 m may be the result of intermittently gaining and losing a LoS path as the vehicles go in and out of traffic, or go around corners. Figure 3.3 shows the measured values of the Nakagami μ parameter as a function of distance for two combined data sets in a suburban environment (Cheng et al. 2007b). The behavior is captured reasonably well by the trend line shown in the figure and given by

$$\mu = -1.3 \log_{10}(r/d_0) + 3.7, \tag{3.5}$$

where $d_0 = 1$ m is a reference distance. Similar data for the highway and rural environments is not yet available. However, higher-speed roadways tend to have larger open areas and gradual curves, as opposed to the street corners encountered in the suburban environment. Because of these differences, it is anticipated that the fading at long distances would be less severe in these environments than in the suburban case.

3.3.4 Coherence time and Doppler spectrum

As mentioned previously, the details of the temporal fading envelope are determined by the Doppler spreading of the signal spectrum. We define the square root of the second-order central moment of the magnitude of the Doppler spectrum as the *Doppler spread*. Empirically, we find that the Doppler spread has a strong linear dependence on the *effective speed*, defined as

$$v_{eff} = \sqrt{v_t^2 + v_r^2}, \tag{3.6}$$

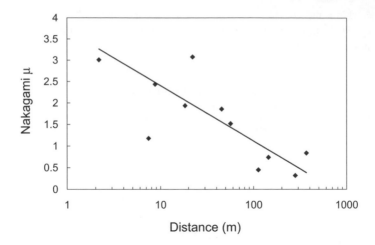

Figure 3.3 Measured Nakagami fading parameter μ along with a fit to the data given by (3.5)

in which v_t and v_r are the speed of a transmitter vehicle and a receiver vehicle, respectively. In terms of the effective speed, the Doppler spread B_D is empirically given by

$$B_D = \frac{a_1}{\lambda\sqrt{2}} v_{\text{eff}} + a_2, \tag{3.7}$$

where the slope a_1 and the offset a_2 for the three environments are given in Table 3.4. As the different frequency components in the Doppler spectrum go in and out of phase, temporal fading of the channel occurs. The *coherence time* is a measure of the time that the channel remains unchanged. We define τ_{90} as the time for which the autocorrelation function falls to 90% of its initial value. Empirical Cumulative Distribution Functions (CDFs) for the Doppler spread and the 90% coherence time are shown in Figure 3.4. The sharp upturn in the τ_{90} CDF at about 35 ms is an artifact caused by the length of the captured time frame for the suburban data set. The length of the frame limited the maximum measured τ_{90} to 40 ms. The coherence time is inversely proportional to the Doppler spread, though the constant of proportionality depends on the precise definitions of these quantities. The relation between τ_{90} and B_D can be extracted from the CDFs as follows:

$$\tau_{90} = CDF_\tau^{-1}(1 - CDF_B(B_D)), \tag{3.8}$$

where $CDF_\tau(\tau)$ and $CDF_B(B_D)$ are the cumulative distribution functions for τ_{90} and B_D, respectively. Using this relation to find (τ_{90}, B_D) pairs allows a determination of the expected value $E(B_D\tau_{90})$. The resulting relationships between B_D and τ_{90} are shown in Figure 3.5 (the stair-shaped nature of the curves results from the binning used in the construction of the CDFs). Part (a) of the figure shows that, for large Doppler spreads, the data is well approximated by the relation $B_D\tau_{90} = 0.14$. However, the expanded view for small Doppler spreads shown in Figure 3.5(b) shows that $B_D\tau_{90} = 0.3$ is a better approximation for the suburban and rural environments for $\tau_{90} > 5$ ms, in agreement with Cheng et al. (2007b). The relative insensitivity of the highway coherence time to Doppler spread for $B_D < 100$ Hz is particularly interesting. The origin of the different behavior for small Doppler spreads (large coherence times) is presently under study.

Table 3.4 Parameters for empirical Doppler spread expressions

Parameter	Suburban	Highway	Rural
Slope a_1	0.428	0.420	0.414
Offset a_2	11.5	0.5	0.2

Data from Cheng et al. (2008a)

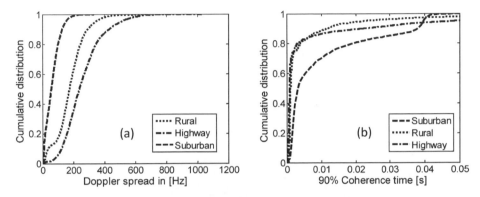

Figure 3.4 Measured CDFs in the three environments: (a) CDF for Doppler spread, and (b) CDF for 90% coherence time (Cheng et al. (2008a) © 2008 IEEE)

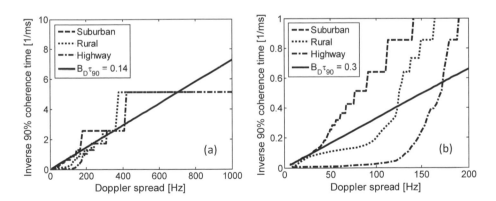

Figure 3.5 Relationships between Doppler spread and 90% coherence time extracted from the cumulative distribution functions

Although the Doppler spread and coherence time are fundamentally related to speed rather than distance, an interesting observation from our measurements is that the effective speed and separation for a V2V link is often correlated owing to driver behavior. This correlation can be visualized with a scatter plot that we refer to as a *speed–separation*

(S–S) diagram. Figure 3.6 shows S–S diagrams for the three experimental environments considered in our measurements. Note that the characteristic patterns are different for each case. For the suburban environment (Figure 3.6a), the S–S diagram shows a strong correlation between effective velocity and separation. Our interpretation of this is that drivers naturally leave more distance between cars at higher speeds. Consequently, in a single-lane road with infrequent passing, we observe a strong correlation between speed and separation. A similar observation can be made for the rural environment (Figure 3.6c), though the S–S diagram has a primary clump at high speeds, representing the open road, and a secondary clump at low speeds resulting from the less-frequent intersections and communities. In contrast, on a multilane highway where passing is common, the effective speed is relatively independent of separation (Figure 3.6b). In each of these cases, the expected Doppler spread for a specific range of separations can be obtained by finding the expected effective velocity from the S–S diagram, and using the relation between effective speed and Doppler spread given by (3.7).

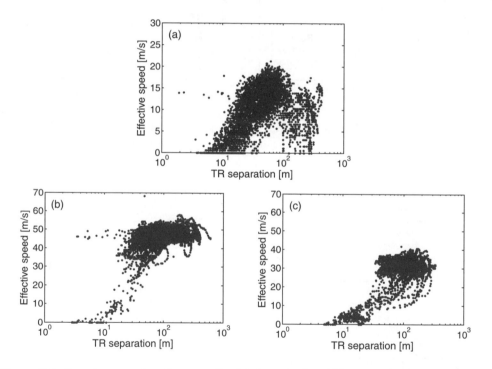

Figure 3.6 Speed–separation diagrams for (a) the suburban (Cheng et al. (2007b) © 2007 IEEE), (b) highway, and (c) rural environments (note the different vertical scale in (a))

3.3.5   Coherence bandwidth and delay spread profile

Time delays between the arrivals of multipath components result in echoes that can cause errors – especially if the delays are large compared with a symbol period. In the frequency domain, the echoes add in phase at some frequencies and out of phase at others, resulting in *frequency selective fades*. Wide bandwidth measurements are needed to resolve the multipath components, or, equivalently, to characterize the frequency selective fading. For these measurements, we used a DSSS signal with a bandwidth of 40 MHz. The *coherence bandwidth* is a measure of the range of frequencies over which the channel can be considered to be approximately constant. We define the 90% coherence bandwidth as the frequency offset for which the magnitude of the autocorrelation function falls to 90% of its peak value.

In the time domain, the multipath components are commonly characterized by the Root Mean Square (RMS) delay spread and the maximum excess delay. The RMS delay spread is the square root of the second central moment of the measured power delay profile, and is discussed by Rappaport (2002). The coherence bandwidth and RMS delay spread are inversely related in a manner analogous to the relationship between the coherence time and the Doppler spread discussed earlier. The maximum excess delay is the time interval between the first time the power-delay profile exceeds a minimum threshold and the last time the threshold is exceeded. It is useful as an estimate of the upper bound of the duration of the channel response. We used a threshold of 15 dB below the strongest peak in our measurements.

The CDFs for the coherence bandwidth and RMS delay spread are shown in Figure 3.7. Using a similar expression to (3.8), the relationship between coherence bandwidth and RMS delay spread can be extracted from the CDFs (shown in Figure 3.7). The result is shown in Figure 3.8. The measured CDFs for the maximum excess delay using a 15 dB threshold are shown in Figure 3.9. The larger values of maximum excess delay for the rural and highway environments when the CDF = 0.9 can be explained by the presence of larger open areas along with wider separation between vehicles, resulting in longer echoes. Similarly, the broader distributions for the rural and highway cases result from the fact that in these environments there are objects both very close (other cars) and far away (hills, trees, etc.). In contrast, the limited distance to buildings in the suburban environment results in a narrower range of echo times.

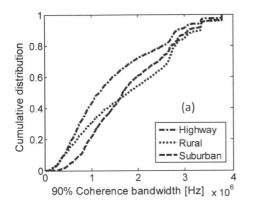 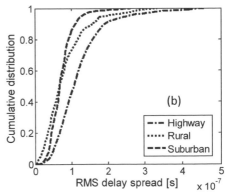

Figure 3.7   Measured CDFs in the three environments: (a) CDF for 90% coherence bandwidth (Cheng et al. (2007a) © 2008 IEEE), and (b) CDF for RMS delay spread (Cheng et al. (2008c) © 2008 IEEE)

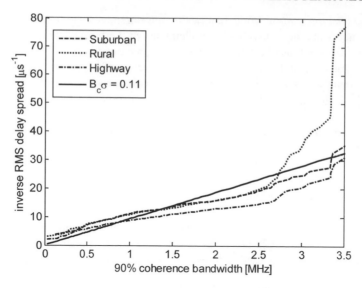

Figure 3.8   Relationships between 90% coherence bandwidth and RMS delay spread extracted from the CDFs

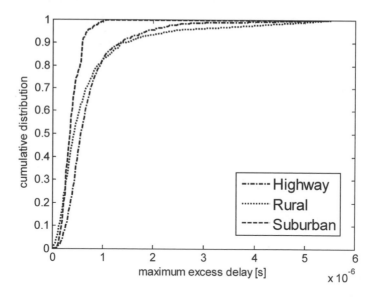

Figure 3.9 Measured CDFs for the maximum excess delay in the suburban, highway, and rural environments (Cheng et al. (2008c) © 2008 IEEE)

3.4   Performance of 802.11p in the Vehicle-to-Vehicle/Infrastructure Channel

The properties of the V2X channel discussed in the previous sections enable estimates of the performance of various types of modulation formats. We now turn our attention to consideration of the performance of scaled versions of the 802.11a/g standard, as proposed for the IEEE 802.11p standard for V2X communications. Our discussion is based on the draft IEEE 802.11p-D1.4 version.

3.4.1 Impact of channel properties on OFDM

The key Physical (PHY) layer parameters for OFDM relating to the channel properties are listed in Table 3.5. Specifically, we are concerned with the duration of the Guard Interval (GI), the carrier spacing, and the interval between channel estimates.

Table 3.5 Key parameters of the 802.11a family of OFDM waveforms and PHY requirements (Cheng et al. (2007a) © 2008 IEEE)

PHY parameter	Relevant channel parameters	Criteria for PHY parameter
Guard Interval	Maximum excess delay (τ_e)	$GI > \tau_e$
Carrier spacing Δf	Coherence bandwidth (B_c), Doppler spread (B_D)	$B_c > \Delta f > B_D$
Interval between channel estimates (packet length T_p)	Coherence time (T_c)	$T_p < T_c$
Pilot spacing Δf_P	Coherence bandwidth (B_c)	$\Delta f_P < B_c$

In OFDM, the carrier spacing is chosen to be the smallest value that results in orthogonality between the carriers over a symbol duration. Also, symbols are transmitted in consecutive non-overlapping time intervals, resulting in orthogonality between them. However, when the symbols are transmitted through a multipath channel, echoes from one may overlap with the next. This destroys the orthogonality between the symbols and the carriers. The orthogonality can be restored by copying a portion of the end of each symbol and inserting it as a prefix on the symbol. Such a *cyclic prefix* or GI is effective provided that it is long enough for the channel echoes to die out before the end of the prefix.

The power-delay profile provides a description of the duration of echoes in the channel. The RMS delay spread is an estimate of the typical duration of the power-delay profile, while the maximum excess delay provides an estimate of the worst-case conditions. Consequently, the 90% maximum excess delay discussed in the previous section provides a conservative estimate of the duration of the power-delay profile. For the GI to provide adequate multipath protection, we therefore desire $GI > \tau_{90}$.

As previously discussed, the delay spread in the channel also causes variations in the channel gain with frequency, resulting in frequency selective fading. For single-carrier systems, tapped-delay line equalizers are needed to correct for this distortion. An advantage of OFDM is that if the coherence bandwidth is large compared with the carrier spacing, each carrier experiences flat fading with no spectral distortion. Thus equalization can be accomplished simply by multiplying by a complex channel gain for each carrier. We conclude that to allow the use of the simpler equalizing scheme we desire the coherence bandwidth to be large compared with the carrier spacing, or $B_c > \Delta f$.

Orthogonality between carriers can also be destroyed by Doppler spreading resulting from motion of objects in the channel. This leads to another constraint on the carrier spacing, namely that the carrier spacing should be much greater than the Doppler spread, or $\Delta f > B_D$.

Finally, the Doppler spread also determines how long a particular channel estimate remains valid. Thus, a particular channel equalization is only effective for time intervals up to the channel coherence time. In the standard 802.11a/g scheme, channel estimation and equalization are performed once at the beginning of each packet, and applied for the duration of the packet. In this case, the packet length must be shorter than the coherence time for the equalization to remain effective, or $T_p < T_c$.

The CDFs discussed in the previous sections are useful for estimating the percentage of time that key parameters are greater than or less than a given value. As an illustration of a conservative system design, let us consider conditions that are valid for 90% of the time. Referring to Figure 3.9, we see that for 90% of the time the maximum excess delays are less than 0.6 μs, 1.4 μs, and 1.5 μs for the suburban, highway, and rural environments, respectively. These as well as the other key parameters determined in similar ways are summarized in Table 3.6. (The maximum Doppler spread is an exception; in this case, the entries in Table 3.6 are simply the maximum observed values, rather than the maximum Doppler spread for 90% of the time.)

Table 3.6   Measured channel parameters (Cheng et al. (2007a) © 2008 IEEE)

Environment	Suburban	Highway	Rural
Maximum excess delay	0.6 μs	1.4 μs	1.5 μs
Minimum 90% coherence bandwidth	750 kHz	410 kHz	420 kHz
Maximum Doppler spread	0.583 kHz	1.53 kHz	1.11 kHz
Minimum 90% coherence time	1 ms	0.3 ms	0.4 ms

For comparison with the measured channel properties, the relevant parameters in the draft standard are summarized in Table 3.7. Comparing the GIs from Table 3.7 with the values of maximum excess delay from Table 3.6, we conclude that the 5 MHz and 10 MHz scaled versions should perform well. In contrast, the duration of the GI for the 20 MHz version is expected to be insufficient for the large multipath delays encountered in the outdoor V2X environment. Comparing the carrier spacings in Table 3.7 with the coherence bandwidth and maximum Doppler spreads given in Table 3.6 shows that the condition $B_c > \Delta f > B_D$ is well satisfied for all scaled versions in all environments.

The primary challenges are associated with channel equalization. To assist with equalization, the standard reserves carriers -21, -7, 7, and 21 for the transmission of pilot signals. To adequately sample frequency selective fading, the spacing between the pilots should be less than the coherence bandwidth of the channel. However, a comparison between the pilot spacings in Table 3.7 and the coherence bandwidths in Table 3.6 shows that the pilot spacings are inadequate for all bandwidth versions in all environments considered.

Further difficulty arises from the short coherence time of the channel. Although the standards do not specify packet durations, they are typically much longer than the measured 0.3–1 ms coherence times (Table 3.6). In fact, the test packet durations at the preferred 5 MHz and 10 MHz bandwidths were longer than the coherence times of any of the three environments. The result of this is that the equalization that is applied at the beginning of the packet expires before the end of the packet, causing increasing bit errors. We conclude that the primary challenges anticipated with the proposed OFDM transmission scheme are

Table 3.7   Scaled OFDM parameters (Cheng et al. (2007a) © 2008 IEEE)

Scale	5 MHz	10 MHz	20 MHz
OFDM GI	3.2 μs	1.6 μs	0.8 μs
Carrier spacing	78 kHz	156 kHz	312.5 kHz
Pilot spacing	1.092 MHz	2.184 MHz	4.375 MHz
Test packet duration	3 ms	1.5 ms	0.75 ms

the inability of the equalization to update rapidly enough to follow changes in the channel, and inadequate pilot carrier spacing to characterize changes in the channel with frequency. The former difficulty is the result of the rapid motion of vehicles in the environment, and the latter is caused by the long delay spreads observed in the outdoor V2V channel.

3.4.2   Potential equalization enhancement schemes

The transmission of pilot symbols at properly spaced intervals in time and frequency is an effective way to obtain the channel state information needed for equalization. The spacing in time should be adequate to Nyquist sample the temporal fading, and the spacing in frequency should be adequate to Nyquist sample the frequency selective fading. In approximate physical terms, this means that the time interval between pilots should be of order the coherence time, and the frequency interval should be of order the coherence bandwidth. As mentioned above, in the 802.11a/g standard, carriers -21, -7, 7, and 21 are used as pilots in each symbol. Since the symbol duration is a few microseconds and the coherence time is of order 1 ms, the temporal channel is oversampled by roughly three orders of magnitude. At the same time, the pilot spacing in frequency undersamples the channel by a factor of up to ten. Using additional carriers as pilots would address this issue, but would also reduce the throughput. However, since the channel is significantly oversampled in time, increasing the time interval between pilot symbols could more than compensate for the extra pilot carriers. Thus it should be possible to select pilot spacings so that the channel state information is complete and the throughput is actually increased.

In addition to more completely characterizing the channel state information, a dynamic equalization scheme is needed to track changes in the channel. Instead of freezing the equalization at the beginning of each packet, the equalization should be updated at intervals not longer than the coherence time of the channel.

A recent overview of channel estimation techniques with applications to OFDM can be found in Ozdemir and Arslan (2007).

3.5   Vehicular Ad hoc Network Multichannel Operation

Up to this point, we have been primarily concerned with a single V2V link. However, realistic applications of DSRC technology generally require interactions between multiple vehicles, forming a Vehicular Ad hoc Network (VANET). We now turn our attention to protocols necessary to coordinate these interactions. Key to our discussion is the medium access protocol for determining when mobile radios are allowed to access the channel.

DSRC technology has the capability of multichannel Medium Access Control (MAC) coordination (regulated in the IEEE 1609.4 standard), whereas the conventional IEEE 802.11 standard does not. The 1609.4 standard is an extension to the proposed IEEE 802.11p DSRC standard that enhances the basic MAC operation to enable coordinated switching between the Control Channel (CCH) and Service Channels (SCHs). In this section we discuss the coordination mechanisms and simulation results as well as other alternative approaches.

3.5.1   Multichannel MAC (IEEE 1609.4)

The PHY of DSRC (specified in IEEE 802.11p 2006) has seven non-overlapping 10 MHz channels around 5.89 GHz. The center channel (178) is the *CCH* and the remaining channels are *SCHs*, including channels 172, 174, 176, 180, 182 and 184. Given that 802.11 is based on a radio model consisting of a single half-duplex transceiver, the DSRC radio can only communicate on one channel at a time. To use multiple channels, the radio must be dynamically switched between them as needed. This channel switching mechanism requires distributed coordination among the radios in the network, so that transmitters and their intended receivers are tuned to the same channel at the same time.

One such protocol providing distributed coordination among radios is proposed for DSRC in IEEE 1609.4 (2006). DSRC Multi-Channel Coordination (DMCC) is a form of *split-phase* channel coordination, where its mechanisms are summarized as follows (Mo et al. 2005):

- Assuming all radios are synchronized to a global time reference (e.g. GPS reference time), at the beginning of every second all radios switch to the CCH and dwell there for some predefined minimum period of time (i.e. the *CCH dwell time*). Once this time expires, any radios that plan to switch to a SCH may do so and stay on the SCH channel for a predefined maximum period of time (i.e. the *SCH dwell time*), after which they must switch back to the CCH again. Thus, the radio resource is split between control channel and service channels.

- Coordination between radios that plan to rendezvous and communicate on a SCH channel is handled by the exchange of management frames on the CCH during the previous CCH interval. In particular, during the previous control channel interval, service providers broadcast *WAVE Service Advertisement (WSA) messages* to announce their services on the CCH channels. For example, in Figure 3.10 there are three service providers, and each of them broadcasts a WSA during the control interval for services on the service channels SCH1, SCH2, and SCH6. Clients that want to use the service must switch to the specific SCH provided in the WSA at the next data phase.

- The CCH is reserved for DSRC management frames and short data messages for safety-relevant applications, while all other data traffic for less critical applications must use the SCH channels. Medium access control on both CCH and SCH uses the basic access protocol (i.e. Carrier Sense Multiple Access, or CSMA) in the 802.11 standard, with the addition of extensions for priority access defined in IEEE 802.11e (2005).

- The *GI* is a period of time during which no radio is allowed to transmit. Its purpose is to account for distributed synchronization errors between radios, and is defined as the sum of the allowable synchronization tolerance between local clocks and the maximum radio channel switch time. In the current version of the IEEE 1609.4 standard, the GI is specified as 4 ms.

Figure 3.10 illustrates the operation of the DMCC MAC protocol. During the control channel interval, providers advertise their services to clients by sending *WAVE Service Announcements* (**w**) containing the channel for their service. Small data frames called *Wireless Access in Vehicular Environments (WAVE) Short Messages* (**d**) may also be exchanged. As seen in this figure, there are a number of *WAVE Service Announcements* (**w** messages) and *WAVE Short Messages* (**d** messages) during the control channel interval. During the service channel interval, providers and clients switch to the selected service channels to exchange data packets. In this figure, during the service channel interval, each provider on SCH1, SCH2, and SCH6 provides its own service and transmits data packets to its clients. At the end of the interval, all nodes must switch back to the control channel. Between channel switches, *GIs* are inserted, during which no node is allowed to transmit, to allow for differences in clock synchronization and channel switching times between nodes.

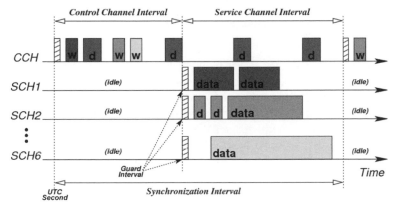

Figure 3.10 Operation of the DMCC MAC protocol

3.5.2 Performance evaluation of the IEEE 1609.4 multichannel MAC protocol

To understand better how the proposed multichannel switching mechanism impacts vehicular communication systems and, in particular, Vehicular Safety Communication (VSC) applications, the proposed IEEE 1609.4 multichannel switching mechanism has been evaluated via QualNet simulations, under the context of Cooperative Collision Warning (CCW) safety applications (Holland et al. 2008).

Simulation setup

In the simulations, 1920 vehicles were simulated on an eight-lane straight freeway stretch of 1.6 km in length, with four lanes in each direction and no entries or exits. Vehicle speeds varied within the same lane and across lanes following a Gaussian distribution. The average speeds in the four lanes were assumed to be 33.0 km/h (slowest lane), 37.8 km/h,

42.6 km/h, and 47.5 km/h (fastest lane). Each simulated vehicle contained a single DSRC multichannel radio configured to match settings and parameters from measurements of hardware prototypes and CCW application requirements (Bai and Krishnan 2006): the data rate was 6 Mbit/s; the transmission power was 9 dBm, which was selected because it presented the best trade-off between effective range and interference for CCW; and the receiver sensitivity was −95 dBm. The wireless propagation model that was used is described in detail in Yin et al. (2006). The size of the message was 100 bytes, which is roughly the estimated packet length for safety messages in WAVE Short Message Protocol (WSMP) format. The frequency at which the server broadcasted safety messages was 10 times per second.

The CCW application consists of client and server processes. The server process is responsible for broadcasting short messages bearing the latest status information (e.g. location, velocity, control settings) for the vehicle. The client process is responsible for receiving and logging all such messages that it receives. Each vehicle had both a client and a server process.

Channel congestion caused by Channel-Switching-Induced Broadcast Synchronization (CSIBS)

With extensive simulation studies, it was found that the proposed DSRC multichannel coordination protocol is able to share the control and service channels correctly among multiple DSRC radios to provide a tolerable level of system performance to users. Nonetheless, at the same time, it was discovered that the packet success probability is significantly reduced, to only 60%, once IEEE 1609.4 multichannel MAC mechanisms are applied.

A careful investigation revealed that the primary cause of packet loss in the study was not queue overflow caused by limited buffer size but channel congestion introduced by the problem of CSIBS (Holland et al. 2008): as DSRC radios serve the SCH during the SCH interval, CCW safety applications still generate safety messages and other management frames and then queue these packets in the radio's MAC buffer. Once the CCH becomes available again, there is a burst of broadcast transmissions as the MAC on each of the vehicles attempts to empty its queued packets as quickly as possible. However, MAC mechanisms are unable to schedule the shared channel efficiently (a common problem among contention-based CSMA-derived MACs), resulting in a high percentage of packet losses due to transmission errors (e.g. collisions and interference). Essentially, the requirement that a CCH interval must occur at the beginning of each cycle creates an artificial broadcast synchronization among vehicles, causing the queued packets to collide and get dropped. This congestion can be seen in Figure 3.11, which shows a snapshot of the network activity for a channel switch into and out of a CCH interval. Here, we simulated the simple scenario consisting of 50 vehicles positioned randomly within a 2 km × 500 m area. The application broadcast interval t was 50 ms, and the dwell time was 90 ms. The switch starts at time 19.8 seconds and ends at 19.89 seconds. At the bottom of the figure, the scatter plot shows the Receive Signal Strength Indication (RSSI) values for the packets that were received by the applications during this interval. The curve at the top of the figure shows the histogram of the number of packets dropped, averaged ('binned') over an interval of 1 ms.

The burst of packets can be seen at the beginning of the interval (in the scatter plot at the bottom of Figure 3.11), as the vehicles attempt to unload all of the packets queued by the application during the previous SCH interval, resulting in a large number of dropped packets (shown by the curve at the top of Figure 3.11). Note that even though the packet transmissions

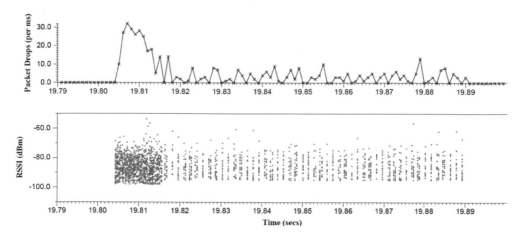

Figure 3.11 Snapshot of network activity for a channel switch into and out of a CCH interval: dwell time = 90 ms, application broadcast interval $t = 50$ ms, and network size = 50 vehicles

are generated at randomly spaced intervals across the vehicles in the network, the switch induces a form of global broadcast synchronization. As previously mentioned, we call this *Channel-Switching-Induced Broadcast Synchronization*. With 50 vehicles transmitting at 50 ms intervals (on average), there are roughly 90 packets queued at the start of the CCH interval (plus four more after the GI). Thus, at the start of the CCH the network utilization is at its maximum as all of the vehicles have a backlog of packets to transmit, which the MAC is unable to schedule adequately, resulting in a high percentage of dropped packets. However, after the queues are drained (around time 19.87 seconds), the congestion dissipates and the number of dropped packets falls.

3.5.3 Other solutions for multichannel operations

Given that research on multichannel MAC operation is still in its early stages, we rely on lessons learned from Mobile Ad hoc Networks (MANETs) to give a brief discussion on alternative solutions. A comprehensive comparison of these protocols in the context of MANETs was presented by Mo et al. (2005), but the protocols, metrics, and network scenarios used are not appropriate for IEEE 1609.4 standards or vehicular applications/networks.

In addition to the *time split* approach that is regulated by the IEEE 1609.4 standard, there are two other major variations on multichannel MAC coordination protocols, as summarized by Mo et al. (2005):

- *Dedicated control channel:* Every vehicle is equipped with two radios. One radio is dedicated to the control channel and the other radio can tune to any other service channel. Uncoordinated vehicles use the control channel as a common rendezvous place to exchange control messages, making agreements on message transmission schedules among them. Based on these agreements, the other radio on each vehicle switches to an appropriate service channel at the scheduled time.

This method – using a dedicated control channel and a dedicated control radio – does not require that synchronization-based rendezvous always happen on the same channel, as the time split approach mandates (IEEE 1609.4 2006). It reduces the complexity of system implementation, with an increased cost of providing a second radio.

- *Multiple rendezvous:* Both the time split and the dedicated control channel approaches are categorized as *single rendezvous* systems, since the exchange of control messages is only allowed to occur on one channel at a time. In contrast with these, multiple rendezvous protocols allow multiple pairs of radios on different vehicles to make agreements simultaneously on distinct channels.

This multiple rendezvous approach is mainly proposed to overcome the scheduling bottleneck caused by a single control channel. On the other hand, since there are multiple rendezvous channels, it is necessary to develop sophisticated coordination mechanisms among vehicles so that two devices can rendezvous on the same channel (which may not be the control channel). One straightforward solution is for each idle device to follow a predetermined hopping sequence and for the sending device to transmit on that channel to find the intended receiver (Kyasanur and Vaidya 2005; Mo et al. 2005).

3.6 Vehicular Ad hoc Network Single-hop Broadcast and its Reliability Enhancement Schemes

The research community is currently evaluating the use of DSRC for V2V and V2I communication applications (Chen and Cai 2005; USDOT 2005; Xu et al. 2004; Yin et al. 2004). Among them, many are based on single-hop broadcast communication. For this basic type of communication, the IEEE 802.11 MAC boils down to a simple CSMA scheme, without using the Request to Send/Clear to Send (RTS/CTS) procedure and Acknowledgment scheme. In other words, the IEEE 802.11 MAC does not provide extra reliability guarantees, beyond the physical-layer best-effort service.

The reliability and characteristics of DSRC wireless communication are examined via empirical experiments by Bai and Krishnan (2006). Such an effort facilitates efficient design and rigorous evaluation of the VSC applications. To achieve this objective, the effort is focused on systematically analyzing both DSRC communication reliability and VSC application reliability: the major aspect is the reliability of DSRC wireless communication itself; at the same time, however, end users (drivers) mainly care about whether VSC applications can provide a reliable and trustable application service.

3.6.1 Reliability analysis of DSRC single-hop broadcast scheme

Experimental setup

Extensive experiments were conducted to collect real-world data by using three vehicles equipped with the VSC system. Each of the three experimental vehicles was equipped with a DSRC radio, an omnidirectional roof-mounted antenna, and a GPS receiver. In addition to the DSRC wireless communication system hardware, the vehicles were equipped with a number of VSC applications capable of providing driver assistance information. To execute the VSC applications, each vehicle periodically broadcasted its current GPS position, velocity, heading and other sensor information (e.g. braking status, acceleration, etc.) so that all the neighboring vehicles within the transmission range (typically, 300m) would receive the message. The periodic DSRC message broadcast rate from each vehicle was 100 milliseconds. The transmission power and transmission rate of DSRC communication was set to 20 dBm and 6 Mbit/s, respectively.

One set of experiments was conducted on a realistic freeway environment of a typical metropolitan/suburban area for five hours. A number of walls, tunnels and overhead bridges are present along this section of the freeway, which represents a harsh environment for wireless signal propagation. Another set of experiments was conducted on test tracks for two hours, representing an ideal open field environment without any hostile environmental and traffic factors affecting the signal propagation. In each set of experiments, the three vehicles equipped with experimental platforms were driven at the drivers' free will in order to emulate normal driving behavior.

Reliability of DSRC wireless communication

Realizing that there exist differences between communication reliability and application-level reliability, Bai and Krishnan (2006) believe that it is necessary to isolate them.

First, the study is focused on analyzing the reliability of wireless DSRC in terms of *average packet drop rate* and *packet drop pattern* under various traffic environments. The former metric captures only average behavior, while the latter examines the detailed distributions of packet drops. Figure 3.12 illustrates the packet delivery ratio of DSRC under the freeway environment and the open field environment. First, we observe that the packet delivery ratio decays with increased distance between vehicles. Moreover, we also notice that the packet delivery ratio decays much faster in the freeway environment than in the open field environment. For example, packet delivery ratios are 93% (open field) and 91% (freeway) at 100 meters, 86% (open field) and 78% (freeway) at 200 meters, 88% (open field) and 67% (freeway) at 300 meters, and 76% (open field) and 58% (freeway) at 400 meters. This observation suggests that fading effects of the DSRC wireless channel on the metropolitan freeway are much more severe than in an open field test track environment. Generally speaking, for a benign traffic environment such as the open field test track, the reliability of DSRC seems to be quite satisfactory, from an engineering point of view. However, even in potentially harsh traffic environments the reliability of DSRC communication still seems to be adequate.

The detailed distribution of packet drops is also investigated. Among others, the metric that is of interest for VSC applications is the probability distribution of consecutive packet drops. This metric describes whether packets dropped over the DSRC wireless channel occur in bursts or not. The less frequently the phenomenon of bursty packet drop occurs, the more reliable the wireless channel. Figure 3.13 gives the probability distributions of consecutive dropped packets in the freeway and open field environments, respectively. It is observed that the majority of packet drops are either single-packet drops (about 90% at 0–25 meters, to 55% at 200–225 meters) or double-packet drops (about 5% at 0–25 meters, to 15% at 200–225 meters), under the less reliable freeway environment. Even for long distance scenarios (200–225 meters), the case that more than five packets drop consecutively occurs rarely (less than 2%) in these experiments. This observation strongly indicates that packet drops during inter-vehicular DSRC wireless communication do not occur in long bursts

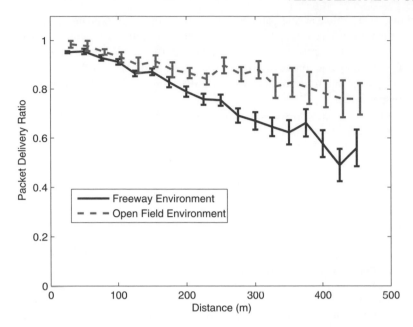

Figure 3.12 Packet delivery ratio vs. distance between vehicles for the freeway and open field environments: data from Bai and Krishnan (2006)

most of the time, suggesting that packet drops seem to be independent of each other in our experiments. This is consistent with our measurements of channel coherence time discussed in previous sections. Bursts of multiple dropped packets would suggest that the channel remained poor for the duration of several packets, while our measurements of coherence time indicate that the channel typically does not remain constant for the duration of even one packet.

3.6.2 Reliability analysis of DSRC-based VSC applications

After analyzing the reliability of DSRC communication, we now investigate the reliability of VSC applications from the viewpoint of end users.

For DSRC-based VSC applications, end users will not experience undesired effects when one or two individual packets are sporadically lost during the periodic routine broadcasts. As long as (at least) one packet from the neighboring vehicle is successfully received within a tolerance time window T, the receiver vehicle should be able to predict and update the neighbor vehicle information with sufficient accuracy for VSC application processing.[2] We thus propose to make use of a novel reliability metric – *T-window reliability*, which is defined as: for each given time t_0, if one packet (or more than one packet) is received during time interval $[t_0 - T, t_0]$, the VSC application is claimed to be reliable at time t_0;

---

[2]T is the maximum tolerance time window. T is determined by the requirements of specific VSC applications, varying from application to application. It is shown that the value of T for the majority of VSC applications falls into the range [0.3 sec, 1.0 sec].

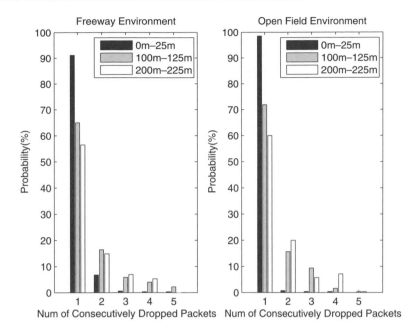

Figure 3.13 Probability distribution function of consecutive packet drops in the freeway and open field environments: data from Bai and Krishnan (2006)

otherwise, the VSC application is said to be unreliable at time t_0.[3] As an example, the average application-level T-window reliability metric is plotted as a function of distance in Figure 3.14 for the freeway traffic environment. The tolerance time window T is the key parameter for application-level reliability and varies with different VSC applications. To illustrate the effect of T on the application-level T-window reliability metric, the impacts of several reasonable T values ($T = 0.3$ sec, $T = 0.5$ sec and $T = 1.0$ sec) were considered. As shown in Figure 3.14, most of the application-level reliability values are more than 97% up to 300 meters (the maximum range for our applications), which leads to the conclusion that the reliability of DSRC-based VSC applications is quite satisfactory.

3.6.3 Reliability enhancement schemes for single-hop broadcast scheme

One-hop broadcast communication of DSRC itself does not provide reliability guarantees beyond best-effort service. To render reliable communication service and reliable safety applications, several schemes are proposed, as follows.

---

[3]In the published literature on networking, the packet delivery ratio (or packet error ratio) is commonly used to describe both communication reliability and application-level reliability. However, in vehicle-safety-oriented VSC communication, a packet with fresh information 'overwrites' the previous packet with stale information. Because of this memoryless property, VSC applications may not be affected even though several packets are lost, once a fresh packet is received.

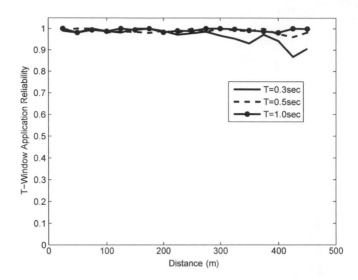

Figure 3.14 *T*-window application reliability vs. distance between vehicles in the freeway environment: data from Bai and Krishnan (2006)

- *Repetitive broadcast:* In VSC applications, vehicles periodically repeat their broadcast messages to compensate for packets being dropped (Bai and Krishnan 2006). With the memoryless nature of VSC application requirements and the non-bursty characteristic of packet loss patterns, the design philosophy of repetitive broadcasts helps provide a reasonably satisfactory reliability even under harsh fading environments.

  The disadvantage of this approach is that it generates a significant number of overhead packets, leading to channel saturation and packet collisions. Therefore, it is suggested that the repetitive broadcast scheme be combined with adaptive transmit power and rate control to achieve the optimal performance (Hartenstein and Laberteaux 2008).

- *Reliable broadcast:* The reliable broadcast protocol has been proposed to create a deterministic, reliable delivery semantic under highly mobile MANET scenarios (Pagani and Rossi 1997). Essentially, the proposed protocol uses clustering algorithms to establish and maintain a routing tree structure among mobile nodes, on which data messages and acknowledgment are exchanged. Once the mobility is deemed too high, the protocol is automatically switched to flooding mechanisms to combat unpredictable topology dynamics.

  This proposed protocol is appropriate for low- or medium-mobility scenarios, partly due to its cluster-based tree structure scheme. However, in the face of high-mobility scenarios, the proposed protocol frequently resorts to flooding mechanisms and thus generates a significant number of overhead packets, degrading the overall performance.

- *Anonymous gossip:* Unlike a deterministic (i.e. all-or-nothing) guarantee, anonymous gossip protocols guarantee reliable packet delivery with a certain probability (Chandra et al. 2001). In other words, anonymous gossip protocols render more reliable communication than *best-effort* service, though they do not provide deterministic guarantees of reliable communication. Anonymous gossip protocols are so-called added-on services that could be built upon any communication service (such as broadcast or multicast) or routing protocol.

Anonymous gossip protocols have two phases of operation. In the first phase, an information source S unreliably broadcasts data messages to its neighbors. In the second phase, one of the receivers A randomly picks another potential receiver B as its *gossip* partner. These two nodes then begin to exchange information and compare whether data messages received at both of them are different. If a difference is detected, one gossip partner will send the lost packets to the other party.

It has been shown that anonymous gossip protocols significantly improve the reliability performance of underlying communication services (Chandra et al. 2001). However, the 100% reliability guarantee, which is mandated under certain mission-critical scenarios, is not achieved using this probabilistic approach.

3.7 Vehicular Ad hoc Network Multi-hop Information Dissemination Protocol Design

Beyond single-hop broadcast communication, multi-hop information dissemination protocols (including broadcast protocols) are also of interest to the research community because they enable a class of public safety applications such as Remote Emergency Warning/Alert.

With multi-hop broadcast protocols, drivers are aware of remote road conditions and the driving status of other vehicles up to a few kilometers away, enabling them to make smart driving decisions well ahead of time. This type of application tends to be highly time critical, mandating an intelligent broadcast mechanism to distribute the warning message. To design a multi-hop broadcast protocol for VANET, one must consider two extreme regimes of vehicle traffic:

- Under dense vehicle traffic scenarios, VANETs experience the broadcast storm problem. A broadcast storm occurs when multiple vehicles attempt to transmit at the same time, thereby causing several packet collisions and extra delay at the MAC layer.

- Under sparse vehicle traffic scenarios, VANETs experience the disconnected network problem. Networks become disconnected, or partitioned, when there are not many nodes in the area to help disseminate the broadcast message.

A brief overview of these two regimes is given in the following sections (Wisitpongphan et al. 2007a,b).

3.7.1 Multi-hop broadcast protocols in dense VANETs

When the traffic density is above a certain value, one of the most serious problems is the choking of the shared medium by an excessive number of instances of the same safety message, broadcast by several consecutive cars. Because of the shared wireless medium, blindly broadcasting the packets may lead to frequent contention and collisions in transmission among neighboring nodes. This problem is sometimes referred to as the *broadcast storm problem* (Ni et al. 1999).

   While multiple solutions exist to alleviate the broadcast storm problem in a usual MANET environment, only a few solutions exist for resolving this issue in the VANET context (Hu et al. 2003; Korkmaz et al. 2006, 2004; Ni et al. 2001, 1999; Wisitpongphan et al. 2007b). Wisitpongphan et al. (2007b) (a) explore how serious the broadcast storm is in a VANET using a case study for a four-lane highway scenario; and (b) propose three contention-based, lightweight broadcast techniques to mitigate the problem: weighted p-persistence, slotted 1-persistence, and slotted p-persistence. The proposed techniques can provide nearly 100% reachability in a well-connected network and up to approximately 70% reduction in the broadcast redundancy and packet loss ratio on a well-connected vehicular network. The proposed schemes are distributed and rely on GPS information (or received signal strength when the vehicle cannot receive GPS signals), but do not require any other prior knowledge about network topology. Specifically, Figure 3.15 shows three contention-based schemes (Wisitpongphan et al. 2007b): weighted p-persistence broadcasting, slotted 1-persistence broadcasting, and slotted p-persistence broadcasting.

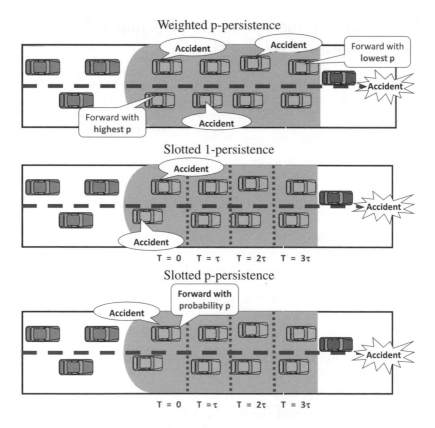

Figure 3.15 Broadcast suppression techniques (Tonguz et al. (2007) © 2007 IEEE)

   The basic broadcast techniques follow either a 1-persistence or a p-persistence rule. Despite the excessive overhead, most routing protocols designed for multi-hop ad hoc wireless networks follow the brute-force 1-persistence flooding rule, which requires that all nodes rebroadcast the packet with probability 1 because of the low complexity and high packet penetration rate. A gossip-based approach, on the other hand, follows the p-persistence rule, which requires that each node re-forwards with a pre-determined probability p. This approach is sometimes referred to as probabilistic flooding (Haas et al. 2002). Figure 3.16 shows the main results obtained with the three schemes designed. Observe that the slotted p-persistence scheme can substantially reduce the packet loss ratio at the expense of a slight increase in total delay and reduced penetration rate.

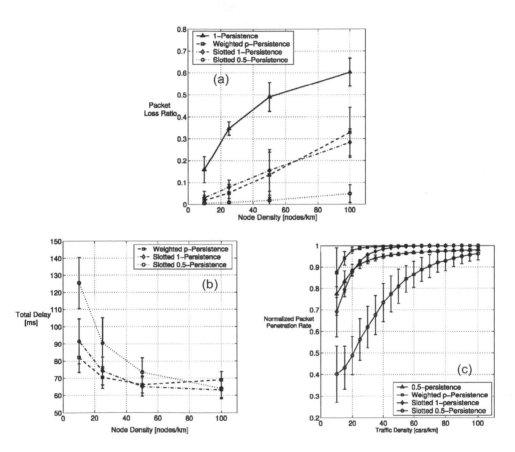

Figure 3.16   Broadcast statistics at various traffic densities; all results shown with 95% confidence intervals: (a) Packet Loss Ratio in VANET, (b) Time required to disseminate the broadcast message to nodes that are 10 km away, (c) Normalized Packet Penetration Rate (Tonguz et al. (2007) © 2007 IEEE)

3.7.2   Multi-hop broadcast protocols in sparse VANETs

The other extreme scenario, which is very troublesome for conventional routing protocols, is the case where there are not many vehicles on the road, as illustrated in Figure 3.17. At certain times of the day (e.g. between midnight and 4 am) the traffic density might be so low that multi-hop relaying from a source (the car trying to broadcast) to the cars coming from behind might not be plausible because the target node might be out of the transmission range (relay range) of the source. To make the situation worse, there might be no cars within the transmission range of the source in the opposite lane either – see Figure 3.17c. Under such circumstances, routing and broadcasting becomes a challenging task. While there are several routing techniques that address the sparsely connected nature of mobile wireless networks, for example Epidemic routing (Vahdat and Becker 2000), Single-copy (Spyropoulos et al. 2004), and Multi-copy 'Spray and Wait' (Spyropoulos et al. 2005), there are only a few that consider a VANET topology (Korkmaz et al. 2006, 2004; Zhao and Cao 2006).

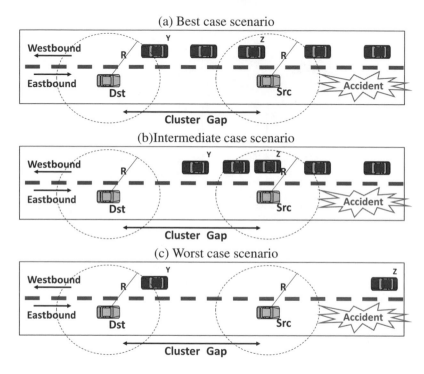

Figure 3.17 Illustration of disconnected VANETs: (a) best case scenario, in which a packet can immediately be relayed to the target vehicles via vehicles in the opposite traffic; (b) intermediate case scenario, in which vehicles in the opposite direction are responsible for relaying packets to the target vehicles using a store–carry–forward mechanism; (c) worst case scenario, in which a packet cannot immediately be relayed to vehicles in the opposite direction (Tonguz et al. (2007) © 2007 IEEE)

In Wisitpongphan et al. (2007a), the authors propose to cope with such extreme cases via the so-called *store–carry–forward* mechanism (Briesemeister and Hommel 2000). Simulation results show that, depending on the sparsity of vehicles or the market penetration rate of cars using DSRC technology, the delay that is incurred in delivering messages between disconnected vehicles can vary from a few seconds to several minutes. This time delay is referred to as the 'network re-healing time'. This suggests that, for vehicular safety applications, a new ad hoc routing protocol will be needed, since conventional ad hoc routing protocols such as Dynamic Source Routing (DSR) or Ad hoc On Demand Distance Vector Routing (AODV) will not work with such long re-healing times. In Figure 3.18 we give both simulation and analytical results obtained via this store–carry–forward approach (Wisitpongphan et al. 2007a).

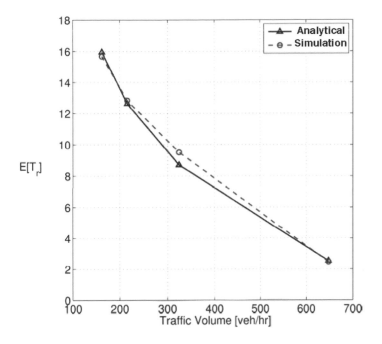

Figure 3.18 Average per-gap re-healing time: simulation results (dashed lines) and analytical results (solid lines) (Tonguz et al. (2007) © 2007 IEEE)

3.8 Mobile IP Solution in VANETs

While safety applications are the main focus of DSRC technology, mobile infotainment applications are expected to provide comfort and convenience during driving. Among these infotainment applications, many require the mobile IP solution and Internet Engineering Task Force (IETF) Network Mobility (NEMO) protocols (IETF 2007; Perkins 1997).

3.8.1 Mobile IP solution

The mobile IP solution is designed to allow mobile device users to move from one network to another while maintaining a permanent IP address. Each mobile node is identified by its home address regardless of its current location in the Internet. While being at remote locations, a mobile node is associated with a Care-of Address (CoA), which gives information about its current location. Using mobile IP, mobile devices still communicate without interruption while roaming, and this allows mobile nodes to connect seamlessly to the Internet.

Each mobile node has two addresses: a permanent home address, which is used when the mobile node is on its home network, and a CoA, which is used in foreign networks. On the home network, a Home Agent (HA) stores information about mobile nodes whose permanent home address is in the home agent's network. In foreign networks, a Foreign Agent stores information about mobile nodes visiting its network. The mobile node's home agent recognizes the node's movement and maintains the CoA of the mobile node in its foreign network.

Another node intending to communicate with the mobile node uses its permanent home address as the destination address for sent packets. Once the home agent receives data packets for the mobile node, it redirects these packets toward the foreign agent using the home agent's lookup table. This is done by tunneling the packets to the mobile node's CoA. The packets are decapsulated at the end of the tunnel and are delivered to the mobile node.

When acting as sender, the mobile node simply sends packets directly to the other communicating node through the foreign agent, without going through the home agent.

Built upon the mobile IP concept, NEMO supports mobility for *entire mobile networks* that move and attach to different points in the Internet. As with the mobile IP solution, NEMO shields the nodes in the mobile network from the movements, and this enables compatibility with devices that are not provided with any mobility support. Moreover, the NEMO technology aggregates the handover signaling procedures for all nodes, resulting in reduced administration overhead.

3.8.2   Mobile IP solution tailored to VANET scenarios

Though mobile IP and NEMO have been extensively studied in generic wireless networks, the study of mobile IP solutions in VANET scenarios is still at an early stage. In a high-level challenging paper, Baldessari et al. (2007) combine the idea of VANET and NEMO and propose a deployable system architecture called *VANET and NEMO (VANEMO)*. The authors then take a holistic approach to analyze the feasibility of this system architecture from economic, functional, and performance perspectives. It is concluded that the VANEMO deployment approach meets the functional and performance requirements better than the pure NEMO-driven one does.

Another open question attracting the attention of the research community is the development of reliable and fast IP address acquisition protocols. It has been shown that neither address configuration protocols in the Internet nor the mobile IP solution for MANETs can be directly applied to VANET scenarios (Nesargi and Prakash 2002; Sun and Belding-Royer 2004). For example, Bychkovsky et al. (2006) revealed that in city driving environments, after a vehicle associates with an Access Point (AP) and acquires an IP address, common connection times range from 5 to 24 seconds. However, the Wi-Fi Dynamic Host Configuration Protocol (DHCP) often requires 2 to 5 seconds once association is complete. In other words, DHCP can consume up to 100% of a vehicle's available connection time.

Fazio et al. (2006) propose to exploit the predictable topology of VANETs and use dynamically elected leader vehicles to run distributed DHCP protocols. This way, these leader vehicles can provide unique IP addresses to moving vehicles and reduce the frequency of IP address reconfiguration. It is shown that this proposed scheme only requires a reasonably low configuration time to acquire an IP address. Arnold et al. (2008) propose an alternative solution to allow precedent vehicles to pass their own IP addresses to the following vehicles in the same geographic region. As an example, as node A leaves an AP's coverage area, node B, which is behind node A, will reuse node A's IP address to access the Internet via the same AP. It is shown that the proposed protocol will significantly improve efficiency, reduce latency, and increase vehicle connectivity.

3.9 Future Research Directions and Challenges

Although recent research has contributed significantly to our understanding of V2X communication systems, much remains to be done. In this section, we briefly discuss remaining challenges and promising future directions.

3.9.1 Physical layer perspective

As discussed in this chapter, the analysis of existing measurement data sets has enabled the statistical characterization of representative V2X channels and has provided insights into the distinguishing characteristics of these environments. However, to verify the widespread applicability of these insights, measurements are needed in a wider range of geographical settings. For example, do suburban channels vary significantly in different parts of the United States, or in different parts of the world? There is a particular need for urban mobile channel measurements with accurate position information. Determining accurate mobile node position in urban canyons is more difficult owing to limited views of the sky and the associated lack of reliable GPS coverage.

An area of research that has received relatively little attention to date but that offers the potential for significant performance improvement is *antenna design*. Use of optimized antennas could improve received signal levels by an order of magnitude or more, or, alternatively, could reduce the required transmit power by a similar factor. Unlike conventional mobile telephony, V2X communications are not omnidirectional but instead require the most gain in the forward and reverse directions with relatively modest gain from side to side. Further, it is desirable for the main antenna beams to point in the *horizontal* direction rather than slightly upward as is the case in the common monopole or collinear antenna above a ground plane.

Finally, the equalization techniques developed for indoor, low-mobility applications with IEEE 802.11a/g are not adequate for the V2X channel. Owing to the longer delay spreads, the pilot carrier spacing is larger than the coherence bandwidth, preventing adequate sampling of the frequency selectivity of the channel. Also, owing to the high mobility of the V2X channel, a single equalization at the beginning of a packet is inadequate. An optimal solution would provide dynamic tracking of the channel, and would likely require changes in the 802.11p standard to capture the channel state information fully. On the other hand, significant performance improvements should be possible without changing the standard, if the receiver updates the equalization at intervals not to exceed the coherence time of the channel (typically of order 1 ms).

3.9.2 Networking perspective

The technical approach defined by the DSRC standard (IEEE 802.11p and IEEE 1609.x) and VANET technology (multi-hop routing and mobile IP) seems to provide a feasible technical solution for active safety applications during the initial phases of deployment, although improvements are needed in the areas of *reliability, scalability, robustness* under various environments, and *flexibility* to support a wide variety of V2X communication applications (such as traffic efficiency management, commercial and infotainment applications).

To further improve communication reliability, besides PHY-layer efforts (antenna design and modulation improvement), the development of MAC-layer error-handling protocols will also help increase the reliability of V2X applications. Localized error-handling mechanisms are able to compensate for dropped packets through implicit handshaking procedures in a hop-by-hop fashion. Specifically, standalone ACK/NACK acknowledgment beacons are used for unicast messages, while a Piggybacked implicit ACK (PACK) is incorporated into the data packets of broadcast messages for vehicular safety applications.

One key challenge of designing VANET multi-hop routing protocols is maintaining scalability and robustness under highly dynamic environments. Under such environments, position-based routing protocols (i.e. Greedy Perimeter Stateless Routing, or GPSR, and geocast) seem to outperform topology-based routing protocols (i.e. DSR and AODV), because it is less costly to maintain network topology using position-based routing protocols. Two other mechanisms are also critical in supporting the operation of multi-hop routing protocols in a vehicular ad hoc network. To enable efficient packet routing among thousands of vehicles, assigning network addresses to vehicles is essential so that destination vehicles can be identified for packet routing. Using geographic information to assist in the dynamic allocation, maintenance and termination of IP addresses seems to be a practical solution for vehicular networks, partly due to its robustness to high vehicle mobility. In addition, autonomous organization of the network structure is required to facilitate better management of a large-scale network. Since highly mobile environments have a negative impact on network hierarchical structures, an on-demand, flat, localized network structure is being considered as the primary method of resolving network scalability issues in large-scale vehicular ad hoc networks.

The time-split multichannel switching protocol enables the flexibility of VANET systems to support both safety and non-safety applications. However, a multichannel switching mechanism is dependent on the availability of GPS for time synchronization and the wide deployment of RSUs for centralized coordination. Unfortunately, ubiquitous availability of GPS and wide deployment of RSUs cannot realistically be assumed, especially during the initial deployment of V2X systems. Supplemental time synchronization and distributed coordination systems will be needed to support scenarios where GPS is unavailable or a centralized RSU is not present. For example, a localized time-synchronization system based on the Automatic Self-Time-Correcting (ASTC) wireless beacon is one method of providing a temporary synchronization mechanism when GPS is unavailable. Moreover, coordination among vehicles in a targeted Service Zone (SZ) can be achieved in a distributed peer-to-peer mode instead of a centralized fashion when an RSU does not exist.

References

Acosta, G., Tokuda, K. and Ingram, M.A. (2004) Measured joint Doppler-delay power profiles for vehicle-to-vehicle communications at 2.4 GHz. *Proceedings of the Global Telecommunications Conference*, vol. 6, pp. 3813–3817.

Acosta-Marum, G. and Ingram, M.A. (2007) Six time- and frequency- selective empirical channel models for vehicular wireless LANs. *IEEE Vehicular Technology Magazine* 2(4), 4–11.

Akki, A.S. and Haber, F. (1986) A statistical model of mobile-to-mobile land communication channel. *IEEE Transactions on Vehicular Technology* **VT-35**, 2–7.

Arnold, T., Lloyd, W., Zhao, J. and Cao, G. (2008) IP Address Passing for VANETs. *The Sixth Annual IEEE International Conference on Pervasive Computing and Communications, 2008 (PerCom 2008)*.

Bai, F. and Krishnan, H. (2006) Reliability analysis of DSRC wireless communication for vehicle safety applications. *Proceedings of the IEEE International Conference on Intelligent Transportation Systems (ITSC)*.

Baldessari, R., Festag, A. and Abeille, J. (2007) NEMO meets VANET: A Deployability Analysis of Network Mobility in Vehicular Communication. *Proceedings of 7th International Conference on ITS Telecommunication (ITST 07)*, pp. 1–6.

Briesemeister, L. and Hommel, G. (2000) Role-based multicast in highly mobile but sparsely connected ad hoc networks. *Proceedings of ACM Mobihoc*, pp. 45–50.

Bychkovsky, V., Hull, B., Miu, A., Balakrishnan, H. and Madden, S. (2006) A measurement study of vehicular Internet access using in situ Wi-Fi networks. *ACM MobiCom*, pp. 50–61.

Chandra, R., Ramasubramanian, V. and Birman, K. (2001) Anonymous gossip: Improving multicast reliability in mobile ad hoc networks. *Proceedings of 21st International Conference on Distributed Computing Systems (ICDCS)*, pp. 275–283.

Chen, W. and Cai, S. (2005) Ad Hoc Peer-to-Peer Network Architecture for Vehicle Safety Communications. *IEEE Communications Magazine*, 100–107.

Cheng, L. (2008) Physical Layer Modeling and Analysis for Vehicle-to-Vehicle Networks. Ph.D. Dissertation, Carnegie Mellon University.

Cheng, L., Henty, B., Bai, F. and Stancil, D. (2008a) Doppler Spread and Coherence Time of Rural and Highway Vehicle-to-Vehicle Channels at 5.9 GHz. *Proceedings of the Global Telecommunications Conference, 2008. IEEE GLOBECOM 2008*, pp. 1–6.

Cheng, L., Henty, B., Bai, F. and Stancil, D. (2008b) Highway and rural propagation channel modeling for vehicle-to-vehicle communications at 5.9 GHz. *Proceedings of the Antennas and Propagation Society International Symposium*, pp. 1–4.

Cheng, L., Henty, B., Cooper, R., Stancil, D. and Bai, F. (2007a) A Measurement Study of Time-Scaled 802.11a Waveforms Over The Mobile-to-Mobile Vehicular Channel at 5.9 GHz. *IEEE Communications Magazine* **46**(5), 84–91.

Cheng, L., Henty, B., Cooper, R., Stancil, D. and Bai, F. (2008c) Multi-Path Propagation Measurements for Vehicular Networks at 5.9 GHz. *Proceedings of the Wireless Communications and Networking Conference, 2008. WCNC 2008*, pp. 1239–1244.

Cheng, L., Henty, B., Stancil, D., Bai, F. and Mudalige, P. (2007b) Mobile vehicle-to-vehicle narrow-band channel measurement and characterization of the 5.9 GHz dedicated short range communication (DSRC) frequency band. *IEEE Journal on Selected Areas in Communications* **25**, 1501–1516.

Davis, J. and Linnartz, J. (1994) Vehicle to vehicle RF propagation measurements. *Twenty-Eighth Asilomar Conference on Signals, Systems and Computers*, vol. 1, pp. 470–474.

Fazio, M., Palazzi, C.E., Das, S. and Gerla, M. (2006) Automatic IP address configuration in VANETs. *VANET '06: Proceedings of the 3rd International Workshop on Vehicular ad hoc Networks*, pp. 100–101. ACM, New York, NY, USA.

Gans, M.J. (1972) A Power Spectral theory of Propagation in the Mobile Radio Environment. *IEEE Transactions on Vehicular Technology* **VT-21**, 27–38.

Haas, Z., Halpern, J. and Li, L. (2002) Gossip-based ad hoc routing. *Proceeding of Infocom*, pp. 1707–1716.

Hartenstein, H. and Laberteaux, K. (2008) A tutorial survey on vehicular ad hoc networks. *Communications Magazine, IEEE* **46**(6), 164–171.

Holland, G., Bai, F. and Krishnan, H. (2008) Analysis of DSRC Multi-Channel MAC Performance Public Presentation in IEEE 1609.x Working Group Meeting. San Francisco.

Hu, C., Hong, Y. and Hou, J. (2003) On mitigating the broadcast storm problem with directional antennas. *Proceedings of ICC*, pp. 104–110.

IEEE 1609.4 (2006) *Wireless Access in Vehicular Environments (WAVE)*. IEEE Press chapter P1609.4/d08 multi-channel operation (draft).

IEEE 802.11e (2005) *Std 802.11 Information Technology Telecommunications And Information Exchange Between Systems, Local and Metropolitan Area Networks, Specific Requirements, Part 11: Wireless LAN Medium Access Control (MAC) and Physical Layer (PHY) Specifications*. IEEE Press chapter Amendment 8: Medium access control (MAC) quality of service (QoS) enhancements.

IEEE 802.11p (2006) *Std 802.11 Information Technology Telecommunications And Information Exchange Between Systems, Local and Metropolitan Area Networks, Specific Requirements, Part 11: Wireless LAN Medium Access Control (MAC) and Physical Layer (PHY) Specifications*. IEEE Press chapter Amendment 3: Wireless access in vehicular environments (WAVE).

IETF (2007) http://www.ietf.org/html.charters/OLD/nemo-charter.html.

Korkmaz, G., Ekici, E. and Ozguner, F. (2006) An efficient fully ad-hoc multi-hop broadcast protocol for inter-vehicular communication systems. *Proc. of IEEE ICC*, pp. 423–428.

Korkmaz, G., Ekici, E., Ozguner, F. and Ozguner, U. (2004) Urban multi-hop broadcast protocol for inter-vehicle communication systems. *Proc. of ACM International Workshop on Vehicular Ad hoc Networks*, pp. 76–85.

Kyasanur, P. and Vaidya, N. (2005) Capacity of multi-channel wireless networks: Impact of number of channels and interfaces. *Proceedings of the ACM International Conference on Mobile Computing and Networking (MOBICOM)*, pp. 43–57.

Liberti, J. and Rappaport, T. (1996) A geometrically based model for line-of-sight multipath radio channels. *Proceedings of the 46th IEEE Vehicular Technology Conf.*, pp. 844–848.

Masui, H., Kobayashi, T. and Akaike, M. (2002) Microwave path-loss modeling in urban line-of-sight environments. *IEEE Journal on Communications* **20**, 1151–1155.

Maurer, J., Fugen, T. and Wiesbeck, W. (2002) Narrow-band measurements and analysis of the intervehicle transmission channel at 5.2 GHz. *Proceedings of the 55th IEEE Vehicular Technology Conf.*, vol. 1, pp. 1274–1278.

Maurer, J., Fugen, T., Schafer, T. and Wiesbeck, W. (2004) A new inter-vehicle communications (IVC) channel model. *Proceedings of the 6th IEEE Vehicular Technology Conf.*, pp. 9–13.

Mo, J., So, H.S. and Walrand, J. (2005) Comparison of multi-channel MAC protocols. *Proceedings of the ACM International Symposium on Modeling, Analysis, and Simulation of Wireless and Mobile Systems (MSWiM)*, pp. 209–218.

Molisch, A.F., Tufvesson, F., Karedal, J. and Mecklenbrauker, C.F. (2009) Propagation aspects of vehicle-to-vehicle communications: an overview. *Proceedings of the IEEE Radio and Wireless Symposium (RAWCON 2009)*.

Nesargi, S. and Prakash, R. (2002) MANETconf: Configuration of Hosts in a Mobile Ad Hoc Network. *Proceedings of INFOCOM*, pp. 1059–1068.

Ni, S., Tseng, Y. and Shih, E. (2001) Adaptive approaches to relieving broadcast storms in a wireless multihop mobile ad hoc network. *Proc. IEEE 21st International Conference on Distributed Computing Systems*, pp. 481–488.

Ni, S., Tseng, Y., Chen, Y. and Sheu, J. (1999) The broadcast storm problem in a mobile ad hoc network. *Proceedings of the ACM International Conference on Mobile Computing and Networking (MOBICOM)*, pp. 151–162.

Ozdemir, M.K. and Arslan, H. (2007) Channel Estimation for Wireless OFDM Systems. *IEEE Communications Surveys & Tutorials* **9**(2), 18–48.

Pagani, E. and Rossi, G.P. (1997) Reliability broadcast in mobile multihop packet networks. *Proceeding of Mobicom*, pp. 34–42.

Patel, C., Stuber, G. and Pratt, T.G. (2005) Simulation of Rayleigh-faded mobile-to-mobile communication channels. *IEEE Transactions on Communications* **53**, 1876–1884.

Perkins, C. (1997) *Mobile IP: Design Principles and Practices*. Wireless Communications Series. Addison-Wesley.

Punnoose, R.J., Nikitin, P.V., Broch, J. and Stancil, D.D. (1999) Optimizing wireless network protocols using real-time predictive propagation modeling. *Proceedings of 1999 IEEE Radio and Wireless Conference (RAWCON)*, pp. 39–44.

Rappaport, T.S. (2002) *Wireless Communications: Principles and Practice*, 2 edn. Prentice Hall.

Sen, I. and Matolak, D. (2008) Vehicle–vehicle Channel Models for the 5 GHz Band. *IEEE Transactions on Intelligent Transportation Systems* **9**, 235–245.

Spyropoulos, T., Psounis, K. and Raghavendra, C. (2004) Single-copy routing in intermittently connected mobile networks. *The Proceedings of the 1st IEEE Comm. Society Conf. (SECON'04)*, pp. 235–244.

Spyropoulos, T., Psounis, K. and Raghavendra, C. (2005) Spray and wait: An efficient routing scheme for intermittently connected mobile networks. *WDTN*, pp. 252–259.

Sun, Y. and Belding-Royer, E.M. (2004) A study of dynamic addressing techniques in mobile ad hoc networks. *Wireless Communications and Mobile Computing*, 315–329.

Tonguz, O., Wisitpongphan, N., Bai, F., Mudadlig, P. and Sadekar, V. (2007) Broadcasting in VANET. *IEEE MOVE Workshop 2007*, pp. 7–12.

USDOT (2005) Vehicle Safety Communications Project, Task 3 Report, Identify Intelligent Vehicle Safety Applications Enabled by DSRC. http://ntl.bts.gov/lib/29000/29500/29505/CAMP3scr.pdf.

Vahdat, A. and Becker, D. (2000) Epidemic routing for partially connected ad hoc networks. Technical report, Duke University.

Wang, L.C. and Cheng, Y.H. (2005) A statistical mobile-to-mobile Rician fading channel model. *Proceedings of the 61st IEEE Vehicular Technology Conf.*, pp. 63–67.

Wiesbeck, W. and Knorzer, S. (2007) Characteristics of the mobile channel for high velocities. *Proceedings of the 2007 Int. Conf. Electromagnetics in Advanced Application*, pp. 116 – 120.

Wisitpongphan, N., Tonguz, O., Bai, F., Mudalige, P. and Sadekar, V. (2007a) On the routing problem in disconnected vehicular ad hoc networks. *Proc. IEEE INFOCOM*, pp. 2291–2295.

Wisitpongphan, N., Tonguz, O., Parikh, J., Mudalige, P., Bai, F. and Sadekar, V. (2007b) Broadcast storm mitigation techniques in vehicular ad hoc networks. *IEEE Wireless Communications* **14**(6), 84–94.

Xu, Q., Mak, T., Ko, J. and Sengupta, R. (2004) Vehicle-to-Vehicle Safety Messaging in DSRC. *1st ACM Workshop on Vehicular Ad hoc Networks (VANET)*, pp. 19–28.

Yin, J., ElBatt, T.A., Yeung, G., Ryu, B., Habermas, S., Krishnan, H. and Talty, T. (2004) Performance evaluation of safety applications over DSRC vehicular ad hoc networks. *Proceedings of the 1st ACM International Workshop on Vehicular ad hoc Networks*, pp. 1–9.

Yin, J., Holland, G., ElBatt, T., Bai, F. and Krishnan, H. (2006) DSRC channel fading analysis from empirical measurement. *Proceedings of the 1st IEEE International Workshop on Vehicle Communications and Applications (Vehiclecomm)*.

Zajic, A., Stuber, G., Pratt, T. and Nguyen, S. (2009) Wideband MIMO Mobile-to-Mobile Channels: Geometry-Based Statistical Modeling With Experimental Verification. *Vehicular Technology, IEEE Transactions on* **58**(2), 517–534.

Zhao, J. and Cao, G. (2006) VADD: Vehicle-assisted data delivery in vehicular ad hoc networks. *Proc. IEEE INFOCOM*, pp. 1–12.

4

Communication Systems for Railway Applications

Benoît Bouchez and Luc de Coen

Bombardier Transportation France

Until recent years, communication between trains and ground did not involve the train by itself as a communication component. New requirements from operators for flexibility and customer services are now transforming the train into communicating equipment that must couple to fixed installations for persistent data exchanges. The goal of this chapter is to examine railway-specific communication requirements and compare them with existing and future wireless communication systems.

4.1 Evolution of Embedded Computers and Communication Networks in Railway Applications

Compared with their use in other industries, embedded computers have only recently been introduced into rolling stock, due to the harsh railway environment (e.g. presence of high voltages and currents, strong magnetic fields, wide temperature range, etc.).

This introduction is partitioned into four different periods (Bouchez 2008).

The *first generation* computers in trains (1980–1990) were completely independent from each other. Each of these computers was dedicated to one task (e.g. propulsion, brakes, etc.). At this time the functions of these computers were related only to train control, with some maintenance functionalities but without any 'passenger comfort' functions. Direct communication between train computers and the ground was not possible, apart from some very specific driverless train control.

Vehicular Networking   Edited by Marc Emmelmann, Bernd Bochow, C. Christopher Kellum
© 2010 John Wiley & Sons, Ltd

The *second generation* architecture of train computers (1990–2000) was still based on dedicated computers for each function, but new functionalities for passenger information were introduced, using LCD and LED displays to provide travel information. Embedded networks, such as Token Ring Network Alsthom Device (TORNAD) networks on TGV trains in France, were introduced in order to connect the different computers together. The lack of direct communication between the train and the ground made it impossible to display dynamic information (e.g. train delays). However, some safety-related systems introduced in this period were able to transmit information punctually from ground to train. Some experiments were also performed to transmit information using a radio modem with a bit rate lower than 1200 bps (for the maintenance link on the Régie Autonome des Transports Parisiens (RATP) MP-89 rolling stock). The KVB system in France is an example of a ground to train communication system of this generation.[1]

Automatic Train Control (ATC) systems can be seen as a specific case, since they have been used in commuter trains (metros) since 1967 (e.g. the RATP PA-BF system), and the first implementations were strictly hardware-based with no software. It should also be noted that fully bidirectional communication within ATC applications appeared only with the RATP MP-89 SAET system, which was developed in 1989 and implemented in 1992.

The *third generation* architecture of train computers (2000–2005) differs from the second generation mainly in its systematic use of communication systems between trains and the ground for normal train operation. The second generation architecture used wireless communication as a backup connection, and only for maintenance purposes. Third generation train to ground communication links are mainly used on these trains for predictive maintenance. Types of service include a journaling report sent each day by trains to a maintenance server, updates of the service database or timetables, and even seat reservation information. In this architecture, a given computer is often shared between multiple functions.

Third generation trains mainly use Global System for Mobile Communication (GSM)-based communication systems. Thus, it is generally not feasible to maintain communication during the whole train service. In most cases, communication is established for a limited time, typically for file transfers between the train and the ground.

The *current generation* architecture of train computers (the fourth generation) began to be implemented on trains in 2005. Compared with the third generation, the most significant difference is the use of distributed systems, which results in a lower number of computers that are connected together with high-bandwidth networks such as Ethernet. The second major difference is the integration of new services based on a network (e.g. digital audio, digital video, and Internet access). This generation targets a cyclical or permanent connection between trains and the ground, making the train a part of a global communication scheme.

4.2 Train Integration in a Global Communication Framework

Train integration in a global communication framework is currently required for multiple kinds of services, especially traffic control and supervision services, those addressing interoperability issues, and dynamic passenger information services.

---

[1]KVB stands for 'Controle de Vitesse par Balise', i.e. speed control by *balise*. A *balise* is an electronic beacon or transponder placed between the rails of a railway as part of an automatic train protection system. The French word *balise* is used to distinguish these beacons from other kinds of beacon, and in French railways operator terminology the letter 'K' is used instead of 'C' to stand for 'Control', in order to avoid confusion with the word 'Command'.

Traffic control and supervision is required on train lines with very high service requirements. In this case, a train can no longer operate autonomously, with actions on the train being made by the engineer only. The Le Réseau Express Régional (RER) line A in Paris is one of the most famous examples, since this line cannot be operated under nominal load without the automated exploitation, maintenance, and driving assistance known as Système d'aide à la conduite, à l'exploitation et à la maintenance (SACEM), since the time between trains is too short for the use of classical signaling.

Interoperability issues are addressed by projects like the European Rail Traffic Management System (ERTMS), which also integrates the train into a global communication framework. For example, Thalys high speed trains operating between Paris, Brussels, and Amsterdam can meet up to six different security systems during a complete service (up to two security systems per country). The two basic components of ERTMS, the European Train Control System (ETCS) and GSM-Rail (GSM-R), are the keys for integration into a global communication framework.

Dynamic information is required by travelers who consider that simple information such as train destination and scheduled arrival time displayed on internal screens is far from sufficient. For these travelers, the train should also display dynamic information such as the next station name, delay estimations, and causes of delays. By integrating the train into a global communication framework, the train itself can receive information directly from the ground, including the reason for a red signal and any potential delay. Various opinion polls taken by train operators have shown that such dynamic information gives a pleasant feeling to travelers even if delays are occurring, because the travelers are 'kept in the loop'.

4.3 Communication Classes and Related Communication Requirements

4.3.1 Real-time data

Real-time data are used for information that is valid only for a limited period of time. Typically, values that can constantly evolve during the normal behavior of the system such, as motor current or door status, are considered real-time data. This class of data is used for unidirectional exchanges between one source and one or many receivers, using multicast or broadcast addressing.

Real-time data are exchanged using cyclic communication, to keep latency as lowest as possible. In this case, exchange reliability is ensured by cyclic repetition of the messages, and it is generally assumed that the loss of a given number of consecutive messages from one source is acceptable, since the data image will be refreshed by the next message. This is especially important when radio links are involved, where messages can be lost due to bad communication conditions or when radio link availability must be raised for other communication classes (Figure 4.1).

Real-time data can be exchanged using underlying protocols that guarantee delivery reliability. However, the maximum jitter and delay of the network must be guaranteed, since an acknowledgment mechanism will be used on the return path. Excessive jitter has a negative impact on sampled systems since it leads the real-time tasks to be processed out of synchronization. Excessive delay leads processing system to consider the data to be outdated.

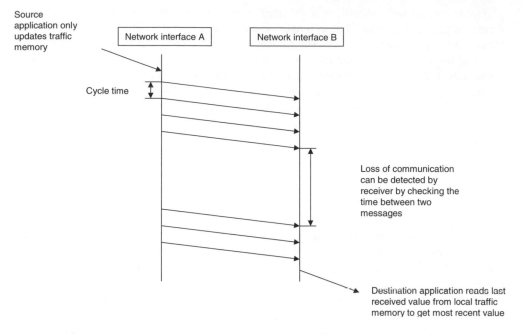

Figure 4.1 Real-time data transmitted by cyclic messages

The receiver can detect loss of communication with the transmitter by measuring the time between two messages. However, it should be noted that the reception of real-time data telegrams is not sufficient to guarantee correct system activity. In most multiprocessor and/or multithreaded systems, the communication and computation tasks use separated resources and exchange data through shared memory and not by events. The cyclic transmission only confirms that the communication resources are active. It is highly recommended that each cyclic telegram contain a dynamic component, or 'life sign', that is computed not by the communication task but rather by the application itself. A life sign value that does not change within a given time (generally the same amount of time as the cyclic timeout) indicates that data contained in the telegram are not valid.

4.3.2   Non-real-time message data

Specific protocols must be used when message delivery across the network must be guaranteed. Command messages typically use this class of data communication, since the sender must be informed of the transaction result. Such data exchanges imply a form of acknowledgment; thus a single command results in multiple messages on the network. In most cases the delays induced by the network lead to unpredictable latency times; therefore, the term 'non-real-time' is used.

Non-real-time messages are also to be used when evolution of data as a function of time is considered sporadic. In this case non-real-time data exchange is a way to limit the bandwidth usage, since no communication takes place when data are not evolving.

The non-real-time message data class is typically used for 'write messages' and 'read messages'.

A 'write message' is formed from two parts: the data message, which contains the data to send to the destination, and the acknowledge message, which is sent from the data receiver to the data sender and confirms to the sender that the message has been received and taken properly into account (Figure 4.2). An acknowledge message is issued by the receivers to confirm the correct message delivery, and triggers a message retransmission by the sender in the case of a network failure.

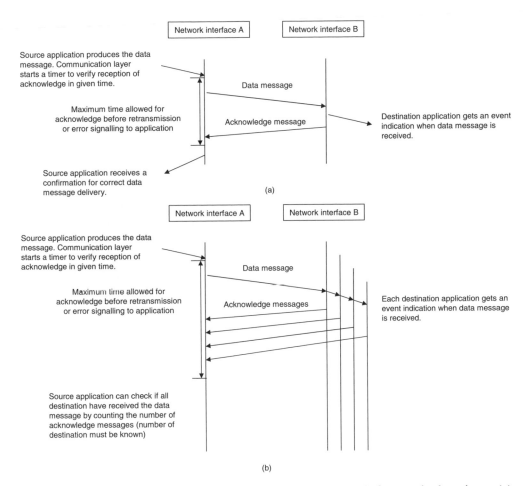

Figure 4.2 A 'write message' being used to transmit commands from a single point to (a) one or (b) multiple destinations, depending on the addressing method

A 'read message' is also formed from two parts: the request message that indicates the requested data to read, and the reply message that contains the requested data. The lack of a reply allows the requester to detect a communication problem (Figure 4.3).

It should be noted that the addressing method has an impact on this class of messaging. A unicast command will produce only one acknowledgement message. A command message being multicasted will produce multiple acknowledgement messages. Thus, it may be

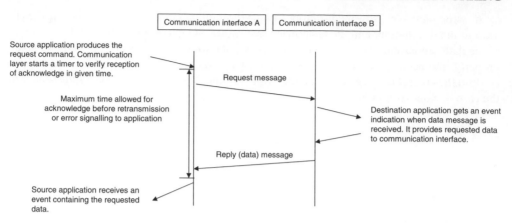

Figure 4.3 A 'read message' being used by a transmitter to retrieve values from remote stations

necessary for the sender to know how many destinations are supposed to answer in order to check that all of them have produced the expected acknowledgement message.

Broadcast addressing is not considered to be compatible with the acknowledgement mechanism since the number of destinations is generally not known.

4.3.3   Streaming data

The streaming data class is quite similar to the real-time data class. However, streams are linked to a command that allows for starting and stopping them. The stream data may then be absent from the network without this being considered a fault.

The jitter and delay requirements for the streaming data class are generally less stringent than for real-time data, since streams are used mainly for audio and video transport.

The loss of packets in audio streams leads to 'click' noises where extreme situations may affect the message intelligibility. In the case of video streams, most compression schemes are based on previous message content. The loss of a single message may affect the whole stream for a given time. The network reliability requirements for the streaming data class are then more stringent than for real-time data.

Streaming data exchange is often seen as unidirectional. However, many applications require a control stream (e.g. Real-Time Control Protocol, or RTCP) that informs the sender of packet delivery at the receiver. The control stream normally uses only a few percent of the throughput of the main stream.

4.4   Expected Services from a Railway Communication System and the Related Requirements

4.4.1   Automatic Train Control

Automatic Train Control (ATC) covers various applications, from train protection – also called Automatic Train Protection (ATP) – to full automatic train driving (Gabillard 1987).

Train protection primarily uses ground to train communication to report ground signalling and speed limitation information into the locomotive cabin.

Full ATC (automatic pilot) is also based on ground to train communication. Older systems often use localized bidirectional communication systems in order to give the train the ability to report information to the ground (e.g. door status). Recent systems require permanent bidirectional communication links.[2]

ATC systems are based on the real-time data class. Bidirectional systems can also use a specific communication class, inherited from the message data class, called a hybrid message exchange. In this case, each message (train to ground and ground to train) contains data. The communication initiator, for example the ground controller, sends a message containing commands to a train within its control range. The receiver then answers with its own data packet. This message reply (Figure 4.3) is then considered by the sender as the acknowledgement message.

ATC applications require the communication to guarantee absolute data integrity (performed most often in OSI layer 2 and/or layer 3). ATC applications using open communication links (i.e. radio) must also be protected against masquerading (by use of authentication), eavesdropping (by use of encryption) and denial of service (by detection of flooding).

The amount of data exchange required for ATC is very low (most often a few kilobits per second). Latency for ATC applications is required to be low to medium, depending on the application profile (one second is generally considered as a maximum).

4.4.2 Passenger Information System

The requirements for passenger information presented in this section are based on audio and display systems.

Passenger visual displays

LED matrix displays require a low amount of data (i.e. typically a few hundred bytes, with a delay between messages of larger than one minute). TFT displays using graphics, however, can require a substantial volume of data, between one and two megabytes, when the display content is updated. The updates result in message bursts on the network.

It is quite rare to update Passenger Information Systems (PISs) dynamically from the ground. In most cases, display content is managed from an on-board server that is updated offline during maintenance phases. The display content is usually related to the mission, and is defined by engineer manual entry.

Seat Reservation Displays (SRDs) are the exception to this last statement. These displays are used on train lines where seat reservation is not mandatory (i.e. travelers have to pay to ensure that they get a seat). When making a seat reservation, travelers receive a voucher indicating the coach and the seat number. Inside the trains, a reservation tag or an electronic display above the seat indicates the reservation and the connection for which it applies. SRDs rely heavily on train integration into the communication framework since they require content update at every station. However, data volume is generally quite low even if a high number of seats have to be updated.

---

[2]For example, the Système Automatisé d'Exploitation des Trains (SAET) system on Paris Metro Line 14.

PIS systems tolerate high latency in most applications, but they require precise synchronization. The synchronization is generally handled by the on-board server, which uses a multicast command to trigger display update.

Passenger audio

Audio streaming is used for multiple purposes with different requirements:

- Public Address (PA) for diffusion of audio from one point to the whole train;

- intercom for communication between two points;

- entertainment: distribution of music or news, either to seats or by public broadcasting;

- communication between train and ground.

The PA application is sensitive to latency, which must be kept less than 50 ms if the source is located in the same train. The echo effect makes it almost impossible to talk intelligibly if people speaking can hear themselves with longer delays. PA with a remote source (typically ground to train or train to ground) tolerates much higher latency.

Bandwidth used for audio streaming can vary from a few kilobits per second up to 500 kilobits per second, depending on the stream encoding. Applications like PA and intercom are generally not compatible with high compression algorithms (such as MP3), since they introduce high latency. Such applications require the use of an uncompressed format, resulting in larger bandwidths.

Intercom applications between on-board points or between ground and train tolerate a higher latency than does PA (up to 250 ms) if echo suppression techniques are used for each source. In this case, it can be highly profitable to use compression systems to lower the required bandwidth. This is especially helpful for train to wayside communication, where the available bandwidth is more restricted than for on-board networks.

Audio streaming applications also require high reliability when they are used for emergency calls.

4.4.3 Video

Video streaming is used for multiple purposes, each with different requirements:

- Closed Circuit TV (CCTV) for security-related recording;

- retrovision, used by the engineer to verify passenger flow without the need to look through cabin windows;

- entertainment: distribution of video (e.g. news or films), either to seats or by public broadcasting;

- platform control from the train.

The main challenge with video streams is the amount of data required. In all cases video compression algorithms must be used, introducing delays on the signal path. Modern compression algorithms such as MPEG-4 allow a $4 \times$ CIF (4CIF) picture (i.e. 704 by 576 pixels) at 25 frames per second to use an average bandwidth of 1 Mbit/s with bursts of up to 2 Mbit/s. It must be noted that some compression algorithms produce much higher burst rates than average bit rates. The choice between any two algorithms must take into account both the average bandwidth used by each stream and the risk of video bursts that may lead to network saturation and packet loss.

Real-time video streaming applications tolerate latencies of up to one second for real-time display, but are less sensitive to packet loss. Recording applications for security tolerate even higher latencies, but there is often a requirement that they must not suffer from any packet loss.

Packet loss for real-time video streaming applications is often critical, since the compression algorithms used rely on previous packets to compute new frames. The loss of a single packet can produce useless video pictures for seconds, which may not be acceptable, especially for security-related recordings.

4.4.4 Maintenance

Maintenance data are stored into local train memory that is transferred to the ground only when the train returns to the depot, or upon receiving a specific request from the ground. Since it is important to guarantee that all maintenance data are transferred properly, an acknowledged message data class must be used. On the other hand, the maintenance data use supports very high latency data transfers.

This class of application data also includes train software updates and train parameters updates, such as changes to timetables. These types of application data have the same requirements as do on-board maintenance data for reliable transfer.

The amount of data to be transferred is extremely variable, but it can easily reach hundreds of kilobytes or even megabytes. This implies a requirement for high bandwidth data communication links between train and ground.

4.4.5 On-board Internet access

Internet access is a specific class of data to be exchanged between train and ground, since it is related to travelers. Access to the Internet for train operation is considered within the data classes previously discussed.

The most common way for travelers to connect to the Internet is through the use of Wi-Fi Access Points located in each car.

Preliminary case studies have shown that most travelers only use the on-board Internet access to read and send e-mails. Only a small proportion currently try to use Internet access for browsing or other high bandwidth usages.

The most critical point about on-board Internet access is its security. It must be ensured that travelers cannot reach embedded operational networks from their computers. The minimum safety requirement is the use of gateways to isolate the travelers' network from the train operational networks. Some train manufacturers prefer to use two different networks, each having its own link to the ground, in order to guarantee that travelers cannot gain access, either accidentally or intentionally, to the operational networks.

4.5 Qualitative and Quantitative Approach for Dimensioning Wireless Links

4.5.1 Environmental influence

Trains encounter many different environments during a typical service. For example, a train that leaves a city's central station, surrounded by high concrete buildings, can then pass into a tunnel to join a track that is located on an open ground. Each of these environments has a different propagation model for wireless communication.

Each environment has its own propagation conditions that affect the radio signal. The same apparent conditions may give different results due to changes in the hidden environmental structures. For example, the wave propagation in a tunnel is affected not only by the concrete walls but also by the underlying metallic structures.

Consequently, it is impossible to predict the wave propagation for a given environment. The only practical way to determine the propagation conditions in a given location is through the use of dedicated measurement tools, such as spectrum analyzers and power meters.

Another environmental influence that must be taken into account is spectrum pollution, especially if the communication system uses frequency bands that are shared between multiple users, such as the Industrial, Scientific, and Medical (ISM) bands. Usage of these bands is often coupled with significant limitations in transmission power, which introduces another limitation. The spectrum pollution can also come from unexpected sources, like microwave ovens, that produce interference in the 2.4 to 2.5 GHz ISM band. In the USA, the situation is also critical for the 901 to 928 MHz ISM band (which is reserved for GSM in Europe), where the interference level is very high.

The only way to protect an installation against interference is to specify the exclusive use of frequency bands as a primary user in a licensed spectrum.

4.5.2 Global propagation model

The Equivalent Isotropic Radiated Power (EIRP), expressed in dBm, is one of the most important parameters to take into account. This value reflects the output power of the transmitter. A higher EIRP gives a wider transmission range, but the maximum EIRP for a given band is most often restricted by local regulations.

The propagation model is influenced by the frequency used for the transmission. High bandwidth radio links require higher frequency carriers and/or more complex modulation schemes. A higher frequency is more easily attenuated by distance for the same propagation conditions, but it also yields better directional propagation characteristics. It must be noticed that antenna gain relative to antenna size is better for higher frequencies, which partially counterbalances the attenuation effect.

Material absorption is also higher for higher frequencies. Absorption is dependent on the material on the wave path and is also related to wave frequency. As an example, a concrete wall with a 25 centimeter thickness produces an attenuation of around 10 dB at 2.4 GHz and 15 dB at 5 GHz. The atmosphere produces a negligible absorption for frequencies up to 10 GHz.

Obstacles met by the waves affect the propagation in three ways: a part of the wave energy is absorbed by the obstacle, a second part is reflected, and the last part is propagated across the obstacle. Since the reflected part can combine with the direct signal and the propagated signal, it produces negative modulation effects. Some modulation schemes are consequently more easily affected by obstacles.

The diffraction phenomenon must also be taken into account when obstacles with dimensions larger than the wavelength are placed on the wave path. Reciprocally, when obstacles with dimensions smaller than the wavelength are placed on the wave path, the scattering phenomenon must be taken into account.

4.5.3 Train motion influence

The handover capability of a radio system can become critical with fast moving trains. Let us consider, for example, a radio system with access points every 500 meters. A train moving at 100 km/h (27.8 m/s) will pass from one access point to another every 18 seconds. If the radio system takes one second to negotiate the transfer from one point to another, the global availability is decreased by 5%. Such a decrease can have a very high impact on safety-related systems, such as ATC.

The Doppler effect may also need to be taken into account, depending on the carrier frequency and train speed, since it may affect the signal (from the receiver point of view) in terms of phase and frequency.

4.5.4 Regulation and licensing

Each country defines the frequency spectrum usage through a regulatory body (e.g. the Federal Communications Commission, or FCC, in the USA and the European Conference of Postal and Telecommunications Administrations, or CEPT, and the European Telecommunications Standards Institute, or ETSI, in Europe). The different national bodies communicate on a regular basis through the World Radiocommunication Conference (WRC) under the International Telecommunication Union (ITU).

The regulatory bodies are responsible for frequency band allotment. Each band can be licensed for exclusive use or for license-exempt use, depending on local regulations. Some shared bands, such as the ISM bands, can be used without a preliminary individual declaration to the regulation body. However, these bands are very restricted in terms of emitted power (i.e. EIRP) in order to achieve interference mitigation.

Critical applications for which interference risk must be minimized, or where the required EIRP exceeds the value allowed for license-exempt use, must then rely on an allotted frequency.

4.6 Existing Wireless Systems Applicable to Railway Communication Systems

4.6.1 Magnetic coupling technology

Magnetic coupling technology is based on induction loops installed along the track. Detectors ('magnetic heads') installed on the train receive the magnetic field produced by induction loops and transmit the signal to on-board decoders (Figure 4.4). A similar system can be installed on board to transmit signals from train to ground.

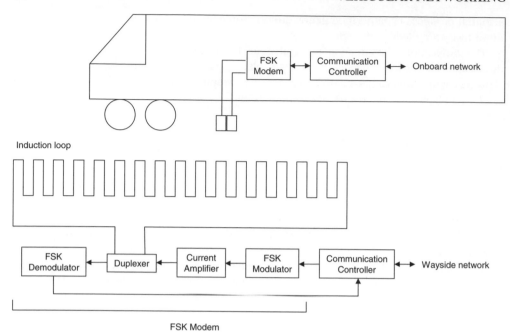

Figure 4.4 The magnetic coupling technology, based on an induction loop running along the train track

In some ATC systems, the induction loop shape is also arranged to provide distance or speed information. A dual-phase system is used on board to detect transmissions on the loop and compute the equivalent relative distance.

Track induction loops have high inductance value, due the great length of cable. The carrier bandwidth is then limited to a few tens of kilohertz. The direct consequence is a limitation on data rate between train and ground. Magnetic coupling technology generally limits the data rate to less than 5 kilobits per second. It should also be noticed that the magnetic loop is quite fragile and must be protected by a container, which has a cost impact when the system must be deployed on the whole track length.

Magnetic coupling technology is often used for ATC due to the low bandwidth requirements of these applications. It is also compatible with maintenance-related communications and Passenger Information Systems.

Unidirectional ATC applications can use punctual magnetic couplers (generally one per signal, used mainly for repetition), but other application profiles require magnetic units to cover the whole track length, making the technology quite expensive. This last kind of application profile is then restricted to limited length lines, typically commuter train lines.

4.6.2   WLAN/WMAN technologies

IEEE 802.11 (Wi-Fi)-based systems

The IEEE 802.11 working group has defined multiple Wireless Local Area Network (WLAN) standards (IEEE 802.11 2007), based mostly on the ISM bands (2.4 GHz and 5 GHz). Each standard uses a different modulation scheme, making the various standards incompatible with one another. User data rates vary from 900 kbit/s (IEEE 802.11) to 54 Mbit/s (IEEE 802.11agh).

The main goal for IEEE 802.11 working groups has not been oriented towards mobility. As of now, there is no official IEEE 802.11 standard dedicated to mobile users, and only a few tests have been performed to check the ability of the various standards to support communication with mobiles.

Experimentation has shown that most Wi-Fi standards, as they were originally designed, do not support communication with mobiles moving at more than 80 km/h, which is extremely restrictive for most train applications. Consequently, Wi-Fi connections are not well suited for moving train applications and should be reserved for fixed connections (typically in stations), since handover delays and the related communication losses quickly become significant.

In 2004 the IEEE launched a working group to develop the IEEE 802.11p standard. This version aims for the capability to support a 6 Mbit/s data rate with mobiles moving at speeds of up to 250 km/h. The IEEE 802.11p will use a licensed 5.9 GHz carrier frequency with OFDM/PSK modulation.

The IEEE 802.11 standards are affected by the contention access mechanism: every station that wants to communicate with the Access Point (AP) competes for the AP's attention on a random interrupt basis. Closer stations get higher priority, since distant stations are more likely to be interrupted. Streaming applications (audio and video) are not well supported by Wi-Fi since they are sensitive to communication loss.

IEEE 802.16 (WiMAX)-based systems

The IEEE 802.16 working group is currently developing a new set of standards known as Worldwide Interoperability for Microwave Access (WiMAX). IEEE 802.16 (2007) aims for wide-range communication with high bandwidth, where IEEE 802.11 aims for short-range communication.

A typical WiMAX system is able to cover up to 30 km using licensed frequencies with a Line-of-Sight (LoS) installation and frequencies from 10 to 66 GHz. Non-LoS installations in the 11 GHz frequency band have lower ranges of up to 10 kilometers, depending on propagation conditions.

The IEEE 802.16d specification describes a communication standard based on Orthogonal Frequency Division Multiplexing (OFDM) technology, supporting data rates up to 100 Mbit/s using a frequency band between 2 and 11 GHz. Preliminary tests have shown that this standard is able to cover some mobile applications when the speed is less than 80 km/h; however, performance with mobile stations quickly diminishes as speed increases.

The IEEE 802.16e standard (IEEE 802.16e 2005) has been issued to address specific mobile requirements. The data rate is less than that of IEEE 802.16d links but it supports communication with mobiles at speeds up to 150 km/h in the 5 GHz band. This standard is based on a different access method, named Scalable Orthogonal Frequency Division Multiplex Access (SOFDMA), dynamically sharing the available radio bandwidth between all users.

A significant advantage of WiMAX over Wi-Fi is the access scheduling algorithm, in which the stations compete to get access only when they want to join the network (Wi-Fi subscribers use contention access). When a subscriber has been granted an access slot, it keeps it until it leaves the network. The access slot time can be enlarged or reduced depending on channel load and Quality of Service (QoS) settings, yielding better bandwidth efficiency.

Another advantage over Wi-Fi is the use of a stream-based QoS, which insures better behavior for streaming applications, such as voice or video. Wi-Fi uses a packet-based QoS tagging mechanism.

4.6.3 Cellular technologies

Cellular networks are made of small cells arranged in a honeycomb manner, each handling a given number of simultaneous connections.

The main advantage of public cellular technologies is that they are now widely distributed, offering the capability of connection almost everywhere around main train tracks. The cellular infrastructure can then be shared between multiple users, avoiding the need for the train operator to install dedicated infrastructure.

GSM/GSM-R

The GSM system is the most common cellular system across the world. It offers the interesting capability of supporting both voice and data transmission. GSM also offers encryption capabilities via various algorithms (e.g. A3, A5, and COMP128), but it has also been demonstrated that some of these have been compromised.

The transmission rate between a Base Station (BS) and a Mobile Station (MS) is 270.833 kbit/s. This bandwidth is shared between eight time slots (TSs), giving each of them a raw bandwidth of 33.8 kbit/s. The resulting usable bandwidth for each TS is 24.7 kbit/s, and this is shared between voice and data channels.

Depending on the propagation conditions, the bandwidth available for the data channel can change from 300 bps to 9.6 kbit/s. This restricts the use of GSM to low bandwidth application profiles, typically PIS updates. Data can be exchanged either by the transparent service or by a non-transparent service. The first has a fixed delivery delay, but data integrity is not guaranteed. The non-transparent service guarantees data integrity through a repetition upon request mechanism (ARQ), which results in a variable delivery delay.

Since a GSM network is shared between multiple users, there is a non-negligible risk that a BS reached by the train has already allocated all available channels. The risk is lowered by the fact that a MS always tries to gain contact with three BSs. It normally selects the nearest one in order to lower power consumption, but it can also connect with a more distant one if no more channels are available on the nearest one, which should lower the risk of communication loss.

The risk of communication loss is generally higher with trains running outside city areas, where the number of BSs is lower. A good rule of thumb for maximum distance between BS and MSs is around 35 km. City areas are normally covered by a higher number of BSs, but cell size is smaller. This has a low impact on slow-moving MSs, but trains are impacted by the need to change the BS to which they are connected more often. Communication availability can be affected by the multiple negotiation phases. In order to avoid the public GSM limitations for train applications, the GSM-R version has been designed by the Union Internationale des Chemins de Fer (UIC). GSM-R is a part of the ERTMS European project (GSM-R 2006).

GSM-R has the same capabilities (voice and data communication) and intrinsic limitations as standard GSM, but it uses a reserved frequency range. The uplink (MS to BS) uses the 876 to 880 MHz band, while the downlink (BS to MS) uses the 921 to 925 MHz band. The entire bandwidth is available for the railway operator, since it is not shared with public users. GSM-R also provides specific functionalities for railway operators that are not available to standard MSs. These specific functionalities are described in the UIC-EIRENE (European Integrated Railway Radio Enhanced Network) document (GSM-R 2006).

The UIC recommends that GSM-R BSs are installed along the rail tracks every three to four kilometers, in order to keep the data rate as high as possible and lower the risk of communication loss when the MS switches from one BS to another.

GPRS

The General Packet Radio Service (GPRS) has been developed over GSM standards in order to enhance data transmission capabilities. It is often termed '2.5G', since its capabilities are between those of the standard GSM (2G) and the UMTS/EDGE Third Generation (3G) ones.

The main difference from GSM is that data are exchanged by packets, and only when needed. GSM uses a 'circuit' mode that is established for the entire communication time even when no data are exchanged. The resources allocated for data transfer under GPRS are released as soon as the packets have been delivered.

The other difference from GSM is the capability of using multiple TSs for each channel, where GSM is restricted to using only one. The number of TSs allocated to each MS, between two and eight, is defined by the BS depending on its availability and saturation.

Each TS can use one of four different coding schemes, depending primarily on the distance between the MS and the BS. Each coding scheme has a speed limit (9.05 kbit/s, 13.4 kbit/s, 15.6 kbit/s, and 21.4 kbit/s). The maximal attainable data rate is then 171.2 kbit/s (8×21.4 kbit/s). However, the average data rate is around 50 kbit/s with a typical maximum data rate of 110 kbit/s.

The ability to run GSM and GPRS in parallel depends on the protocol capability of the devices. Class A devices are able to use both GSM and GPRS links in parallel. Class B devices are the most common. They can use both GSM and GPRS, but have to switch automatically from one mode to another (GSM mode stops the GPRS mode and vice-versa). Class C devices can only use one of GSM or GPRS. GPRS modems are often class C devices.

GPRS is a 'best-effort' system. It does not guarantee any quality of service nor delivery delay. The resulting latency can be very high (up to one second), which makes GPRS unsuitable for the real-time data class.

EDGE

Enhanced Data Rates for GSM Evolution (EDGE) has been a first step to third generation cellular systems in order to reach the higher data rates required for multimedia applications. Since it still uses the 2G base architecture but is part of ITU's 3G specification, EDGE is sometimes referred to as '2.75G'.

As with GPRS, EDGE systems are able to use multiple channels in parallel. Each channel can use one of nine different coding schemes: four coding schemes using Gaussian Minimum Shift Keying (GMSK) modulation (like GPRS) plus five new coding schemes based on 8 level PSK (8-PSK). The 8-PSK modulation schemes allow transmission of three bits per symbol (as opposed to one bit per symbol with GMSK), giving a capability of 59.2 kbit/s per channel.

The maximum theoretical data rate is then 473.6 kbit/s when the eight slots are used with MCS-9 encoding; however, the ITU limits the usable data rate to 384 kbit/s. The typical end-to-end latency with an EDGE link is around 150 ms.

EDGE is, however, characterized by a greater sensitivity to propagation conditions. The data rate capabilities are drastically reduced when the MS is far from the BS (leading to the use of MCS-1 encoding). Train applications using EDGE must take this risk into consideration in order to avoid data link saturation on distant links.

EDGE has an important advantage over Universal Mobile Telecommunications System (UMTS) since it is far less expensive. It can be seen as an alternative to UMTS for low population density regions, where UMTS operators will most likely not upgrade existing networks to 3G. Therefore, compatibility with EDGE should be considered as a requirement when designing on-board communication modules for trains.

UMTS

The Universal Mobile Telecommunications System (UMTS) is a part of the Third Generation (3G) of mobile phone standards. This new generation has been designed specifically to support high bandwidth data rates in order to be able to provide video-related services.

UMTS devices use a different coding method from that used by GSM or EDGE, called Wideband Code Division Multiple Access (WCDMA). This technology is based on a spread-spectrum approach that uses completely different radio subsystems. UMTS is thus unable to use the existing GSM/GPRS radio access, and also requires the deployment of an enhanced network infrastructure of gateways and service nodes.

WCDMA-FDD and WCDMA-TDD are normally able to provide up to 1.9 Mbit/s throughput, but the available throughput varies drastically depending on mobile speed and the distance between the MS and the BS. A distant connection in a high speed train can reduce the available bandwidth to less than 150 kbit/s.

As with GPRS for GSM, there is specific support for data transfer using UMTS. This takes the form of two protocols, named High Speed Downlink Packet Access (HSDPA) and High Speed Uplink Packet Access (HSUPA). Note that Third Generation Partnership Project (3GPP) specifications use the name Enhanced Up Link (EUL) in place of HSUPA. Both HSUPA and HSDPA are asymmetric, and do not provide the same data rate in both directions.

HSDPA is able to provide a downlink speed of 14.4 Mbit/s in theory (limited to 3.6 Mbit/s in UMTS release 5, and 7.2 Mbit/s in release 6). The uplink uses the Dedicated Channel (DCH) channel, which limits the speed to 128 kbit/s (384 kbit/s in release 6).

HSDPA does not support a soft handover mechanism. When the MS moves from one cell to another, it switches into a specific mode ('Compressed Mode') with a communication drop that lasts for a few seconds. During this time, the MS performs measurements to identify the best BS to join. This communication loss can induce high traffic penalties for fast moving trains.

HSUPA provides uplink speeds between 730 kbit/s and 11.5 Mbit/s, with a downlink speed of 14 Mbit/s similar to that of HSDPA.

Compared to GSM, GPRS and EDGE, UMTS networks provide better data rate capabilities, but they also have some serious weaknesses. The 3G license fees are extremely high and UMTS requires deploying a completely new architecture. Currently, this has resulted in a weaker coverage compared to GSM-based technologies. For now, UMTS is mainly limited to urban areas.

4.6.4 Satellite link technologies

Satellite data links for trains are described here only for illustration purposes. This technology has been experimentally used on Thalys trains, where data rates of 4 Mbit/s for the downlink and 2 Mbit/s for the uplink were achieved.

Satellite data links are capable of much higher data rates than those achieved by other technologies, but the train motion makes the dynamic antenna alignment a complex task.

Satellite links are also extremely expensive, and most railway operators reject this solution due to its cost. Experiments conducted on Thalys trains used a specific method to share the satellite link bandwidth between all trains in the fleet, which can be a solution to reduce the overall costs.

It must also be noticed that satellite links are lost when trains are operated in tunnels. An automatic switch-over must be used in this case in order to maintain the data link with the ground, using cellular communication, for example.

4.7 Networks for On-board Communication and Coupling with the Wayside

4.7.1 Multifunction Vehicle Bus

The Multifunction Vehicle Bus (MVB) is a serial communication bus with a bit rate of 1.5 Mbit/s, and is part of the International Electrotechnical Commission (IEC) Train Communication Network (TCN) standard (IEC 61375-1-3 2007). This bus was designed specifically for harsh electrical environments, such as power plants and trains, but it is currently used only on trains.

MVB is based on a Master/Slave communication scheme (the Master station being called the Bus Administrator). Each bus can support up to 4096 devices, with up to 255 devices being able to become Bus Administrator. It supports both real-time and non-real-time communication classes.

The real-time communication is handled by so-called Process Data (PD) ports. A single MVB bus supports up to 4096 PD ports, each of them containing 16, 32, 64, 128 or 256 bits. The main distinctive feature of MVB is that it uses fixed cycle times between master and slave devices. The MVB bus administrator triggers an exchange with each device in its management list on a cyclical basis (with a cycle time of between 4 and 1024 ms). The slave device must always respond to this exchange. If no answer is received from the slave, an error is signaled and the master device will not retry to send data until the next cycle. Consequently, MVB is a 'hard real-time' bus, since it is possible to guarantee that data are exchanged between master and slave devices in a given period.

The non-real-time communication class is called 'Message Data', and it allows the exchange of up to 64 KB of data. Due to the cyclical structure of the MVB message handling, these messages are split into small chunks each containing fewer than 256 bits. The Message Data protocol uses an underlying acknowledgement mechanism for reliable transfer of these chunks. Due to the huge fragmentation of messages, MVB is not very efficient for this class of messaging.

4.7.2   Wire Train Bus

The Wire Train Bus (WTB) is a serial communication bus that is also part of the IEC TCN train communication standard (IEC 61375-1-4 2007). It has been developed specifically to interconnect train consist (i.e. an ordered train segment in which multiple engines or train cars are coupled as a service unit) through couplers, and it supports dynamic reconfiguration. The WTB bus uses a 1 Mbit/s bit rate with High-Level Data Link Control (HDLC) (ISO 13239 2002) encoding, and supports physical media redundancy with automatic switching between lines.

Like MVB, WTB supports two communication classes (real-time data and non-real-time data), but access to the bus uses a different algorithm, which cannot guarantee the bus cycle time as with MVB. Therefore, WTB is considered 'soft real-time'.

Each message on the WTB bus can contain up to 1056 bits. The message for PD between two WTB nodes is restricted to 1024 user bits per node.

When WTB is used to transfer non-real-time data, messages are split into smaller elements that can fit into the 1056-bit HDLC frame. The TCN stack is responsible for reconstructing the complete message on the receiver side.

4.7.3   Ethernet

Due to their limited bandwidth, MVB and WTB appear to be unable to cover most of the application profiles described previously. In most cases, they are used for process communication (control/command and diagnostics). They can also be used for some PISs when the amount of data is limited (character strings for displays).

The growing need to carry digital audio and video over on-board networks has led major train manufacturers to use Ethernet (IEEE 802.3 2005) in order to be able to meet these new needs. Currently, there are two existing implementations: Profinet, which comes from industrial automation and is promoted by Siemens, and IPTrain, which is designed by Bombardier Transportation specifically for train communication.

IPTrain uses a double network structure. The first one is used for component-level communication and is based on a redundant network ring. The second one is a redundant bus used to couple consists together. IPTrain and Profinet have been designed to support both real-time and non-real-time communication classes.

4.7.4   Coupling on-board communication with wayside communication

On-board networks have specific characteristics that make them hardly compatible with existing wireless solutions. The cyclical nature of MVB and WTB buses must be taken into account when it is intended to exchange their data over links that are not able to respect the timing constraints. The problem is similar with Ethernet-based networks when they are used to exchange cyclical messages or streaming data.

Train manufacturers have developed specific gateways in order to be able to establish punctual or permanent radio links between the train and the ground. For example, the Bombardier MITRAC Mobile Communication Gateway (MCG) is able to establish links between train and ground using both GPRS and Wi-Fi technologies (Figure 4.5). This gateway is connected to both MVB and Ethernet networks and covers PIS, maintenance, and audio and video communication requirements.

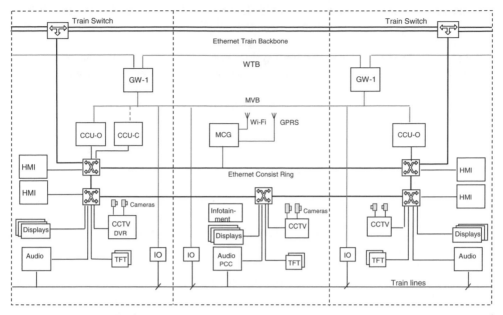

CCU-O : Central Computer Unit for Operation
CCU-C : Central Computer Unit for Comfort
GW : TCN gateway
MCG : Mobile Communication Gateway
DVR : Digital Video Recorder
HMI : Human Machine Interface

Figure 4.5 The different communication requirements in a modern train, demanding use of multiple networks in order to provide high bandwidth, interoperability with other consists, and real-time data exchange (© Bombardier Transportation)

4.8 Integration of Existing Technologies for Future Train Integration in a Global Communication Framework

4.8.1 European Rail Traffic Management System

The European Rail Traffic Management System (ERTMS) is already being used by various railway operators in Europe. However, it is still considered a future technology, since the last operation level is still under development, and current use is considered experimental. ERTMS is based on two wireless transmission technologies: magnetic coupling and GSM-R for train to ground data exchange. The use of GSM-R depends on the ERTMS integration level.

ERTMS has been designed in order to solve the signaling systems incompatibilities between European countries. More than 20 different systems have been developed by the various countries, each of them being incompatible with the others. Without using ERTMS, each train must be equipped with as many country-specific ATC systems as the train will meet during its journey. Thanks to ERTMS, international trains are able to cross borders without needing to change the locomotive and/or switch the ATC system from one standard to another.

ERTMS is designed to cover train/wayside audio and ATC application profiles (see Section 4.4.1). It is based on two major components: GSM-R (see Section 4.6.3) and the ETCS.

The ETCS part is based on *Eurobalise* and *Euroloop* devices. These devices use magnetic coupling technology to transfer their information from ground to train.

ERTMS Level 1 does not use the GSM-R part. It only uses Eurobalise devices, which report signal data to the train. They are also used in some cases to resynchronize the on-board train position measurement. The Eurobalises are sometimes replaced by Euroloop devices where trains require ground to train communication over longer distances.

ERTMS Level 2 uses both the Eurobalise/Euroloop and the GSM-R components. The main difference is that Eurobalise components are not used to transmit signal data to the train, but are used to synchronize the embedded odometry system with the ground. The signal data are transferred from ground to train using the GSM-R link. The train is also able to report its position continuously to the ground using GSM-R.

ERTMS Level 3 is still under development. The difference from ERTMS Level 2 is that train localization information is given by the trains themselves, and train detection by ground systems is no longer used. Within ERTMS Level 3, Eurobalise devices are used only to synchronize embedded odometry systems.

4.8.2   MODURBAN Communication System

The MODURBAN Communication System (MODCOMM) is a subproject of the European Modular Urban Guided Rail Systems (MODURBAN) project (MODURBAN Project 2009). This project aims at the definition of a standard architecture for the next generation of urban guided transport systems (tramways and commuter trains). The MODCOMM subproject has defined various requirements for the communication system between train and ground.

MODCOMM is the first project for which all communication services, including video, have been included in the requirements. Communication with stations is also part of the MODCOMM concept. All communications within MODCOMM are handled by a global system called the Data Communication System (DCS). The MODCOMM documents do not specify how the DCS is supposed to work internally, but specify the system operation. It is possible to create multiple implementations of a MODCOMM-compliant DCS depending on the needs of each train operator.

The MODCOMM group issued the first reference documents for implementation in Q4 2007, and first preliminary tests started in 2008. It was expected that the first MODCOMM practical implementations would start in 2009.

Both proprietary (such as Alstom's Informatisation et Automatisation par Guide d'Onde, or IAGO, or Siemens' *Airlink*) and open wireless technologies were evaluated by MODCOMM. Even if MODCOMM specifications do not require the use of a given wireless technology, the various requirements can be met currently only by the use of IEEE 802.11 and/or IEEE 802.16 technologies. Other technologies existing at the date of MODCOMM meetings were unable to provide all required functionalities.

4.9 Conclusion

Train communication is currently facing two challenges. The first one is the need for high bandwidth embedded communication links that are able to support new services such as video. The second one is the need for high bandwidth wireless communication between train and ground.

The first challenge is currently achieved with the use of Ethernet-based communication. However, the current Ethernet technology does not cover all communication requirements, especially those related to safety, due to the non-deterministic behavior of Ethernet. Specific networks still need to be used in parallel to cover these specific requirements. Coupling new technologies with already existing trains still using older trainwide networks (like WTB or TORNAD) also requires the coexistence of multiple different networks.

To solve this problem, there are current developments to create a future network that will be able to cover both high bandwidth and security and safety requirements. The coupling issue is addressed by the use of specific communication gateways (this solution is currently being used on recent trains), but solutions involving wayside are also being investigated.

Train to ground communication is the biggest challenge, since existing solutions require a choice between mobility capacity (also known as 'roaming' capacity) and available bandwidth. Wi-Fi solutions are currently only applicable to local communications with stopped or slowly moving trains. WiMAX-based solutions seem very promising, but the current experience with mobile applications is still very limited. Cellular technologies can give the desired mobility, but available bandwidths are still very limited, especially when distance between stations is taken into account.

The current state of the art in mobile technologies is to share the data between multiple communication channels. However, railway operators and train manufacturers are still expecting a wireless communication system that can support both high bandwidth and high speed roaming capabilities. These requirements will grow continuously in the coming years, as train integration in a global communication scheme becomes the standard.

References

Bouchez, B. (2008) Evolution des systèmes informatiques embarqués pour une intégration dans l'architecture ferroviaire globale. Colloque GEII – University of Lille.

Gabillard, R. (1987) Method of and apparatus for controlling the movement of a plurality of vehicles travelling on one track in a closed circuit. Patent FR2151440 and related Agence Nationale de Valorisation de la Recherche (ANVAR), Neuilly-sur-Seine, Hauts-de-Seine, FR.

GSM-R (2006) EIRENE Specification for GSM-R. Union Internationale des Chemins de Fer (UIC).

IEC 61375-1-3 (2007) Electric Railway Equipment – Train Bus – Part 1: Train Communication Network – Part 3 – Multifunction Vehicle Bus.

IEC 61375-1-4 (2007) Electric Railway Equipment – Train Bus – Part 1: Train Communication Network – Part 4 – Wire Train Bus.

IEEE 802.11 (2007) IEEE Standard for Information technology – Telecommunications and information exchange between systems – Local and metropolitan area networks – Specific requirements – Part 11: Wireless LAN Medium Access Control MAC and Physical Layer PHY Specifications. IEEE Std 802.11-2007 (Revision of IEEE Std 802.11-1999).

IEEE 802.16 (2007) Draft Standard for local and metropolitan area networks – Part 16: Air Interface for Broadband Wireless Access Systems (Revision of IEEE Std 802.16-2004 and consolidates material from IEEE Std 802.16e-2005, IEEE Std 802.16-2004/Cor1-2005, IEEE Std 802.16f-2005 and IEEE Std802.16g-2007). IEEE Unapproved Draft Std P802.16 Rev2 D2.

IEEE 802.16e (2005) Unapproved Draft Amendment to IEEE Standard for Local and Metropolitan Area Networks – Part 16: Air Interface for Fixed and Mobile Broadband Wireless Access Systems – Amendment for Physical and Medium Access Control Layers for Combined Fixed and Mobile Operation in Licensed Bands (Amendment and Corrigendum to IEEE Std 802.16-2004). IEEE Std P802.16e/D9.

IEEE 802.3 (2005) Carrier sense multiple access with collision detection (CSMA/CD) access method and physical layer specifications IEEE Std 802.3-2005 Part 3 (Revision of IEEE Std 802.3-2002 including all approved amendments).

ISO 13239 (2002) ISO 13239 High-level data link control (HDLC) procedures. International Organization for Standardization.

MODURBAN Project (2009) Homepage of the Modular Urban Guided Rail System (MODURBAN) Project. http://www.modurban.org.

5

Security and Privacy Mechanisms for Vehicular Networks

Panos Papadimitratos

Ecole Polytechnique Fédérale de Lausanne

The recent increase in interest and developments in the area of Vehicular Communication (VC) systems indicate that the technology could become broadly available in the near future. At the same time, authorities, industry, and researchers in academia agree that security and privacy enhancing mechanisms are a prerequisite for the acceptance and deployment of VC technology. A number of concerted efforts have been undertaken, with various projects providing significant results. As these approaches have common elements, this chapter surveys the state of the art in security and privacy enhancing methods for VC systems. A considerable distance has already been covered, but there is some uncertainty on how exactly VC systems are to be instantiated. Even though there is convergence in terms of securing communication, there are still challenges regarding other system aspects. A discussion on steps towards deployment and on the future landscape concludes the chapter.

5.1 Introduction

Intelligent Transportation System (ITS) and related technologies have been deployed in recent years for toll collection, fleet logistics and management, anti-theft protection, pay-as-you-go insurance, traffic information, and active road-side signs. These systems, relying on various communication technologies, continue to evolve and proliferate. More recently, a new trend has emerged: on-board computing units (On-Board Units, or OBUs) and short-range high bit-rate radios (in addition to cellular network transceivers) are integrated into vehicles, and dedicated (likely sparse) road-side infrastructure is expected to be deployed. Vehicle-to-Vehicle (V2V) and Vehicle-to-Infrastructure (V2I) communication, or, in general, Vehicular

Vehicular Networking   Edited by Marc Emmelmann, Bernd Bochow, C. Christopher Kellum
© 2010 John Wiley & Sons, Ltd

Communication (VC) would enhance transportation safety and efficiency and would support various other applications. V2V communication would enable real-time safety applications, extending the driver's horizon, while both V2V and V2I communication would enhance, for example, the distribution of environmental and traffic conditions information and in-vehicle entertainment.

The unique features of VC are a double-edged sword: a rich set of tools would be offered to drivers and authorities but a formidable set of abuses and attacks would become possible if the appropriate safeguards were not in place. An attacker could 'contaminate' large portions of the vehicular network with false information, announcing, for example, non-existent dangerous or congested road conditions, and thus mislead drivers and cause traffic jams. Alternatively, drivers could purchase software or hardware VC system 'hacks', just as they now often purchase police radar detectors or modify their cars for additional horsepower. Such VC system modifications could, for example, allow private vehicles to transmit messages as if they were an emergency vehicle (e.g. ambulance, police patrol, or road maintenance vehicle), or they could have unsuspecting drivers notified by their OBUs to slow down and yield, and in this way offer fast movement for some vehicles even in traffic jams. From a different point of view, receivers deployed in a city center, at highway exits, or even in a celebrity's neighborhood could record transmissions from passing vehicles to be used later in tracing their location and inferring private information about their passengers.

Privacy Enhancing Technologies (PET) are needed, especially because attacks are relatively easy to mount. To begin with, VC relies on a variant of the widely adopted IEEE 802.11 wireless communication technology. Besides this, attackers could use any low-cost computing platform, such as palmtop or laptop computers, or wireless local area network Access Points (APs); for example, a wireless network operator, licensed to provide services unrelated to VC systems, could 'tune' its APs to intercept VC traffic. Finally, VC equipment can be left unattended for long periods, increasing the likelihood of physical compromise. Overall, without security, VC systems could make antisocial and criminal behavior easier, in ways that would actually jeopardize the benefits of their deployment.

The awareness of the need to secure VC has spurred a number of projects to design VC security architectures: the Secure Vehicular Communication (SeVeCom) project and the IEEE 1609.2 working group are two prominent efforts that seek to secure communication and protect private user information. The envisioned systems rely on multiple Certification Authorities (CAs), with each CA managing identities and credentials for nodes (vehicles and Road-Side Units, or RSUs) registered within its *region* (e.g. national territory, district, or county). Each node is uniquely identified and holds one or more private–public key pairs and certificates, thus digitally signing messages it transmits.

In this chapter we survey security and PET solutions for vehicular networks. With significant commonalities among approaches to secure communication, we ponder whether these aspects of the overall problem of securing VC systems are addressed. More generally, are there research challenges to be addressed, or is it rather clear which VC security architecture would be deployed? We reflect on these questions, and discuss alternative viewpoints and non-technical factors that are likely to influence the deployment of secure VC systems.

In the rest of the chapter we first discuss threats and security requirements, followed by an overview of a secure VC and a set of basic system assumptions in Section 5.4. We present secure communication schemes that also enhance privacy in Section 5.5.1. Then we discuss approaches for data-centric security, notably secure localization with the help of Global Navigation Satellite Systems (GNSS) (Section 5.7.1), and data trustworthiness (Section 5.7.2). After that we briefly discuss considerations on design choices and challenges to deployment, before giving our conclusion.

5.2 Threats

VC systems comprise network nodes, or, in other words, wireless-enabled computing platforms mounted on vehicles and RSUs. Their complexity varies from relatively powerful devices (e.g. vehicle OBUs or servers run by authorities) to relatively simple ones (e.g. alert beacons on the road-side). These VC entities can be *correct* or *benign* (i.e. comply with the implemented protocols) or they can be *faulty* or *adversarial* (i.e. deviate from the protocol definition).

Faults might not be malicious; for example, the communication module of a node may discard or delay messages or set packet fields to inappropriate values. We do not consider here benign faults, such as communication errors, or message delay or loss, which can occur either under normal operational conditions or due to equipment failure. Malicious behavior can result in a much larger set of faults. Papadimitratos et al. (2006a) provide a detailed discussion of faults and adversary models, aspects germane to VC systems, and models used in other types of distributed systems.

The behavior of adversarial nodes can vary widely, according to the implemented protocols and the capabilities of the adversary. The incentive of the adversary may be its own benefit, or it may be malice. Active adversaries can meaningfully *modify* in-transit messages, beyond what the protocol definitions allow or require them to modify. More generally, they can *forge* and *inject* messages, based on their own prior observations (messages they received) and the protocol they are attempting to compromise. An active adversary may also *jam* communications (i.e. interfere deliberately and prevent other devices within its range from communicating). Alternatively, it can *replay* messages it has received that were previously transmitted by other system entities. In contrast, *passive adversaries* only gather information about system entities and cannot affect or change their behavior.

The adversary can be *external* but still be able to influence the protocol execution, by jamming communications and replaying the messages of correct nodes. Alternatively it can be *internal*, with cryptographic keys and credentials to participate in the execution of the protocol(s). Even though the VC implementations would be proprietary, standards, needed for interoperability, would provide extensive information on the VC protocol stack. Attackers could in principle clone its functionality, build their own rogue protocols, and modify the functionality of VC system nodes. If they obtain compromised cryptographic keys, by physically extracting them from a node for example, then they can act as internal adversaries. In fact, a node holding multiple such keys can appear as multiple nodes.

More generally, many adversarial nodes can be present. Often, they can act *individually*, but they might also act in *collusion*, coordinating their actions. It is, however, likely that colluding adversaries are unwilling to share their private keys and allow other nodes to fully impersonate them (and obtain, for example, their access rights). Over time, the number of adversaries can change, depending on the type of compromise and the defensive reaction of the system. It is reasonable to expect that at any point in time a small fraction of the network nodes are adversaries. At any time and location, only a few adversaries are likely to be physically present. This does not preclude a group of adversarial nodes surrounding an honest one, which should be a rare situation.

A rather peculiar type of adversary is relevant to VC systems: an *input-controlling adversary* that alters (sensory) inputs to VC protocols, rather than compromising the protocols. Such an adversary is weaker than an arbitrary internal adversary, because it cannot induce arbitrary behavior. But it would often be much easier to affect inputs, or compromise sensors or sensor-to-OBU connections, than to compromise the OBU itself. Such an adversary is weaker than an internal one: controlling inputs alone cannot induce arbitrary behavior if self-diagnostics and other controls are available and out of reach of the adversary.

5.3   Security Requirements

In general, we seek to secure the operation of VC systems, or, in other words, to design protocols that mitigate attacks and thwart deviations from the implemented protocols to the greatest possible extent. Each protocol has its own specifications, but instead of requirements per protocol and application, stand-alone security requirements have been considered, largely independently of specific applications and protocols.

Message authentication and integrity mechanisms protect messages from alteration and allow receivers to corroborate the node that created the message. If necessary, *entity authentication* can provide evidence of the sender *liveness* (i.e. the fact that the sender generated a message recently). To prevent a sender from denying having sent a message, *non-repudiation* is needed. Furthermore, *access control* and *authorization* can determine what each node is allowed to do in the network, in terms of the implemented system functionality. *Confidentiality* can keep message content secret from unauthorized nodes.

Privacy and anonymity are required, at least at the level of protection achieved before the advent of VC systems. In general, VC systems should not allow disclosure of private user information. In particular, the identity of a vehicle performing a VC-specific action (e.g. transmitting a message) should be concealed. Anonymity, with respect to an observer, depends on the set of involved vehicles: an observer cannot determine, among all vehicles in a set, which vehicle performed an action. Moreover, any two actions by the same vehicle cannot be linked. But under specific circumstances an observer could consider a vehicle more likely to perform an action.

Rather than seeking *strong anonymity*, along with authentication and other security properties, less stringent requirements are considered. Cryptographically protected messages should not allow for the identification of their sender, and two or more messages generated by the same vehicle should be *difficult to link* to each other. More precisely, messages produced by a node over a protocol-selectable period of time τ can be linked, but messages m_1, m_2 generated at times t_1, t_2 such that $t_2 > t_1 + \tau$ cannot. The shorter τ is, the fewer the linkable messages are, and the harder it becomes to trace a node.

Beyond security and anonymity, *availability* is also sought, so that VC systems remain operational even in the presence of faults, and resume normal operations after the removal of the faulty nodes. Another significant dimension is that of *non-cryptographic security*, including the determination of data *correctness* or *consistency*. Traditionally, if the sender of a message is trusted, then the content of the message is trusted as well. This notion is valid for long-lived, static trust relationships. But in VC systems there are often no grounds for similar approaches. It is thus necessary to assess the *trustworthiness of data* per se, as obtained by other nodes in the VC system.

These general requirements can be mapped to specific VC-enabled applications. Ideally, one could argue that all requirements must be satisfied for all applications (Papadimitratos et al. 2006a). But the relative importance of security requirements can differ across applications. General application characteristics and security requirements were assessed for a large number of VC applications; for example, for a roadwork zone warning application, it may be relatively less important to rigidly determine the recency of its messages than for a collision avoidance application (Kargl et al. 2006; Papadimitratos et al. 2008a). Of course, for both applications it is critical to ensure that no message content can be fabricated by an attacker. Privacy protection is not required for infrastructure- or public vehicle-sent messages (e.g. the work zone and emergency vehicle warnings).

5.4 Secure Vehicular Communication Architecture Basic Elements

Efforts are being made in academia and industry with the aim of providing adequate security solutions: for example, the IEEE 1609.2 trial-use standard (IEEE 1609.2 2006), the Network On Wheels (NOW) project (NoW 2007), and the SeVeCom project (Kargl et al. 2008a; Papadimitratos et al. 2008a; SeVeCom 2009). Essentially, security architectures first seek to address two fundamental issues: (a) *identity, credential, and key management*, and (b) *secure and privacy enhancing communication*. The focus is primarily on securing the wireless part of the VC system and enhancing the privacy of its users, in order to satisfy the requirements outlined in the previous section. Additional aspects, such as the *in-car system protection* and the *data trustworthiness*, have received relatively less attention. In this section we present an overview of the basic elements of secure VC systems. Then, in the sections that follow, we give a more detailed presentation of secure communication, revocation, and mechanisms for data trustworthiness.

5.4.1 Authorities

Authorities are trusted entities responsible for the issuance and management of *identities* and *credentials* of the parties involved in the vehicular network operation. In general, authorities can be *multiple* and *distinct* in their roles and have a subset of network parties in their jurisdiction. We denote the set of system entities, S_X, registered with an authority X determined by geographical, administrative, or other criteria, as the *domain* of X. All parties in S_X trust X by default. The presence of online authorities is not required, as connectivity and communication, especially over a wireless medium, with an authority may be intermittent. Nodes can, in general, establish two-way communication with the authorities, even though one-way communication (from an authority towards the nodes) can also be meaningful in general.

In the context of secure VC, we interchangeably call authorities *Certification Authorities (CAs)*. Each CA is responsible for a *region* (national territory, district, county, etc.) and the identities and credentials of all the nodes registered from that region. To enable interaction between nodes from different regions, CAs provide certificates for other CAs (cross-certification) or provide *Foreigner Certificates (FCs)* to vehicles that are registered with another CA when they cross the geographical boundaries of their region (Section 5.6).

5.4.2   Node identification

At a basic level, we consider a network node, a vehicle or an infrastructure node, as: (a) a unique identity, V, (b) a public/private key pair K_V, k_V, (c) a module implementing the networking and the overlying application protocols, and (d) a module providing communication across a wireless network interface.

Each node is registered with only one CA, and has a unique *long-term* identity and a pair of *private* and *public* cryptographic keys. Accordingly, it is equipped with a long-term *certificate*. A list of *node attributes* and a *lifetime* are included in the certificate that the CA issues upon node registration and upon certificate expiration.

The binding of K_V to V and the binding of K_V to other data or *attributes* pertinent to V can be achieved by an *identity certificate* or an *attribute certificate*, respectively. We denote a certificate on K_V issued by an authority X as $Cert_X\{K_V, A_V\}$, with A_V being a possibly void attribute list. Similarly, infrastructure nodes have a unique identity, I, and k_I and K_I private and public keys, with $Cert_Z\{K_I, A_I\}$ a certificate issued by an authority Z for I with attribute list A_I.[1]

Note that infrastructure nodes are not necessarily static. Vehicles can be grouped into two categories, *public* and *private*. The former can include vehicles related to public safety (e.g. highway assistance or firefighting, and police vehicles or helicopters), or public transportation vehicles (e.g. buses or trams). Public vehicles, like infrastructure nodes, are considered more trustworthy, and they can be used to assist security-related operations.

The CAs are also responsible for the *eviction* of nodes or the *withdrawal* of compromised cryptographic keys. This is achieved by revoking the corresponding certificates. In all cases, the interaction of nodes with the CAs is infrequent and intermittent, with the road-side infrastructure acting as a gateway to and from the vehicular part of the network, and the use of other infrastructure (e.g. cellular) also being possible.

5.4.3   Trusted components

Nodes are equipped with Trusted Components (TCs) (i.e. built-in hardware and firmware) that basically have two types of functionality: *cryptographic* operations and *storage*, in order to protect the vehicle's cryptographic material data (usable for liability identification). The TCs enforce a policy on interaction with the on-board software, including access to and use of the securely stored keys, credentials, and secrets. Access (read or write) to any information stored in the TCs and modification of their functionality is possible only through the interface provided by the TCs. The TCs should be *tamper-resistant*, in order to provide enhanced protection of the cryptographic material and other data.

A specific example of a TC is that of the Hardware Security Module (HSM) that the SeVeCom architecture envisions for vehicles and RSUs. The HSM stores and physically protects sensitive information (primarily private keys for signature generation) and it provides

[1]Users of VC systems can accordingly have a unique identity, U, and they can be bound to their credentials and secrets. They can be owners and/or the driver or passengers of vehicles, associated in general in a *many-to-many* manner. This chapter does not elaborate further on the user–vehicle interactions.

a secure time base. If a HSM were to be tampered with, for example to extract private keys, the physical protection of the unit would ensure that the sensitive information would be erased so that the adversary could not obtain it. Moreover, since all private key cryptographic operations are performed in the HSM, sensitive information never leaves the physically secured HSM environment. Essentially, the HSM is the basis of trust; without it, private keys can be compromised and their holders can masquerade as legitimate system nodes.

5.4.4 Secure communication

The basic way for nodes to undertake *secure communication* is for them to sign messages digitally, after attaching a time-stamp and the signer's location and certificate to the message. This way, modification, forgery, replay, and relay attacks can be defeated. The latter relate to secure neighbor discovery, which is possible since safety beacons include the time and location at the point they are sent across the wireless medium, as explained in Section 5.5.2. Signatures can be applied in different ways, to beacons or to multi-hop flooded and position-based multi- or uni-casted messages, not only by the message originator but also by relaying nodes (Section 5.5.3).

A conceptual view of a node is given in Figure 5.1. To provide both security and a degree of anonymity, long-term keys and credentials are *not* used to secure communication. Rather, the approach of *pseudonymity* or *pseudonymous authentication* is used. Each vehicle is equipped with multiple certified public keys (pseudonyms) that do not reveal the node identity. It obtains these pseudonyms via a trusted third party, a Pseudonym Provider (PNP), by proving it is registered with a CA. Then the vehicle uses each pseudonym and private key for at most τ seconds (the pseudonym lifetime), before it switches to another, not previously used, pseudonym. Messages signed under the same pseudonym can be trivially linked, but messages signed under different pseudonyms cannot (Sections 5.5.1 and 5.5.4).

5.5 Secure and Privacy-enhancing Vehicular Communication

5.5.1 Basic security

Periodic single-hop broadcasting, *beaconing*, is typically used for the so-called cooperative awareness applications: beacons, typically sent γ times per second, contain information on the sender's status such as vehicle position, speed, and heading. The frequency of beacons is expected to range from 10 Hz to 1 Hz. Beacon messages are digitally signed, and the signer's certificate is attached. More precisely, after the beacon message assembly is complete and before submitting a message m to the data link layer for transmission, the sending node (V) calculates a signature.

Rather than using its long-term cryptographic material, each node V is equipped with a set of *pseudonyms* (i.e. *public keys* that do not carry any information that identifies V). For the ith pseudonym K_V^i for node V, the CA provides a certificate $Cert_{CA}(K_V^i)$ that is simply a CA signature on the public key K_V^i (unlike, for example, the more complex X.509 certificate). The node uses the private key k_V^i corresponding to the pseudonym K_V^i to digitally sign messages. We term this approach the Baseline Pseudonymous Authentication (BPA) scheme, since it is essentially the same across different projects. Note that we should consider

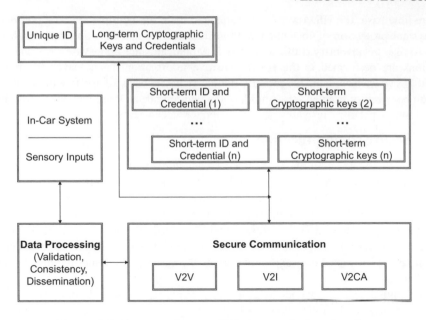

Figure 5.1 Conceptual secure VC view: node functionality

only vehicles when using this approach, as the privacy of RSUs or other infrastructure does not need to be protected.

With $\sigma_{k_V^i}$ denoting V's signature under its ith pseudonym and m the message payload, the message format is m, $\sigma_{k_V^i}(m)$, K_V^i, $Cert_{CA}(K_V^i)$. Upon receipt of this, a node, with the public key of the CA assumed available, validates $Cert_{CA}(K_V^i)$. It makes use of a *Certificate Revocation List (CRL)*, also assumed to be distributed to vehicles, as discussed in Section 5.6. If successful (i.e. the K_V^i is not included in the CRL and the CA signature on K_V^i is valid), the node validates $\sigma_{k_V^i}(m)$. The CA maintains a map from the long-term identity of V to the node's set of pseudonyms, $\{K_V^i\}$. If presented with a signed message, the CA can perform the inverse mapping and identify the signer.

Each pseudonym is used for at most a period τ and is then discarded. A number of possible implementation aspects should be considered: the dynamic adaptation of the period of pseudonym usage; the number of pseudonyms (and the corresponding certificates and private keys) with which to 'pre-load' V; the frequency of such refills; other policies for pseudonym change, such as factors rendering a pseudonym change unnecessary (e.g. a TCP connection to an access point); and interactions of pseudonym changes with the network stack (Papadimitratos et al. 2007).

5.5.2 Secure neighbor discovery

Vehicles transmit safety beacons at high rates, and in this way they obtain a frequently updated view of other vehicles nearby (i.e. *physical neighbors*). This is the essence of cooperative awareness, which allows vehicles to have an up-to-date view of their neighborhood for transportation-related actions (e.g. avoidance of collisions).

However, it is often important that vehicles also discover other nodes (vehicles or RSUs) that are directly reachable (i.e. their *communication neighbors*) (Papadimitratos et al. 2008d). Typically, it is assumed that if two nodes are communication neighbors then they are physical neighbors, and vice versa, but this is not always the case, because adversaries mount *relay attacks*, by receiving and quickly retransmitting (replaying) messages from remote nodes.

The inclusion of sender time-stamp and location, along with authentication, enables the system to perform provably secure neighbor discovery against *external* adversaries (Poturalksi et al. 2008a,b). The basic idea is to estimate the sender–receiver distance based on a node's coordinates, the location in the received message, and the time-of-flight (difference between the node's time and the message time-stamp). Of course, to obtain precise results it is necessary to have time-stamps added by the hardware, and to have a cryptographic protection for the time-stamp calculated exactly at that point when the beacons are outgoing from the sender's transmitter. For a protocol-selectable acceptable neighbor range, the receiving node accepts the sender as a communication neighbor when the two distance estimates are equal and the sender is authenticated. As a result, vehicles can be assured that their *neighbor table* includes only nodes that are indeed communication neighbors.

5.5.3 Secure position-based routing

In order to disseminate data to a geographically defined destination, position-based communication benefits from location-aware nodes. Nodes maintain the locations of their neighbors, and forward data to the closest neighbor to the destination region. This approach is well suited for VC systems, given that location information is assumed to be available (e.g. by GNSS, for example the GPS). However, it can be abused by adversaries. As a basic security measure for both position-based routing and message distribution, source nodes sign created messages and attach the corresponding certificate, which is similar to the basic security functionality. Moreover, forwarding nodes can also sign packets they relay, so that these can be authenticated by the next-hop relay (Harsch et al. 2007). This way, only qualified network participants can create messages that are accepted by others, and message integrity is protected towards the destination. Replay and neighbor discovery attacks can be prevented, as discussed in the previous subsections. However, the location information in safety beacons can be forged by adversaries that seek to attract traffic illegitimately. A position verification scheme based on plausibility heuristics is capable of detecting such position falsifications (Leinmüller et al. 2006). More generally, tests that take into account constraints of the transportation network and the vehicle dynamics, and of course secure neighbor discovery, can be used for nodes to maintain a truthful neighborhood view (Festag et al. 2009).

5.5.4 Additional privacy-enhancing mechanisms

As the BPA scheme requires that the short-lived certificates are preloaded to the vehicle, a series of problems arise. What should a vehicle do when all of its pseudonyms are used up? How should the refilling of pseudonyms be designed and secured? What are the storage requirements, in terms of disk space and security? As the lifetime of a pseudonym is inversely analogous to the degree of message unlinkability, the stronger the protection needed the higher is the number of temporary identities and keys (pseudonyms) per node. For large-scale systems, this can be a significant burden.

To enhance system usability and efficiency, a method that allows nodes to self-generate (i.e. self-certify their own pseudonyms) was proposed by Calandriello et al. (2007) and Papadimitratos et al. (2008b). This approach extends the BPA, offering significant advantages: vehicles do not need to be sidelined or to compromise their users' privacy if a 'fresh' pseudonym is no longer available, no 'over-provisioning' in the supply of pseudonyms is necessary, and the cost of obtaining new pseudonyms over an 'out-of-band' channel is avoided.

This can be achieved with the use of *anonymous authentication* primitives, notably *Group Signature (GS)* as described in Section 5.5.4. As the practicality of GSs in the VC context is limited by their overhead, in terms of computation and communication, we propose in Section 5.5.4 the *Hybrid Pseudonymous Authentication (HPA)* scheme that allows on-the-fly generation of pseudonyms by combining the BPA and GS approaches. This alleviates the management overhead of the BPA; but in principle it is more costly than BPA. The mechanisms discussed in Section 5.5.5 reduce the cost of the HPA scheme to be roughly the same as that of BPA and increase the robustness of any pseudonymous approach.

Anonymous authentication: Group Signatures

Each node V is equipped with a secret *group signing key* gsk_V; the *group* comprises as members all vehicles registered with the CA. A *group public key* GPK_{CA} allows for the validation (by any node) of any *group signature* $\Sigma_{CA,V}$ generated by a group member. Intuitively, a group signature scheme allows any node V to sign a message on behalf of the group *without* V's identity being revealed to the signature verifier. Moreover, it is impossible to link any two signatures of a legitimate group member. Note that no public key or other credentials need to be attached to an anonymously authenticated message; the format is: $m, \Sigma_{CA,V}(m)$. Group signatures were introduced by Chaum and van Heyst (1991), with numerous subsequent works (e.g. Ateniese and Tsudik 1999; Boneh et al. 2004; Brickell et al. 2004; Syverson and Stubblebine 1999). If the identification of a signer is necessary, the CA can perform an *Open* operation and reveal the signer's identity (Bellare et al. 2003, 2005).

Hybrid Pseudonymous Authentication

The combination of the BPA and GS schemes is the basic element of the proposal of Calandriello et al. (2007) and Papadimitratos et al. (2008b). Each node V is equipped with a group signing key gsk_V and the group public key GPK_{CA}. Rather than generating group signatures to protect messages, a node generates its own set of pseudonyms $\{K_V^i\}$. As in Section 5.5.1, a pseudonym is a public key without identification information, and $\{k_V^i\}$ is the set of corresponding private keys. In this case, the CA does not provide a certificate on K_V^i, but V uses gsk_V to generate a group signature $\Sigma_{CA,V}()$ on each pseudonym K_V^i instead.

This way, nodes generate and 'self-certify' K_V^i on the fly, producing $Cert_{CA}^H(K_V^i)$. The H superscript denotes the HPA scheme and differentiates this certificate from that of the BPA approach. The CA subscript denotes that the certificate was generated by a legitimate node registered with the CA. V attaches the $Cert_{CA}^H(K_V^i)$ to each message, and signs with the corresponding k_V^i: $m, \sigma_{k_V^i}(m), K_V^i, Cert_{CA}^H(K_V^i)$.

When receiving an HPA message, the group signature $\Sigma_{CA,V}(K_V^i)$ is verified, using the GPK_{CA}. If this is successful, the receiver infers that a legitimate system (group) member generated pseudonym K_V^i. We emphasize that, as per the properties of group signatures, the receiver/verifier of the certificate *cannot* identify V and *cannot* link this certificate and pseudonym to any prior pseudonym used by V. Once the legitimacy of the pseudonym is established, the validation of $\sigma_{k_V^i}(m)$ is identical to that of the BPA message. To identify the signer of message, an *Open* on the $Cert_{CA}^H(K_V^i)$ group signature is necessary; the message m is bound to K_V^i via $\sigma_{k_V^i}(m)$, and K_V^i is bound to V via $\Sigma_{CA,V}(K_V^i)$.

5.5.5 Reducing the cost of security and privacy enhancing mechanisms

Mechanisms have been proposed in the literature to reduce overhead (Mechanisms 1, 2, and 4 below) and enhance robustness (Mechanism 3). They are all applicable to both BPA and HPA schemes. To reduce overhead, Calandriello et al. (2007) and Papadimitratos et al. (2008b) propose *not* to attach certificates to all messages, but rather to do so for one in every α successive beacons; they also propose certificate caching to reduce the verification processing overhead. Moreover, Kargl et al. (2008b) propose to avoid attaching certificates to beacons *unless* a change in the vehicle neighborhood is detected.

Mechanism 1

At the sender side, the $Cert_{CA}^H(K_V^i)$ is computed only once per K_V^i, because $Cert_{CA}^H(K_V^i)$ remains unchanged throughout the pseudonym lifetime τ. Note that the notation here does not distinguish which method is used for the certificate generation. For the same reason, at the verifier's side the $Cert_{CA}^H(K_V^i)$ is validated upon the first reception and stored, even though the sender appends it to multiple (all) messages. For all subsequent receptions, if the $Cert_{CA}^H(K_V^i)$ has already been seen, the verifier skips its validation. This optimization is useful because $\tau \gg \gamma^{-1}$ (i.e. the pseudonym lifetime is much higher than the beaconing period).

Mechanism 2

The sender appends its signature $\sigma_{k_V^i}(m)$ to all messages, but it appends the corresponding K_V^i, $Cert_{CA}^H(K_V^i)$ only once every α messages (termed the *certificate period*). The message structure is $m, \sigma_{k_V^i}(m)$. To make the choice of the right K_V^i to verify such an incoming message easy, all messages signed under the same pseudonym can carry a short key ID field. When a pseudonym change occurs, the new tuple $\sigma_{k_V^{i+1}}(m), K_V^{i+1}, Cert_{CA}^H(K_V^{i+1})$ must be computed and transmitted. V will sign messages with the new k_V^{i+1} corresponding to K_V^{i+1} from then on.

Mechanism 2 can affect the protocol robustness, if the message that carries K_V^{i+1} and $Cert_{CA}^H(K_V^{i+1})$ is not received. Then, nodes in range of V must wait for α messages before the next pseudonym transmission, while being unable to validate *any* message from V. This can be dangerous if vehicles are close to each other and/or moving at high relative speeds.

Mechanism 3

To address the aforementioned issue with Mechanism 2, the transmission of K_V^{i+1}, $Cert_{CA}^H(K_V^{i+1})$ is repeated for β consecutive messages when K_V^{i+1} is issued, with β denoted as the *push period*.

Mechanism 4

The K_V^{i+1}, $Cert_{CA}^H(K_V^{i+1})$ is transmitted (repeated) only when V detects a new neighbor, so that transmissions from V can be validated by the 'newcomer'. Mechanism 3 can be combined with Mechanism 4 to enhance reliability.

Cryptographic overhead and system performance

Cost constraints in today's car manufacturing make it hard to equip vehicles with powerful state-of-the-art desktop processors. Instead, relatively inexpensive and energy-saving embedded processors are used. But cryptographic operations create a significant overhead, both in terms of processing and of communication bandwidth, especially because vehicles send information frequently (e.g. position and environment conditions) – typically one beacon per 100 milliseconds (i.e. $\gamma = 10$).

Without ignoring other factors, the computational security overhead is due to the generation and verification of packet signatures and certificates. The communication security overhead is due to signatures and certificates attached to packets. Each safety beacon has to be signed, and each vehicle has to validate (for example, every 100 milliseconds) beacons from *all* neighboring vehicles in range, which, do not forget, may also change their identity (pseudonym) in the meantime.

Cryptographic and communication security overhead can affect VC applications in multiple ways. The first dimension of the problem is communication reliability: increased beacon size contributes to interference. In principle, the higher the offered load, with the number of transmitters in the area, the beaconing rate, and the message overhead, the worse the channel performance. The second dimension is processing overhead. Attention should be paid to the cost of verifications: safety beaconing, for example, entails that each node verifies at least one signature for each received packet within the beaconing period, with all its neighbors sending one beacon, and it has to generate one signature during the same period.

The mechanisms discussed above can significantly reduce both communication and processing overhead. This can be beneficial in any case, as long as the supported applications are not affected. Any safety-related warning can be trusted only if it can be cryptographically validated. For example, a safety beacon can be validated only if the corresponding short-term certified public key (pseudonym) of the signer was previously verified. A specific safety application is considered by Calandriello et al. (2007) and Calandriello et al. (2009); Papadimitratos et al. (2008b): an emergency braking warning application operating on top of secure and privacy-enhancing communication, which can be almost as effective as the same application operating without any security (or related overhead).

5.6 Revocation

All projects on security architectures for VC consider the eviction of faulty or illegitimate nodes. More generally, certificates of faulty or compromised nodes should be revoked. Without valid credentials, faulty or adversarial nodes can no longer damage the VC system. Nodes are revoked in principle in three cases: if they are deemed faulty, if it is detected that their cryptographic keys have been compromised, or for administrative reasons. The CA is responsible for a revocation decision. If the decision is the result of a (detected) faulty operation or key compromise, then the CA should obtain or be presented with evidence. One option, which relieves the CA from operating its own monitoring infrastructure, is to have misbehavior evidence collected by vehicles.

The basic revocation approach, as is the case for other systems beyond VC, is through the distribution of Revocation Lists (RLs) that the CA generates and authenticates. At a first stage, in a system that relies on pseudonyms (short-term identities) and corresponding private keys, the CA (or PNP) would not provide new pseudonyms to a revoked node. Nonetheless, the use of a RL would be necessary for the revocation of the pseudonyms that are still valid. The RL can be a list of certificate identifiers, similar to certificate revocation lists (CRLs) for classic public key cryptography; or it can be a list of elements that allow for the identification of the signer when group signatures are used (e.g. as in a GS scheme, when HPA is used). For the rest of the discussion, we do not dwell on the exact type of RL; a discussion about some quantitative aspects is available in Calandriello et al. (2009).

The basic challenge is to distribute a RL efficiently and effectively across a large-scale multi-domain system, illustrated by Figure 5.2. This can be achieved, in spite of the constraints of the VC environment, by leveraging on a sparse road-side infrastructure. A scheme proposed by Papadimitratos et al. (2008c) achieves that with very low bandwidth used for RL transmissions, on the order of a few kbit/s at each RSU. In practice, with RSUs a few kilometers apart, all vehicles can obtain the latest RL within tens of minutes (e.g. the duration of a commuter journey). *Scalability* can be achieved by keeping RL sizes low and RSU–CA interactions minimal, with no RSU–RSU interactions.

This scheme relies on few basic elements. Thanks to the *collaboration between regional CAs*, RLs contain only regional revocation information and their size is kept low. *Encoding of RLs* into numerous (cryptographically) self-verifiable pieces provides resilience to disconnections, radio impairments, and malicious message injection.

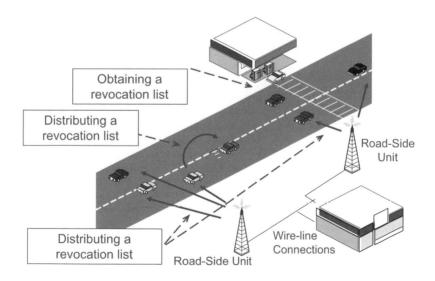

Figure 5.2 Illustration of revocation list distribution

The multi-domain CA structure keeps RL sizes low, but vehicles need revocation information from other regions to validate the certificates of foreigner (visiting) vehicles.

Rather than distributing RLs of other regions, the CA validates certificates of visiting nodes: if they are not revoked in their home region, it issues them short-lived foreign certificates (*FCs*) that they must use in the foreign region. If the certificate of a FC holder is revoked later on, it is included in the RL of the CA that issued the FC, and its actual certificate is added to its home CA's RL.

The encoding of the RL into multiple pieces can be done in different ways, using *Fountain* or *Erasure* codes. The original RL is segmented into M parts and then encoded, with added redundancy. Erasure codes produce $N > M$ *RL pieces*, such that for any M out of N pieces received, the original RL can be reconstructed (Rabin 1989). Fountain codes and among them a special class, Raptor codes, with linear time encoding and decoding complexity, produce for M input pieces a potentially limitless stream of output RL pieces (Shokrollahi 2006). For a protocol-selectable parameter $\sigma > 0$, the original M pieces can be recovered with a high probability from any subset of $M(1 + \sigma)$ RL pieces. The RL version and time-stamp, a piece sequence number, the CA identifier, and a digital signature covering all previous fields, are added to each RL piece, so that each of them can be validated individually.

For areas that are not covered by RSUs, vehicles could undertake the role of distributing the RL in an '*epidemic*' manner (Laberteaux et al. 2008; Papadimitratos et al. 2008a). It would also be possible to use alternative media for such information, such as cellular links (e.g. General Packet Radio Service, or GPRS, or Universal Mobile Telecommunications System, or UMTS), broadcast digital radio, or simply localized wireless or wired links when the vehicle is static (e.g. overnight parking).

If the adversary could not control the communication of the CA and the on-board trusted hardware, all credentials and cryptographic keys could be remotely removed by the CA. This can be achieved by a 'kill' command issued to the HSM: once it authenticates the command, the HSM erases all its private keys (Papadimitratos et al. 2008a; Raya et al. 2007). This essentially prevents the revoked node from participating further in the protocol execution: its messages cannot be signed and validated.

The decision to revoke or not the credentials of a node is made exclusively by the CA. Nonetheless, CRLs would be issued infrequently – for example, once per day or every few days. This would leave a vulnerability window, until a faulty or otherwise compromised node's credentials are revoked. A local reaction mechanism can protect honest nodes from a misbehaving node that is not yet revoked. This, of course, presumes that nodes can reliably detect a misbehavior and attribute it to a node. One option is to have each node that detects a wrongdoer broadcast a warning in its neighborhood. When multiple warnings are issued and corroborated locally for a given node, then newcomers in the vicinity of the detected attacker can ignore its messages (Raya et al. 2007).

Clearly, the redundancy in detection information can be beneficial: it can reduce false positives (i.e. 'branding' a correct or honest node as a misbehaving one). Similarly, an attacker that issues false warnings by trying to exclude honest nodes cannot achieve anything if it acts alone. Or, more generally, multiple attackers would need to be present together in a neighborhood, with correct nodes being the minority, in order to have honest nodes excluded. On the flip side, the distributed computation can be slow to complete, and can be affected significantly by the high mobility, or incur higher communication costs in a dense topology. An alternative method, which is complementary, would be to allow a single node to announce any misbehaving nodes it detects (Moore et al. 2008). This, of course, would be less robust against attackers that abuse the misbehavior reporting mechanism. But, overall, it would allow for a faster reaction in terms of detecting attackers.

5.7   Data Trustworthiness

Vehicular communication systems are *data centric*: (a) the VC-enabled on-board system depends on many sensory inputs; (b) frequent data and event reports are exchanged among vehicles and road-side infrastructure; and (c) the identity of the sender of such data is of lesser importance. The sensory inputs vary in nature, ranging from vehicle-specific data (e.g. motion or temperature sensors) to location and time corrections, provided, for example, by GNSS. Transportation safety and efficiency applications are built on the exchange of data. Safety beacons carry the location (but can also carry other information such as speed and direction) of the transmitting vehicle; dangerous conditions or re-routing warnings can be broadcast by infrastructure and vehicles. For all such messages, the identity of their sender is not important in the way that an IP address is for other networks. On the contrary, information such as the time and location of the reporting node, as well as its attributes (e.g. type of vehicle or RSU), are the important information, along with the reported data *per se*. Finally, privacy-enhancing mechanisms conceal the identity of the vehicles.

Cryptographic protection, including misbehavior detection and node eviction, addresses a significant part of the problem: it prevents external adversaries from injecting bogus data. External adversaries could still affect sensory inputs, with attacks remaining undetected in the absence of additional mechanisms, whereas internal adversaries can inject any bogus data at will. Only when they are detected can they be isolated and eventually evicted (Section 5.6). But interacting with possibly adversarial (faulty) data senders in order to determine their trustworthiness is not easy. Adversaries can intelligently change their attack patterns (e.g. remaining below the 'detection radar' most of the time but still harming the system). Moreover, detection implies a lengthy interaction, which is often impossible to sustain; encounters are in general short lived and without prior association.

Security mechanisms that allow nodes to detect and ignore bogus data are necessary. In general, nodes cannot rely only on their own measurements or have access to trusted data, and data often come from remote sources. Each node should be able to assess their trustworthiness alone. The use of *non-cryptographic* protection mechanisms is paramount. In the rest of this section we consider two cases: location information and data from other VC-enabled applications.

5.7.1   Securing location information

Location information is critical for VC systems, especially for cooperative awareness, vehicle collision avoidance and essentially all safety applications, as well as for position-based information dissemination. GNSS, such as the GPS, its Russian counterpart (GLONAS), and the upcoming European GALILEO system, are the most widely used technologies: GNSS transmit signals bearing reference information from a constellation of satellites; computing platforms, equipped with the appropriate receiver, can decode them and determine their own location. Most importantly, these units are already integrated in vehicles or available in large numbers, as part of commodity devices for navigation.

However, commercial instantiations of GNSS are open to abuse: according to a recent article (Humphreys et al. 2009), software-defined GPS receivers make spoofing (i.e. the injection of forged navigation messages by an adversary, see Figure 5.3) relatively easy: the hardware can be assembled with off-the-shelf components. In the same article, an effective spoofer is presented. With such an ability, the adversary can influence the location information, $loc(V)$, that a node V calculates, and compromise the node's operation. For example, in the case of a fleet management system, an adversary can target a specific truck. First, the adversary can use a transmitter of forged GNSS signals that overwrite the legitimate GNSS signals and are received by the victim node (in this example, the truck) V. This would cause a false $loc(V)$ to be calculated and then reported to the fleet center, essentially concealing the actual location of V from the fleet management system. Once this is achieved, physical compromise of the truck (e.g. breaking into the cargo or hijacking the vehicle) is possible with a reduced or no ability for the system to detect it and react in time.

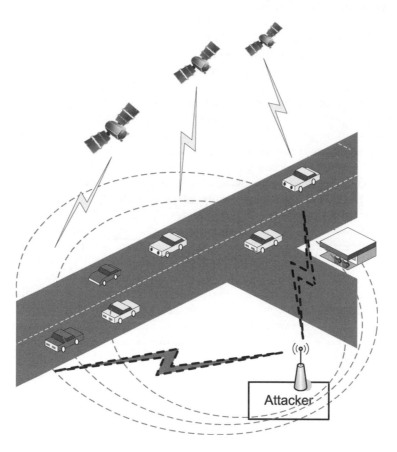

Figure 5.3 Illustration of an attack against GNSS-based localization, in which the attacker forges or replays GNSS signals

It is useful to repeat here a long-known fact: location information in the VC system cannot be considered trustworthy by default. Spoofers fall into the category of input-controlling adversaries (Section 5.2), and such attacks are possible without compromising, physically or otherwise, the GNSS receiver or other on-board equipment or software. A variant of such attacks, termed replay attacks, would be possible even if the GNSS were cryptographically protected, as is expected for the authentication services of the upcoming GALILEO system. Replay attacks can be fine grained, so that the gradual manipulation of the victim's location can remain small and thus hard to detect. But, cumulatively, they can lead to substantial distances between the actual and the perceived (provided by the GNSS) location of the victim nodes (Papadimitratos and Jovanovic 2008b).

Defense mechanisms, complementary or even orthogonal to cryptographic protection, are a promising, low-cost, and effective approach. Three defensive mechanisms that allow receivers to detect forged GNSS messages and to fake GNSS signals have been developed and analyzed by Papadimitratos and Jovanovic (2008a). The countermeasures rely on information that the receiver obtained before the onset of an attack, or, more precisely, before the suspected onset of an attack. This can be (a) a node's own location information, calculated from GNSS navigation messages; (b) clock readings, without any re-synchronization with the help of the GNSS or any other system; and (c) received GNSS signal Doppler shift measurements. Based on these different types of information, it can be detected whether the received GNSS signals and messages are genuine or whether they originate from adversarial devices. If the latter, location information induced by the attack can be rejected, and manipulation of the location-aware functionality be avoided.

5.7.2 Message trustworthiness

The classic approach is to assess the trustworthiness of messages sent by a node (vehicle or RSU) primarily based on the trustworthiness of the sender's credentials. CAs (and nodes when they issue their own pseudonyms) make statements about public keys, identities, and attributes, and special types of data (e.g. revocation lists). Nodes (vehicles and RSUs) make statements about various types of data. At any point in time, messages from any newly encountered nodes are trusted as long as their certificates are valid. These entity-centric trust relations, which are set *a priori*, are useful, but they do not provide the necessary flexibility for the highly volatile and data-centric VC systems.

In VC systems, unlike traditional trust establishment schemes, the data trust level cannot be the same as that of the data-generating node. Clearly, different types of nodes can have by default different levels of trustworthiness; for example, police cars are more trustworthy than private cars. But the trust level of reports from the same type of vehicle can also vary: reports issued at different distances from the corresponding events, in terms of location or time, or the timeliness of the report itself, should have their trust values adapted accordingly. The receiver of such reports could assign the same trust level to those coming from different vehicles but referring to the same event. Moreover, it could use such multiple pieces of evidence to corroborate the event, and then issue its own report. In these cases, it is also clear that the report (data) trust values would differ from those of the reporting vehicle(s).

In such a context, it is far more useful to assess the trustworthiness of data *per se*, rather than to rely on the fixed node trust levels. Raya et al. (2008) present a scheme to instantiate this approach. Default trust levels are used, but only as one factor. In addition, dynamic factors (location, time, number, and type of the statements on data) are used to calculate on the fly the trustworthiness of data, or more essentially the truthfulness of the reported event. This assessment is based on multiple sources of evidence, each of which is assigned a weight calculated via well-established rules. All pieces of weighted evidence and their weights are combined in a decision logic (such as Dempster–Shafer Theory, Bayesian inference, or majority voting) that accounts for the uncertainty of the data.

5.8   Towards Deployment of Security and Privacy Enhancing Technologies for Vehicular Communication

5.8.1   Revisiting basic design choices

Is security always necessary and meaningful?

Is an arguably complex secure VC system actually necessary? This is a legitimate question, given that cellular telephony and nomadic wireless Internet access were deployed without strong security features. They proliferated and continue to do so, in spite of significant security and privacy breaches. But, the significant differences of VC systems from these two systems imply that a different approach is necessary. First, cellular and nomadic network access rely on infrastructure, which simplifies the provision of security; for example, associations (trust) are needed between the mobile node and the infrastructure. Moreover, a compromise would not have costly or even fatal consequences, such as a multi-car accident caused by an attack on a VC system.

The stakes for VC systems are higher: reducing accidents, saving lives, improving transportation. Thus, a single security incident would perhaps suffice for the public to lose confidence in this new technology. Then, assuming sufficient security is in place, one can ponder on the *degree of protection* a security architecture should and can offer. Would strong cryptographic protection of network and application protocols suffice to ensure that false data are not injected into the system? Given the rich location information on VC traffic (e.g. in safety beacons), how easy is it for an adversary to recreate trajectories of vehicles even if their transmissions are fully anonymous (e.g. utilizing GS or BPA, with each message signed under a different temporary identity)? This is feasible in an area that is fully covered by the adversary, but the adversary could also derive and use additional knowledge, so as to be effective even when it cannot intercept VC in some areas (Buttyan et al. 2007). Would anonymous or pseudonymous secure VC ensure location privacy when numerous cameras and optical plate recognition systems are deployed?

VC security does not address problems that are present independently of the use of VC. But it fends off a broad range of exploits that could otherwise wreak havoc with VC and the transportation system. External adversaries or vehicle modifications are mitigated, the results of key compromise are thwarted, and accountability, even if anonymous authentication is used, can lead to the eviction of adversarial nodes. Until this happens, redundancy or the absence of corroborating evidence from other nearby vehicles could enable the degree of truthfulness of received VC application messages to be inferred. Initial results are promising, showing that *data-centric trust establishment* is feasible (Section 5.7.2). Investigations for specific applications and complex environments, as well as measures to thwart determined adversaries, can lead to stronger protection.

Choice of cryptographic tools

The high volatility and large scale of VC systems led to the choice of digital signatures. For the selection of appropriate algorithms, the following basic factors were considered: the processing times for signature generation and verification, the security overhead (public key and signature sizes), the standardization of cryptographic algorithms (and confidence in their strength), and the experience in implementation. Elliptic curve-based algorithms (e.g. Elliptic

Curve Digital Signature Algorithm, or EC-DSA) seem to be preferred, primarily because of low network overhead for strong security. Thanks to usability limitations, approaches such as the HPA (Section 5.5.4) enable on-board, on-the-fly generation of short-term keys and credentials. However, the use of anonymous authentication for this purpose should prevent any abuse: adversaries should not be allowed to use anonymity to influence or control VC protocols and applications. Detection of multiple uses of anonymous transmissions, beyond a threshold within a given period of time, perhaps by revealing the identity of the wrongdoer, is a necessary addition (Camenisch et al. 2006).

Security levels for VC system entities have not been clearly defined yet, but 80-bit or higher security seems to be favored, to prevent practical cryptanalytic attacks. It is important, though, to consider *for which operation* a specific security level for cryptographic primitives is needed. Clearly, CAs and PNPs should have higher security levels. Then high security would be needed for long-term keys and certificates. The lowest level should be assigned to short-term keys, for which security levels below 80 bits could perhaps be considered. Even if such a key, valid for minutes or hours, were broken within weeks or months by a determined adversary, there would be no immediate consequences, only a reduction in overhead. Of course, this would be true if and only if VC traffic were not used in the long run – for example, logged for future liability attribution (Section 5.8.2). In all cases, as discussed in Section 5.5.5, sufficient processing power for the employed cryptographic functionality should be provided to ensure the overall application performance that is sought.

Trustworthy Vehicular Communication equipment?

VC system nodes can often have low physical protection. If vehicular equipment is indeed trustworthy, with the appropriate VC security in place, the overall problem of securing VC systems could be more easily addressed. There are critical resources to be protected: A TC, for example, such as the HSM proposed by SeVeCom, would store private keys and perform private cryptographic operations. With a tamper-resistant TC, extraction of the private keys would be impossible. With a real-time clock and a battery integrated in the TC, the adversary would be unable to feed the TC with fake future time-stamps and obtain falsified cryptographically protected messages. However, cost is a major concern, and making all of the on-board equipment tamper-proof or tamper-resistant would be impractical.

What type of revocation?

The distribution of RLs, discussed in Section 5.6, is the basic approach in VC systems. A RL can be a CRL, if traditional cryptography is used (e.g. for BPA), but it can differ in the case of the HPA scheme. Checking whether a node is in the RL can incur significant processing overhead, primarily because of anonymity mechanisms. For 'classic' cryptography, each pseudonymous certificate should be validated at first reception, but with many pseudonyms per vehicle, the RL would be large. For anonymous credentials, each received message should be checked against the much shorter RL, but each check is orders of magnitude costlier than that for 'classic' cryptography.

The challenge is not the RL distribution but rather its on-board processing cost, which is proportional to the RL length. The natural question to ask is whether it is necessary to ignore messages signed by all nodes in a RL. In fact, this is closely related to the composition of the RL. For example, if a stolen vehicle is in the RL, its VC equipment is not necessarily compromised; thus it would be unwise, in the interests of the safety of receiving vehicles, to ignore its messages. A flexible approach to address the problem could reduce the length of RLs and, thus, the processing overhead: distinct RLs are created, according to the 'urgency' of using them for real-time message validation. The RL of highest priority, processed at all times, can then contain only the truly faulty or compromised nodes. At a lower priority can be a RL with nodes that are possibly faulty, perhaps in different RLs according to the type of fault, and at the lowest priority would be a RL with nodes evicted for other reasons. Lower priority RLs can checked if possible, or if a specific event triggers the need to do so (e.g. suspected faulty behavior by a nearby node).

5.8.2   Future challenges

Introducing (secure) Vehicular Communication to the market

The development of VC systems can significantly influence security solutions and thus the system trustworthiness. The primary question is how the VC deployment would take place. Would it be based on '*all-in-one*' on-board equipment, or would it be based on a sequence of *add-on* components? In other words, would the OBU be one or two powerful, multi-purpose box(es), or perhaps a multi-core processor, running all protocols? Or would it be a set of boxes, each of them added on board gradually, running a single application with just enough processing power for the specific tasks?

   An all-in-one model resembles what has been considered thus far in the development of secure VC architectures. But the add-on approach may be closer to what a strongly market-centric deployment commands, driven by the applications (e.g. entertainment) preferred by consumers. Reflecting the mindset of some stakeholders, the evolutionary deployment would most likely lead to a minimum application-specific security as well as a heterogeneous on-board network. The situation would become more involved if user devices (e.g. Personal Digital Assistants, or PDAs, cell-phones, and home or corporate computers) interact with the OBUs, for example to obtain useful personal information for navigation, to record trip data, or to access physical spaces or digital content. All of these aspects would raise new challenges in terms of security and privacy (Papadimitratos 2008).

Organizational concerns

The reliance on authorities is in line with long-lived approaches in managing vehicles (Papadimitratos et al. 2006b). However, the effort of operating CAs often results in a degree of skepticism, with frequently recurring questions on the operational cost or the difficulties of collaboration among diverse CAs. The alternative, of vehicle manufacturers running their own CAs, is being considered. Nonetheless, this raises concerns about monopoly or oligopoly situations that could be imposed, or even the likelihood that proprietary solutions that do not provide fully fledged security might be adopted. Existing *multi-domain* systems, such as cellular systems, which require access control and accounting, indicate that addressing organizational issues is feasible. In fact, the success story of cellular systems provides useful clues. Numerous distinct providers, each having high numbers of registered clients and devices, all uniquely identified and able to operate in other regions while being billed for network usage via their 'home' providers, is a model that has interesting features and even similarities to ponder.

Legal considerations

User awareness of the offered protection is paramount: the guarantees that the VC system offers, the residual vulnerabilities, and the role of all system entities should be clearly stated in end-user agreements. Analogies can be drawn with existing systems: for example, with recent privacy breaches against cellular telephony perpetrated by insiders. The responsibility of each entity, the users included, should be clear, as this also relates to VC equipment maintenance and accreditation.

The use of VC systems to assist in the attribution of *liability* for transportation incidents is a controversial issue. Clearly, a non-secure VC system would be out of the question for such a task. Strong accountability in secure VC systems, as discussed above, is possible. However, determining which entity can perform this, and under which circumstances, and then through which procedure liability could be attributed, is far from straightforward.

Policies for VC systems would also have to deal with the issue of *voluntary or mandatory* use of the equipment. Would, for example, safety and traffic efficiency functionality be mandatory, in the same way that seatbelts are in many countries nowadays? It is likely that users would have an incentive to run these applications, for example in order to lower their insurance premiums. But if deployment is mandatory, would privacy concerns be fully addressed? Solutions discussed above can indeed achieve this. But users may still raise legitimate arguments in favor of powering off their VC boxes. Or perhaps users may raise the need for distinct secure VC instantiations, for example for government vehicles that do not wish to take any risk of being traced by terrorists.

5.9 Conclusions

Significant progress has been made already towards comprehensive security- and privacy-enhancing solutions for VC systems. Moreover, the design of VC protocols is evolving and standardization efforts are ongoing. The research community has a unique opportunity: to understand the problems at hand deeply, and, at the same time, to design VC protocols and applications by taking into account security and PET mechanisms. For example, OBU characteristics can be set to a certain standard to enable security; or protocol features that enhance performance but allow high-impact attacks can be disabled for enhanced resilience.

At the same time, investigations of VC systems and their security reveal new dimensions and complexity. VC systems would interact with various other systems, essentially forming a wireless system of systems. Moreover, results and insights can be far reaching and applicable to other computing systems. The current extensive interest, rising awareness, and significant results in terms of security for VC systems, along with demonstrations, are most welcome (Ardelean and Papadimitratos 2008; Gerlach et al. 2008; Kargl et al. 2009). Nonetheless, further progress in securing all aspects of the VC system, and extensively evaluating the overall system performance through test beds, is necessary. The objective is to have trustworthy VC systems at the time of their initial deployment, so that societies can reap the benefits of intelligent transportation systems.

References

Ardelean, P. and Papadimitratos, P. (2008) Secure and Privacy-Enhancing Vehicular Communication Demonstration. *IEEE WiVec*, Calgary, AL, Canada.

Ateniese, G. and Tsudik, G. (1999) Group Signatures à la carte. *SODA '99*, Baltimore, MD, USA.

Bellare, M., Micciancio, D. and Warinschi, B. (2003) Foundations of Group Signatures: Formal Definition, Simplified Requirements and a Construction based on Trapdoor Permutations. *Advances in Cryptology*.

Bellare, M., Shi, H. and Zhang, C. (2005) Foundations of Group Signatures: The Case of Dynamic Groups. *CT-RSA*, San Francisco, CA, USA.

Boneh, D., Boyen, X. and Shacham, H. (2004) Short Group Signatures. *Crypto '04*, Santa Barbara, CA, USA.

Brickell, E., Camenisch, J. and Chen, L. (2004) Direct Anonymous Attestation. *CCS '04*, Washington DC, USA.

Buttyan, L., Holczer, T. and Vajda, I. (2007) On the Effectiveness of Changing Pseudonyms to Provide Location Privacy in VANETs. *ESAS*.

Calandriello, G., Papadimitratos, P., Hubaux, J.P. and Lioy, A. (2007) Efficient and Robust Pseudonymous Authentication in VANET. *ACM VANET*, Montreal, Quebec, Canada.

Calandriello, G., Papadimitratos, P., Hubaux, J.P. and Lioy, A. (2009) On the Performance of Secure Vehicular Communication Systems. *LCA-REPORT-2009-006*.

Camenisch, J., Hohenberger, S., Kohlweiss, M., Lysyanskaya, A. and Meyerovich, M. (2006) How to Win the Clone Wars: Efficient Periodic n-Times Anonymous Authentication. *ACM CCS*, Alexandria, VA, USA.

Chaum, D. and van Heyst, E. (1991) Group Signatures. *EUROCRYPT '91*, Brighton, UK.

Festag, A., Papadimitratos, P. and Tielert, T. (2009) Design and Performance of Secure Geocast for Vehicular Communication. *LCA-REPORT-2009-007*.

Gerlach, M., Friederici, F., Ardelean, P. and Papadimitratos, P. (2008) Security Demonstration – C2C-CC Forum and Demonstration, Dudenhofen, Germany.

Harsch, C., Festag, A. and Papadimitratos, P. (2007) Secure Position-Based Routing for VANETs. *IEEE Vehicular Technology Conference (VTC2007-Fall)*, Baltimore, MD, USA.

Humphreys, T., Psiaki, M., Kintner, P., Ledvina, B. and O'Hanlon, B. (2009) Assessing the Spoofing Threat. *GPS World*.

IEEE 1609.2 (2006) Trial-Use Standard for Wireless Access in Vehicular Environments – Security Services for Applications and Management Messages.

Kargl, F., Ma, Z. and Schoch, E. (2006) Security engineering for VANETs. *Proceedings of the Fourth Workshop on Embedded Security in Cars (ESCAR)*, pp. 15–22, Berlin, Germany.

Kargl, F., Papadimitratos, P., Buttyan, L., Müter, M., Wiedersheim, B., Schoch, E., Thong, T.V., Calandriello, G., Held, A., Kung, A. and Hubaux, J.P. (2008a) Secure Vehicular Communications: Implementation, Performance, and Research Challenges. *IEEE Communications Magazine* **46**(11), 110–118.

Kargl, F., Papadimitratos, P., Holczer, T., Cosenza, S., Held, A., Mueter, M., Asaj, N., Ardelean, P., de Cock, D., Sall, M. and Wiedersheim, B. (2009) Secure Vehicle Communication (SeVeCom) Demonstration. *IEEE MobiSys*, Krakow, Poland.

Kargl, F., Schoch, E., Wiedersheim, B. and Leinmüller, T. (2008b) Secure and Efficient Beaconing for Vehicular Networks. *Proceedings of the Fifth ACM International Workshop on Vehicular Ad hoc Networks (VANET)*, San Francisco, CA, USA.

Laberteaux, K., Haas, J. and Hu, Y. (2008) Security Certificate Revocation List Distribution for VANET (short paper). *ACM VANET*, San Francisco, CA.

Leinmüller, T., Schoch, E. and Kargl, F. (2006) Position Verification Approaches for Vehicular Ad Hoc Networks. *IEEE Wireless Communication Magazine* **13**(5), 16–21.

Moore, T., Raya, M., Clulow, J., Papadimitratos, P., Anderson, R. and Hubaux, J.P. (2008) Fast Exclusion of Errant Devices from Vehicular Networks. *IEEE SECON*, San Francisco, CA, USA.

NoW (2007) Network On Wheels. URL: http://www.network-on-wheels.de/.

Papadimitratos, P. (2008) 'On the Road' – Reflections on the Security of Vehicular Communication Systems. *Proc. IEEE International Conference on Vehicular Electronics and Safety ICVES 2008*, pp. 359–363.

Papadimitratos, P. and Jovanovic, A. (2008a) GNSS-based Positioning: Attacks and countermeasures. *Proc. IEEE Military Communications Conference MILCOM 2008*, pp. 1–7.

Papadimitratos, P. and Jovanovic, A. (2008b) Protection and fundamental vulnerability of GNSS. *Proc. IEEE International Workshop on Satellite and Space Communications IWSSC 2008*, pp. 167–171.

Papadimitratos, P., Buttyan, L., Holczer, T., Schoch, E., Freudiger, J., Raya, M., Ma, Z., Kargl, F., Kung, A. and Hubaux, J.P. (2008a) Secure vehicular communication systems: design and architecture. *IEEE Communications Magazine* **46**(11), 100–109.

Papadimitratos, P., Buttyan, L., Hubaux, J.P., Kargl, F., Kung, A. and Raya, M. (2007) Architecture for Secure and Private Vehicular Communications. *Proc. 7th International Conference on ITS Telecommunications ITST '07*, pp. 1–6.

Papadimitratos, P., Calandriello, G., Hubaux, J.P. and Lioy, A. (2008b) Impact of vehicular communications security on transportation safety. *INFOCOM Workshops 2008, IEEE*, pp. 1–6.

Papadimitratos, P., Gligor, V. and Hubaux, J.P. (2006a) Securing Vehicular Communications – Assumptions, Requirements, and Principles. *Workshop on Embedded Security in Cars (ESCAR)*, Berlin, Germany.

Papadimitratos, P., Kung, A., Hubaux, J.P. and Kargl, F. (2006b) Privacy and Identity Management for Vehicular Communication Systems: A Position Paper. *Workshop on Standards for Privacy in User-Centric Identity Management*, Zurich, Switzerland.

Papadimitratos, P., Mezzour, G. and Hubaux, J.P. (2008c) Certificate Revocation List Distribution in Vehicular Communication Systems (short paper). *ACM VANET*, San Francisco, CA.

Papadimitratos, P., Poturalski, M., Schaller, P., Lafourcade, P., Basin, D., Capkun, S. and Hubaux, J.P. (2008d) Secure neighborhood discovery: a fundamental element for mobile ad hoc networking. *IEEE Communications Magazine* **46**(2), 132–139.

Poturalksi, M., Papadimitratos, P. and Hubaux, J.P. (2008a) Secure Neighbor Discovery in Wireless Networks: Formal Investigation of Possibility. *ACM ASIACCS*, pp. 189–200, Tokyo, Japan.

Poturalksi, M., Papadimitratos, P. and Hubaux, J.P. (2008b) Towards Provable Secure Neighbor Discovery in Wireless Networks. *ACM Workshop on Formal Methods in Security Engineering*, Alexandria, VA, USA.

Rabin, M. (1989) Efficient Dispersal of Information for Security, Load Balancing, and Fault Tolerance. *Journal of the ACM*.

Raya, M., Papadimitratos, P., Aad, I., Jungels, D. and Hubaux, J.P. (2007) Eviction of Misbehaving and Faulty Nodes in Vehicular Networks. *IEEE Journal on Selected Areas in Communications, Special Issue on Vehicular Networks*.

Raya, M., Papadimitratos, P., Gligor, V. and Hubaux, J.P. (2008) On Data-Centric Trust Establishment in Ephemeral Ad Hoc Networks. *IEEE INFOCOM*, Phoenix, Arizona, USA.

SeVeCom (2009) Secure Vehicular Communications. http://www.sevecom.org.

Shokrollahi, A. (2006) Raptor Codes. *IEEE/ACM Transactions on Networking*.

Syverson, P.F. and Stubblebine, S.G. (1999) Group Principals and the Formalization of Anonymity. *FM '99*, Toulouse, France.

6

Security and Dependability in Train Control Systems[1]

Mark Hartong

Office of Safety, Federal Railroad Administration, US Department of Transportation

Rajni Goel

Howard University

Duminda Wijesekera

George Mason University

Railroads are a critical component of the US transportation infrastructure, hauling over 1.7 trillion ton-miles of freight per year. Recently they have begun implementing advanced Supervisory Control and Data Acquisition (SCADA)/Digital Control Systems (DCS) (Chai et al. 2007). These network-based systems, known as Communications-Based Train Control (CBTC)/Positive Train Control (PTC) carry control and status messages, providing significant safety enhancements to the traditional methods of railroad operations. However, due to their reliance on wireless communications, they suffer from the same security vulnerabilities

---

[1]The views expressed are those of the authors, and do not necessarily represent those of the United States Government, The US Department of Transportation, or the Federal Railroad Administration, and shall not be used for endorsement or advertising purposes.

Vehicular Networking   Edited by Marc Emmelmann, Bernd Bochow, C. Christopher Kellum
© 2010 John Wiley & Sons, Ltd

inherent in any wireless-based communications system. Countering these security vulnerabilities imposes additional requirements on PTC systems that must be considered throughout
the system development life cycle. Failure to address these vulnerabilities provides mal-
actors (Sindre and Opdha 2005) with another avenue of attack to prevent safe, secure, and
efficient rail operations.

6.1   Introduction

Railroads are a critical component in the US transportation and distribution system. From
humble beginnings in 1827, with a single company, the Baltimore and Ohio Railroad,
operating over thirteen miles of track along the Patapsco River in Maryland, the US rail
system has grown to over 600 companies operating a complex network of over 140,000 miles
spanning the continental United States and Alaska. Freight trains have evolved from single
horse-drawn carts to diesel electric locomotives cable of pulling more than one hundred
286,000 pound cars. Meanwhile the US rail passenger service has changed from 24 people
riding on open platforms at twelve miles per hour to over ten times that number in closed
climate-controlled cars at speeds of up to 150 mph.

Railroads have a significant economic impact on the US economy, and this influence
cannot be understated. With revenues of over 40 billion dollars, the seven largest US
freight railroads alone employ over 167,000 people at an average total compensation of over
94,000 dollars per individual. The diverse mixture of cargos hauled supports all facets of the
US industrial base. From raw materials such as coal, and agricultural products such as wheat,
to finished consumer goods such as automobiles, freight railroads move over 1.7 trillion ton-
miles of freight (AAR 2007b), which equates to 25% of all intercity freight tonnage carried in
the US, and 41% of all ton-miles. Commuter railroads move roughly 1.4 million passengers
per day (APTA 2007) and intercity passenger services another 67,000 (BTS 2007).

Railroads are also responsible for moving significant amounts of hazardous material
created or used in the United States. They operate 30,000 miles of the Department of
Defense Strategic Rail Corridor Network (STRACNET) for the movement of Department
of Defense munitions and other materials (Banks and Barclay 1976). They transport 1.7 to
1.8 million carloads of hazardous material (AAR 2007a), including over 21.6 million ton-
miles of substances that are Toxic by Inhalation (TIH) (FHWA 2006).

6.2   Traditional Train Control and Methods of Rail Operation

Trains are constrained to travel along on a single track, and cannot pass except where there
are sidings, which results in a system with a single degree of freedom. From the beginning in
1827, when multiple trains began to share the same set of tracks, railroads began to formalize
various methods of operations in order to improve efficiency and safety through the reduction
of collisions, derailments, and the associated deaths and property damage. Over the years
these have developed into a complex set of operating rules and practices, often integrated
with various technologies that allow for safe movement, into a complex system of rules
and equipment to control train operations. Today methods for the control of trains range

from verbal authorities to wayside signal indications relayed directly to locomotive cabs, and automatic train stop systems.

6.2.1 Verbal authority and mandatory directives

The traditional means of controlling train operations is through the use of verbal authorities and mandatory directives. A dispatcher who issues the authorities and directives controls train operations. The dispatcher, akin to an aircraft flight controller, takes responsibility for knowing what trains are located where, and ensures that no two trains are issued authorization to occupy the same location of track at the same time. Once the dispatcher issues the orders, the train crew is responsible for ensuring that they obey them.

Verbal authority and mandatory directive operations are generally broken into one of two possible types based on how the railroad tracks are partitioned. In Track Warrant Control (TWC) the instructions are given for the crew to proceed between train stations or specific mileposts. The furthest point along the track to which the dispatcher authorizes a train movement is known as the authority limit. The alternative to TWC is known as Direct Traffic Control (DTC). DTC is similar to TWC in that trains may only operate to the limit of their authority. However, unlike TWC, where the authorities may be issued to different locations by the dispatcher based on the situation, DTC only allows the dispatcher to issue movement authorities in terms of fixed predefined partitions known as 'blocks'. Because all movement authorities are defined in terms of the same set of constantly fixed blocks, it is simpler in execution. The trade-off for this is that the dispatcher does not have the flexibility in the selection of authority boundaries that is provided by TWC.

TWC and DTC are both codified in one of two different rule books. One is the General Code of Operating Rules (GCOR), and other is the Northeast Operating Rules Advisory Committee (NORAC) Rules. The GCOR rules are used primarily by railroads in the western United States, while railroads in the eastern United States use NORAC rules. Railroads may elect to use DTC, TWC, or a combination of both; the decision is made by individual railroads based on what is most efficient for their operations. Roughly 40% of all tracks in the United States are controlled in this manner.

6.2.2 Signal indications

The remaining 60% of track in the USA is controlled through signal indications. Block signal systems were first installed in the USA in 1872. In their simplest form, a section of track, called a block (usually several thousand feet in length), has a low-level electric voltage in it. Each block is separated from the next block by insulated joints between the rails that prevent current from flowing from one block to the next. When a train or other vehicle with steel wheels and a steel axle connects the two energized rails, the circuit is completed between the two rails and the signal system responds by displaying a stop signal at the entrance of that block to following trains.

By 1927 groups of signals and switches could be centrally controlled, where they became known as Centralized Traffic Control (CTC) systems. CTC, also known as Traffic Control System (TCS), uses signals located at predetermined intervals and locations to indicate to the train operator the condition of the track block ahead, and provides authority for train movements. The train dispatcher at the control center determines train routes and priorities, and then remotely operates switches and signals to direct the movement of trains. In addition to track circuits to detect the presence of a train, as well as hazardous conditions such as a broken rail, specially designed devices called switch point detectors are used to determine the position of switches.

Some CTC systems have been further enhanced to provide direct indications of wayside signal conditions to the locomotive engineer inside the cab, in order to provide authority information to the engineer. These cab signal systems provide the on-board display of trackside signal indications through the transmission of signal aspect information in coded pulses along the track. Their function is simply to relay the external signal indications to a visual display inside the cab of the locomotive, making it easier for the crew to note the signal aspect and the associated order that it conveys. In the event of a total compromise of a signal system, the railroad can resort to TWC or DTC mode of operation, with a corresponding reduction in operating efficiency.

6.3   Limitations of Current Train Control Technologies

The main types of train control system technology that are available to reduce the impact of a failure of the train crew to comply with authorizations are Automatic Train Stop (ATS) and Automatic Train Control (ATC). ATS provides enforcement for signal indications. This can be done with or without a cab signals system in place. ATS, however, does not provide speed enforcement. It only enforces the indication provided by the wayside signal in the event that the train crew fails to react. ATC, on the other hand, provides both signal indication enforcement and speed enforcement.

ATS and ATC systems rely on the information relayed through Audio Frequency (AF) current to transmit ATS- or ATC-related information along the track circuit (AREMA 2005). The amount of information that can be transmitted in this way is limited. ATS and ATC systems are both capital intensive. Train location can only be determined to the resolution of the track circuits. Shortening the length of a track circuit increases resolution, but at the cost of more track and wayside hardware. There is a practical (and economic) limit to the number and length of track circuits that a railroad can install. The seven major US railroads spent over $490 million in operations, administration, and maintenance of all types of communications and signaling systems with another $153 million in depreciation of the existing plant (STB 2003) on approximately 65,000 miles of track in 2003. As a result, deployment, not only of ATS and ATC, but also of cab signals, is limited. Consequently less than 5% of route-miles in the US (BTS 2003) have these systems in place.

6.4   Positive Train Control

PTC systems are wireless communication SCADA systems that provide high precision train position information independent of track circuits. They utilize a continuous high bandwidth Radio Frequency (RF) data communications network that allows train-to-wayside and wayside-to-train exchange of control and status information. Wayside, office, and train borne computers process received train status and control data to provide continuous train control. PTC offers protection against failures of existing methods of operation. Basic PTC: (FRA 1994)

- prevents train-to-train collisions (positive train separation);

- enforces speed restrictions, including civil engineering restrictions and temporary slow orders;

- protects roadway workers and their equipment operating under specific authorities.

PTC overcomes one of the fundamental limitations of conventional ATS and ATC, a reliance on track circuits. By substituting wireless communications for encoded track communications, PTC can also increase the amount of data that can be exchanged between the train and the wayside, permitting more effective utilization of the track wayside infrastructure. Additional benefits include the potential for improved reliability and reductions in maintenance costs through a significant reduction in the amount of wayside equipment. PTC can also provide the advantages of signal-based train control systems to non-signalized territory.

6.4.1 Functions

One way of classifying PTC is based on five differing levels of functionality (Table 6.1) (RSAC 1999). Level 0 is the lowest level of functionality, and it provides no protection. Level 4 is the highest, providing not only positive train separation, speed enforcement, and roadway worker protection, but also advanced functionalities that improve the safety and the efficiency of railroad operations. Each higher level of PTC functionality includes the functionality of all lower layers.

Table 6.1 PTC functional levels

PTC Level	Functionality
0	None
1	Prevention of train to train collision
	Enforcement of speed restrictions
	Protection of roadway workers
2	PTC Level 1 +
	Automated digital dispatch of authorities
3	PTC Level 2 +
	Wayside monitoring of the status of all switch, signal, and protective devices in traffic control territory
4	PTC Level 3 +
	Wayside monitoring of all mainline switches, signals, and protective devices
	Additional protective devices such as slide, high water, and hot bearings detectors
	Advanced broken rail detection
	Roadway worker terminals for communications with dispatcher and train

PTC systems are also classified based on whether they augment or replace existing methods of operation. Full PTC systems replace existing methods of operation, while overlay PTC systems provide their functionality while maintaining the existing method of operation. Overlay systems provide increased safety benefits at significantly reduced development and implementation costs relative to full PTC systems, but make potentially unsubstantiated assumptions regarding human reliance on the system and the ability of system operators to recognize failures promptly and take corrective actions (NTSB 2008).

6.4.2   Architectures

The generic PTC architecture is illustrated in Figure 6.1. It consists of three major subsystems interconnected by communications links. Each major subsystem consists of various databases with data communications and information processing equipment. The specific hardware and software configurations of each subsystem vary based on the level of PTC functionality being implemented. The most critical attribute of advanced PTC systems is the integration of precise navigation and location determination technologies, such as those provided by modern GPS along with accompanying office, wayside, on-board and intercommunication technologies that are used to transmit and interpret safety-critical data.

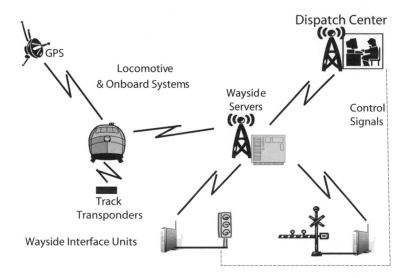

Figure 6.1 Generic Positive Train Control (PTC) architecture

On board

The on-board subsystem consists of locomotives or other on-rail equipment, with their on-board computer, communication, and location determination systems. The on-board subsystem takes data from the office and wayside subsystems, analyzes it, and translates the analyzed information into actions for locomotive systems. It communicates with the office, and possibly directly with the wayside subsystems, as required to report on-board conditions and information received, as well as to request additional operational information. The on-board system monitors its own health, and acts independently when it detects failures and defects. This assures fail-safe system operation even when inter-subsystem communication is lost.

Wayside

The wayside subsystem consists of elements such as highway grade crossing signals, switches, and interlocks. These systems receive commands originating from the on-board system and the office system. Depending on the implementation, the wayside system may communicate directly with the on-board system, providing health and status information. The wayside system might also relay commands from the office subsystem to the on-board system. Alternatively it may communicate only with the dispatch system, communicating with the on-board system via the office. Like the on-board system, the wayside system self monitors for its own internal defects and failures, falling to a safe state if they are detected.

Office

The central control office operates the whole railroad. Through it, the dispatcher controls the behavior of selected wayside units such as signals and switches, prepares and transmits authorities to the on-board system, receives and interprets data received from the wayside and on-board systems, and acts as relay, if required, between the wayside and the on-board systems. The office system is also interconnected with various other IT systems used by the railroad and relays operational data, such as consist weight and length, that is required by the on-board system and the crew for the safe operation of the train.

6.4.3   US communication-based systems

The limited deployment and the inability to use cab signals, ATS, and ATC to prevent human error and effectively incorporate collision and accident avoidance measures with the current methods of operation is the motivation for installation of PTC in the USA. There is, however, significant concern about the rate of deployment. The US National Transportation Safety Board (NTSB) has expressed concern regarding the slow rate of implementation of PTC (NTSB 2004). The US Congress shares similar concerns that the pace of development and implementation of PTC is inadequate, to the extent that in Section 214 of the Federal Railroad Safety Authorization Act of 1994 (Pub. L. No. 103-440) they required the US Secretary of Transportation to report on the status of PTC development and deployment. Progress in installing PTC has been slow, particularly along rail lines that primarily serve freight carriers. Even those lines with significant passenger traffic remain largely unprotected today, some twelve years after positive train control was first placed on the NTSB's Most Wanted list (USS 2003).

Today in the US there are eleven PTC systems either operational or being deployed on over 3000 route miles on eight railroads across 21 states. All of these systems give PTC Level 1 functionality and also provide, to varying degrees, different elements of PTC Level 2, Level 3, and Level 4 functionality. Generally these systems function as designed, although there have been some technical deficiencies encountered. Although the vitality of safety-critical functionalities has not been affected, the deficiencies could potentially have an adverse impact on the efficiency of train operations. Issues encountered include: limited geographical communication coverage; the problems of determining appropriate and accurate location of infrastructure critical points, and appropriate and accurate braking distance prediction; and train tracking.

The eleven systems that are either operational or being deployed for revenue service are the:

- Advanced Civil Speed Enforcement System (ACSES);

- Incremental Train Control System (ITCS);

- Communications-Based Train Management (CBTM);

- Electronic Train Management System (ETMS) Version 1;

- ETMS Version 2;

- ETMS METRA Configuration;

- Vital Train Management System (VTMS);

- Collision Avoidance System (CAS);

- Optimized Train Control (OTC);

- Train Sentinel (TS);

- North American Joint Positive Train Control System (NAJPTC).

Developed for the US Amtrak, ACSES is installed and fully operation on 240 route miles of the North East Corridor (NEC) between Boston, Massachusetts (MA) and Washington, District of Columbia (DC). It supports National Passenger Rail Corporation (Amtrak)'s ACELA, currently the fastest passenger service in the USA, to speeds up to 150 miles per hour. ACSES is a track embedded transponder-based system that supplements the existing NEC cab signal/automatic train control system. Amtrak also operates the ITCS system to support high-speed passenger operations between Niles, Michigan (MI) and Kalamazoo, MI. Operating on 74 route miles, ITCS currently supports speeds up to 95 miles per hour. ITCS is unique among PTC system implementations in that it includes advanced high-speed highway–rail grade crossing warning system activations using radio communication rather than track circuits. Depending on the reports received from the Highway Grade Crossing Warning System (HGCW), the ITCS on board imposes and enforces appropriate speed restrictions. Upon completion of the verification and validation of the software, maximum authorized speeds will be raised to 110 miles per hour.

CSX Transportation (CSXT) is preparing to field test the latest version of their CBTM on approximately 200 route miles of their Aberdeen and Andrews, South Carolina (SC) Subdivisions. Early versions were installed on their Blue Ridge and Spartanburg, SC lines. Current CSXT efforts are focused on harmonization of CBTM with the Burlington Northern Santa Fe (BNSF) Railways ETMS Version 1 and 2, the Union Pacific (UP) Railroad VTMS, and the Norfolk Southern (NS) OTC to support interoperable freight operations.

BNSF Railways has undertaken an extensive PTC development and deployment effort to support their freight operations. ETMS Version 1 for low density train operations has received full approval from the Federal Railroad Administration, and BNSF has started deployment on 35 of its subdivisions. BNSF also has an enhanced version of ETMS, ETMS Version 2, to support high density train operations under active test on its Fort Worth and Red Rock Subdivisions in Texas (TX). A related configuration of ETMS Versions 1 and 2 is under development for the Commuter Rail Division of the Chicago Regional Transportation Authority North East Illinois Commuter Railroad (METRA). Created in response to a series of fatal accidents resulting from excessive train speed through turnouts, the METRA implementation of ETMS is intended to support passenger, as opposed to freight, operations. This system is under deployment on the Joliet and Beverly Subdivisions in Chicago, Illinois (IL).

Unlike the METRA, CSXT, and BNSF variants of ETMS, which are overlays, the UP and NS are developing full (or vital) system variants of ETMS. The UP VTMS has begun test operations on fifteen different UP subdivisions in Washington State in the US Pacific Northwest and the Powder River Basin of Wyoming (WY). The NS OTC variant, which integrates their new NS Computer Aided Dispatch (CAD) System with PTC and other specialized business functionalities, is under test on the NS Charlestown to Columbia, SC Subdivisions. CBTM, ETMS, OTC, and VTMS are all developed by the same manufacturer, and share a common code base. They differ only in their specific hardware configurations.

The Alaska Railroad is undertaking installation of CAS on all 531 miles of its track. Also designed to be a full PTC system, it is built to implement the same PTC functional architecture as other PTC systems, using a completely different hardware and software implementation. CAS enforces movement authority, speed restrictions, and on-track equipment protection in a combination of DTC and signaled territory. All of the wayside and office components have been installed and tested, and on-board system test operations are in progress on the Portage and Whittier Subdivisions outside Anchorage, Alaska.

The Ohio Central Railroad System (OCRS) version of a PTC system is the Train Sentinel (TS). TS is currently in use on various railroads in South and Central America. The OCRS version of TS is based on the TS installation currently operating in mixed passenger and high-speed freight service on the Panama Canal Railroad connecting Balboa and Panama City in the Republic of Panama. The OCRS has completed installation of its office subsystem, and is conducting integrated office, wayside, and on-board subsystem tests between Columbus and Newark, Ohio (OH).

Another system, the NAJPTC System, was removed from service due to technical issues associated with communications bandwidth. The NAJPTC is a joint effort of the Federal Railroad Administration (FRA), the Association of American Railroads (AAR), and the Illinois Department of Transportation, to develop an open industry CBTC standard for high-speed passenger and freight service. The system was relocated to the United States Department of Transportation (DOT) Transportation Technology Center (TTC) Test facility in Pueblo, Colorado (CO), for study and resolution of the communications issues associated with the standard in a controlled environment.

Finally, in the preliminary design stage is the Port Authority of New York and New Jersey (PATH) CBTM. The PATH design is entirely separate and independent from the CSXT CBTM System. The PATH CBTM will provide PTC functionality to the Trans-Hudson River Commuter Rail Line running underground between New Jersey and New York City.

The technical challenges associated with developing all of these complex microprocessor-based software and communication systems are similar to those associated with creation of similar industrial or commercial technology applications. Unlike the past, when the level of technical knowledge possessed by the railroad exceeded that of vendors, today the railroads often lack the level of expertise required to implement and deploy PTC systems independently. Modern microprocessor and software (digital system) design has become such a specialized field that railroads are often at a disadvantage in trying to make sound procurement and acquisition decisions when trying to evaluate detailed technical specifications and system performance. Further, much of the detailed information associated with digital systems is valuable intellectual property owned by vendors, who may not routinely, or ever, release it to the railroads.

Current work by the FRA to implement the new statutory requirements of the Rail Safety Improvement Act of 2008 (RSIA 2008) (USC 2008) involves an extensive effort to facilitate interoperability of PTC system installations mandated by the Act. This effort has involved regulatory reform to encourage the use of standard system designs. Since the RSIA 2008 mandates the installation of interoperable PTC by the end of 2015, the major US passenger, commuter, and freight railroads affected by the statute have informally adopted ETMS- or ACSES-based technology as the standard building blocks for their individual railroads. This has created de facto industry standards that allow the reuse of component designs between different railroads. While the market created by the mandatory installation requirement is large (over 24,000 locomotives will require on-board component installations and over 1.2 million wayside interface units will be required to equip the over 100,000 route miles of affected track), the degree to which the installation will positively impact costs is unknown. The railroads' decision to adopt two very specific proprietary architectures, however, has created an unregulated oligopoly.

Perhaps the most significant non-technical issue in the USA is the transition from traditional prescriptive-based regulations to performance-based regulations by both government and industry. Performance-based standards, by definition, create uncertainty for both regulators and regulated entities with respect to enforcement and compliance issues. Regulators accustomed to enforcing prescriptive standards are frequently uncomfortable with the discretion inherent in loosely specified performance standards. Similarly, regulated entities are uncomfortable with loosely specified performance standards because they believe they give regulators too much discretion when considering enforcement issues or making compliance decisions.

6.5   System Security

The size and scope of the rail network over which PTC may be deployed makes protection of PTC systems from mal-actor exploitation a non-trivial task. PTC has a number of security threats that, if successfully exploited, may have potentially significant consequences for public health and safety.

6.5.1   The security threat

Studies have shown that the threat to railroads is real (Chittester and Haines 2004; GAO 2004; TRB 2003). Worldwide between 1995 and 2005 there have been over 250 terrorist incidents against rail targets (CRS 2005). PTC systems, being based on wireless networks, share some of the same vulnerabilities as other wireless SCADA systems. PTC command protocols run at the Open Systems Interconnection (OSI) application layer, and hence are subject to faults and mal-acts in the lower layers of the protocol or stack as well as those that are in the application layer protocols. Failure of a PTC system to adequately address these issues could allow a mal-actor to exploit them, neutralizing the safety functionality of the system. Regulatory initiatives (CFR 2005) and industry efforts to deploy PTC systems have increased the level of potential risk. Studies by the US National Security Telecommunications Advisory Committee (NSTAC) (NSTAC 2003) and the US Government Accountability Office (GAO) (GAO 2004) have reported both the capability for, and the success of, mal-actors' attacks against control systems.

Although not the result of a compromise of a PTC system, a recent PTC-preventable accident in Graniteville, SC illustrates the potential impact of a successful attack on a PTC system. In this particular accident, the accidental release of chlorine gas after a train-to-train collision killed nine people, injured 554 people, and required the evacuation of 5400 people, with direct and indirect damages exceeding 40 million dollars (NTSB 2005). Fortunately the potential impact of a successful attack via a PTC communications infrastructure can be mitigated through the use of various security controls. Consequently, an understanding of potential attacks and mitigations is critical to effectively implementing and maintaining CBTC safety advantages.

6.5.2 Attacks

Mal-actor resources for conducting wireless communication attacks are significant (ISST 2002). Detailed vulnerability information, tutorials on exploiting the information, and complete attack tools are publicly available and widely distributed via the Internet and World Wide Web. The United States National Security Agency (NSA)-sponsored Information Assurance Technical Framework Forum (IATFF) has classified mal-actors into five general classes (Table 6.2: passive, active, close-in, insider, and distribution – see NSA 2002).

Table 6.2   IATF attack class definitions (NSA 2002)

Attack Type	Definition
Passive	Passive attacks include traffic analysis, monitoring of unprotected communications, decryption of weakly encrypted traffic, and capture of authentication information. Passive interception of network operations can give adversaries indications and warnings of impending actions. Passive attacks can result in disclosure of information or data files to an attacker without the consent or knowledge of the user.
Active	Active attacks include attempts to circumvent or break protection features, introduce malicious code, or steal or modify information. Active attacks can result in the disclosure or dissemination of data files, denial of service, or modification of data.
Close-in	In a close-in attack, individuals attain close physical proximity to networks, systems, or facilities for the purpose of modifying, gathering, or denying access to information. Close physical proximity is achieved through surreptitious entry, open access, or both.
Insider	Insider attacks can be malicious or non-malicious. Malicious insiders intentionally eavesdrop, steal or damage information, use information in a fraudulent manner, or deny access to other authorized users. Non-malicious attacks typically result from carelessness, lack of knowledge, or intentional circumvention of security for benign reasons.
Distribution	Distribution attacks focus on the malicious modification of hardware or software at the factory or during distribution. These attacks can introduce malicious code into a product, such as a back door to gain unauthorized access to information or a system function at a later date.

Passive attacks

The first category of attacks are passive attacks. The surreptitious gathering of information by a mal-actor can leave victims without any knowledge that they are under attack. As long as the mal-actor does not actively transmit or otherwise disturb the victims' transmission, the victims have no cause to take mitigating measures – unless they have information from other sources that they may be under attack. A passive attack may be particularly easy for two reasons: frequently, confidentiality features of wireless technology are not enabled by the user; and the inherent ease of access to wireless signals enables mal-actors to compromise the system at any number of points.

Active attacks

The second category of attacks are active attacks. Mal-actors may select from a broad continuum of attacks depending upon their knowledge of the parameters of the messages being exchanged between the sender and the receiver. One type of active attack is jamming. In jamming the communication a mal-actor uses some mechanism to disable the entire communications channel between the sender and the receiver. With the original sender and receiver unable to recognize transmissions between them, they cannot exchange information and are unable to communicate. The mal-actor requires no knowledge of the parameters of the messages being exchanged between sender and receiver, only a device capable of blocking transmissions over the communications channel.

A more sophisticated method of blocking communications between the sender and receiver by a mal-actor is the Denial of Service (DOS) or Distributed Denial of Service (DDOS) attack. The specific mechanisms of a DOS or a DDOS are dependent on the communications protocol and product implementation, since these attacks exploit weaknesses in both the communications protocol itself and the product's implementation of it. Both the DOS and the DDOS require that the mal-actor have a good understanding of the parameters of the messages being exchanged between sender and receiver. Using this knowledge, the mal-actor floods the communication channel with invalid messages, preventing the exchange of valid messages between the sender and the receiver. The two attacks differ from each other in the source of the invalid messages. DOS attacks originate from only one location, and DDOS from multiple locations.

Other active attacks are based on exploitation of weaknesses associated with the sender (identity theft, where an unauthorized user adopts the identity of a valid sender), with the receiver (malicious association, where an unsuspecting sender is tricked into believing that a communications session has been established with a valid receiver), or with the communications path ('man in the middle', where the attacker emulates the authorized receiver for the sender – the malicious assertion – and emulates the authorized transmitter for the authorized sender – identity theft). These attacks are primarily geared towards disrupting integrity in the form of user authentication (assurance that the parties are who they say they are), data origin authentication (assurance that the data came from where it said it did), and data integrity (assurance that the data has not been changed).

Close in, insider and distribution attacks

The last three categories of attack describe the nature of system access by the mal-actor, as opposed to the nature of the attack itself (passive or active). Close-in, insider, and distribution attacks consist of some form of passive attack, active attack, or combination of the two. The effectiveness of these attacks is enhanced by a higher degree of specialized knowledge and access to the system on the part of the mal-actor.

6.5.3   Required security attributes

The required attack mitigation strategy that must be used by US Government information processing and communications systems is codified into law (USC 2002). While commercial activities, such as railroads implementing PTC, are not bound by law to implement these strategies, the same strategies are still applicable. Specifically these are confidentiality, integrity, availability, authenticity, accountability, and identification.

Confidentiality

Confidentiality is concerned with ensuring that the data and the system are not disclosed to unauthorized individuals, processes, or systems. The enforcement of confidentiality must be considered for two different points in time. One is when the data is moving through the communications channel between sender and receiver, and the other is when data is received and retained by the sender and receiver. Generally the enforcement of confidentiality when data is located at a sender or a receiver is by one of three common strategies: Mandatory Access Control (MAC), Discretionary Access Control (DAC) or Role-based Access Control (RBAC) (DOD 1985; Ferraiolo and Kuhn 1982). Enforcement strategies for confidentiality for data in transmission are usually derived from the enforcement strategy at the sender or the receiver.

Integrity

Integrity ensures that data is preserved with regard to its meaning, completeness, consistency, intended use, and correlation with its representation. Like confidentiality, it can be considered in multiple ways. Another approach further classifies integrity in terms of a hierarchical set of properties (Sandhu 1993). These properties are: consistency (the least restrictive property); no improper modification; no unauthorized modification; and no undetected change. These properties are applicable after receipt at the sender or receiver, and during transmission.

Availability

Availability assures that there is timely and uninterrupted access to the information and the system. Unlike confidentiality and integrity of the data, which are properties of the data or access to it, availability is more a property of the system. The availability of data is a function of the ability of a system to provide users with access to it, rather than whether a system will disclose the data, or whether the disclosed data has been changed.

Authenticity

Authenticity is the ability to verify that a user or process attempting to access information or a service is who or what it claims to be. It, and identification, are the critical components of accountability. The ability to ensure accountability requires not only that a system be able to determine whether entities requesting access are who they say they are, but that such an entity can be uniquely determined after the fact. Authenticity and identification are often not considered as separate objectives. User authenticity and data origin authenticity differ in that the former involves corroboration of the identity of the originator in real time, while the latter involves corroboration of the source of the data (and provides no timeliness guarantees).

Accountability

Accountability enables events to be recreated and traced to entities responsible for those actions. Accountability has two aspects. The first is assignment of responsibility for the use of the resource or information. The second is the assignment of responsibility to the entity for the consequences of its use of the resource or information. Accountability requires the ability to uniquely determine who an entity is (identification) as well as the validity (authentication) of the entity seeking access.

Identification

Identification is the specification of a unique identifier for each user or process. It allows for the assignment of accountability.

6.5.4   Analysis of requirements

The requirements associated with the security of a PTC system can be expressed in terms of the elements necessary to mitigate an attack, or in terms of the elements required for the execution of a PTC function.

Analysis of requirements by attack

Protection against passive attacks is provided by confidentiality. The access control mechanisms to prevent unauthorized users accessing data, services, and resources at the sender or receiver also protect data in transmission. The need, however, for protection against passive attacks with respect to the identity of the communicating parties, or the details of information communicated between them, has not been fully established for PTC systems.

Protection against active attacks can be provided by a combination of confidentiality, availability, accountability, authentication, and integrity mechanisms. The access control mechanism used to enforce confidentiality can prevent a mal-actor from obtaining knowledge of different aspects of the communication protocols, reducing the vulnerability to DOS or DDOS attacks. Capacities used to provide for availability can also provide an effective counter to DOS or DDOS attacks. If the valid message originator, the valid receiver, and the communications channel have sufficient capacity to ensure availability, the mal-actor will be unable to disrupt communications.

Ensuring integrity (and confidentiality) places restraints on availability and also has performance costs. Encrypting agents and messages in transit may impose unacceptable delays in environments where near real-time response is required.

Authentication mechanisms, in providing accountability, counter identity theft, malicious association, and 'man in the middle' attacks. User authentication methods range from time invariant weak authentication methods, such as simple passwords, to time variant strong cryptographic authentication methods. In non-hostile environments, no or weak user authentication may be acceptable, while in hostile environments strong user authentication is essential to provide authenticity.

Data origin authentication provides assurances regarding both message integrity and sender identity. One of the most common techniques for identifying the authenticity of the sender is the use of symmetrical or asymmetrical digital signatures. Cyclic Redundancy Codes (CRCs) can provide message integrity in a non-hostile environment, although it will not do so in a hostile environment. This is the result of the CRC's many-to-one relationship between its input and output. It is possible for multiple inputs to checksum to a single CRC value. As a result, a data substitution can be made with a correct CRC and remain undetected. A cryptographic hash function, where there is a unique one-to-one relationship between input and output, and where only one data input can checksum to one hash value, will provide for integrity. Any change in the input results in a change in the hash value, which is detected at the receiver when the hash calculation is carried out and the received hash value does not correspond to the calculated hash value (Menezes et al. 2001).

Analysis of requirements by function

The consist is the collection of rail cars that make up a train. The 'consist information' for a train refers to data about the length, weight, and distribution of loads and key elements in calculating the safe braking distance used to provide positive train separation, speed enforcement, and roadway worker–train separation. Underestimating the safe braking distance can result in train-to-train collisions if a train is unable to initiate braking to avoid exceeding the limits of its authority. A train exceeding the limits of its authority and violating the limits of another train's authority introduces the potential for collision as the two trains attempt to occupy the same section of track at the same time. Similarly, maintenance worker casualties may occur if the train overshoots the limits of its authority and enters a maintenance work zone prior to the maintenance workers clearing the track.

In addition to determining the stopping distance, the weight and load information are used in calculating brake applications necessary to reduce speed to comply with a temporary or permanent speed restriction. Underestimation of the brake application required to decelerate to the correct speed will result in an 'over speed' condition. If the speed restrictions were applied due to track geometry or track maintenance conditions, an over speed could result in a derailment. Consist length is also used for determining clearances when attempting to park a train in a siding, and a following train uses the length information to determine the trailing end of a leading train when estimating the distance separating the two.

Knowledge by a mal-actor of the specifics of consist information does not create any security issues; hence, confidentiality is not required. However, incorrect consist information that may be the result of mal-actor actions does create security issues. Integrity, to ensure that the consist information has remained unchanged; availability, to ensure that the correct consist information is obtainable; and accountability, to ensure that the consist information provided is from a reputable source, are required.

The knowledge of train location is mandatory for the execution of all PTC functions. It is used in the calculation of speed and subsequent braking distances, and in the determination of train location relative to switches, wayside devices, work zones, speed restrictions, authority limits, or other trains and on-track equipment. Train location information is exchanged between trains, between trains and waysides, between trains and roadway workers, or between trains and the dispatch/control center. As with consist information, the knowledge of a train's location by a mal-actor is generally not an issue; hence, confidentiality is not required. However, because of the uses of train location, it is necessary, as it is with consist information, to ensure the integrity, availability, and accountability of the data. There may also be an exception regarding the non-confidentiality of train location. It may not be advisable to broadcast the locations of certain high value trains, hauling extremely valuable or hazardous materials, openly. Knowledge of the train's location may allow a mal-actor to execute non-communication-based attacks on the train more easily.

The track database defines the railroad as seen by the PTC system. As such, it contains the position description of the track, and its turnouts, permanent speed restrictions, mileage markers, grade crossings, and critical wayside devices. This is augmented with the position of temporary speed limits, work zones, and other movement or operating restrictions. The track database, when coupled with consist position information, defines the exact location of the train on the railroad. Without the information, correct operation of the PTC system is not possible. The data is sent either from the wayside or from the dispatch center to the train. This information is readily available from other sources, so confidentiality is not an issue. However, given the importance the database plays in the correct operation of the PTC system, data integrity and availability are required, as well as accountability.

Authorities provide authorization for a train or roadway worker to occupy a particular segment of track. Generally knowledge of the content of the authorities, as well as the identities of the affected train crews, dispatchers, and roadway workers is not an issue. Confidentiality, therefore, is not important. However, the timely receipt of authorities and wayside information is critical to safe train operations. This information must be from the appropriate source (requiring accountability), unaltered (requiring integrity) and timely (requiring availability).

6.6 Supplementary Requirements

In addition to the traditional security requirements associated with confidentiality, integrity, availability, and accountability are requirements for other functionalities to support successful PTC operations. These include interoperable performance management, configuration management, accounting management, fault management, and security management.

Defining and implementing these requirements is often complicated because railroads may often share trackage rights. The Interstate Commerce Act authorizes the Surface Transportation Board (STB) of the DOT to impose conditions (between railroad companies) on the granting of reciprocal switching, track access provisions, and reciprocal access to facilities under 49 U.S.C. 11323.[2] This requires negotiation between railroads, with respect not only to security requirements, but to supplementary requirements to interact effectively to achieve shared goals (Brodie 1993).

This is currently a difficult situation; there is no guarantee that the on-board PTC subsystems of a tenant railroad will be compatible with the wayside and dispatch subsystems of the host railroad. Even within a railroad, it is unlikely that PTC will be installed across the entire road simultaneously; so not all subsystems will necessarily be available. This leads to a host of significant security issues, both technical and policy related.

6.6.1 Performance management

Performance management measures various aspects of systems performance. There are two competing issues that require different types of data collection, with differing types of data that the performance management system must support. The first is that of obtaining the

---

[2]Title 49: Transportation; U.S.C. §11323: US Code – Section 11323: Consolidation, merger, and acquisition of control.

necessary data to measure system service reliability and the degree to which the negotiated quality of service is met. The second is the optimization of the use of the resources being monitored.

Development of the management system requires its own security to protect and to monitor the management data. Accomplishment of this, without adversely impacting overall system performance, often requires significant financial and technical effort. Establishing the appropriate variables of interest, collecting performance data, and determining a baseline is difficult. Setting appropriate thresholds to prevent false positives, as well as not to miss legitimate attacks, requires training and ongoing tuning.

By establishing a baseline on performance, and through careful selection and monitoring of appropriate variables such as throughput, response times, and bandwidth utilization, it is possible to determine when the performance thresholds are exceeded. This would indicate a possible attack on the communications links.

6.6.2 Configuration management

Implementation of Configuration Management (CM) for US PTC systems is a regulatory requirement, addressed in both 49 Code of Federal Regulation (CFR) 236.18 and 49 CFR 236,907(c)(3). CM studies and tracks system hardware and software configuration and membership, to ensure that only the correct versions of software and hardware are in use. It also helps to prevent the installation, substitution, or use of non-authorized hardware/software in the network of PTC devices, and it provides a record of system components that can be used in troubleshooting security issues.

CM also has some negative sides. A CM system can provide, to a mal-actor, complete information regarding each component in the user's network. This can provide the mal-actor with the information needed to determine the system's exploitable vulnerabilities more easily. The need for greater accountability for CM data therefore arises. The importance of determining who accessed the CM data, when they accessed the data, what data they accessed, and why they accessed the data, becomes increasingly important. Without knowledge of this information, it cannot be guaranteed that tampering or modification has not occurred.

6.6.3 Accounting, fault, and security management

Accounting management, like performance management, provides measures of network behavior in order to evaluate individual PTC systems and their use. It also supports protection against unauthorized use of resources by non-PTC systems, albeit not necessarily in real time. By careful identification and control at the individual PTC system component level, defense against unauthorized use can be provided. Selection and monitoring of appropriate variables such as throughput, response times, and bandwidth utilization allow determinations to be made when resources use limits have been exceeded.

Fault management primarily provides protection against non-malicious technical issues that threaten to affect PTC system network operations. This primarily affects availability of the system. Although the goal of fault management is to detect, log, notify, and fix network problems, fault management is useful in resolution of security issues since it can help determine or rule out issues as potential system problems.

Security management provides and manages the access control mechanisms for PTC systems in accordance with the security policy of the network. Access control is intended to prevent either intentional or unintentional sabotage, and misuse of PTC devices and the network resources. Security management subsystems identify sensitive network resources and determine mappings between these resources and authorized users. They also monitor access points to the network and notify of inappropriate access to sensitive resources.

6.7   Summary

Positive Train Control systems are being designed to decrease railway accidents by electronically enforcing speed restrictions, inter-train separation, and a host of other requirements. Due to potential misuses of such a system as a consequence of communication vulnerabilities, some additional security mechanisms need to be included in various designs.

The recent RSIA 2008 statutory mandate for the installation of PTC by 2015 introduces further challenges to the US railroads and suppliers of PTC systems. Estimates for railroad compliance with the statute based on available technology range from $3 to $6 billion. This requires a significant, new, unplanned capital investment on the part of the affected railroads, which are publicly traded commercial entities. When these unplanned capital requirements are combined with the associated RSIA 2008 requirements for system interoperability and the rapid system build out, it introduces additional new design and implementation constraints that have not been previously considered by either the railroads or their PTC suppliers. Migrating the currently available PTC technologies, which were designed and built assuming a more limited deployment, slower implementation rates, a less hostile security environment, and with less concern for system interoperability and unit cost, to PTC systems that are more secure, have greater interoperability, and support rapid, widespread deployment with significantly lower per-unit costs presents a technical and managerial challenge unlike any previously faced by the US railroad industry.

This chapter has introduced the basics of traditional and positive train control systems in use in the United States, along with their associated security and performance issues. Further effort, both in terms of specific implementation policy and technology, is required to design an efficient, effective, and economical security solution to PTC. If appropriately chosen, and when considered in light of organizational and environmental factors, a combination of managerial, operational, and technical controls can synergistically work together to ensure safe, secure, interoperable, and cost-effective PTC systems.

References

AAR (2007a) Mandatory Hazmat Rerouting. Policy and Economics Department, Association of American Railroads, Washington, DC.
AAR (2007b) Railroad Facts, 2007 Edition. Policy and Economics Department, Association of American Railroads, Washington, DC.
APTA (2007) Commuter Rail Public Transportation Ridership Report, Fourth Quarter 2007. American Public Transportation Association, Washington, DC.
AREMA (2005) Communications & Signaling Manual of Recommended Practices, Volume 4, Part 16.4.50. American Railway Engineering and Maintenance of Way Association, Washington, DC.
Banks, W. and Barclay, R. (1976) An Analysis of a Strategic Rail Corridor Network (STRACNET) for National Defense. Military Traffic Management Command, Washington, DC.

Brodie, M. (1993) The promise of distributed computing and the challenges of legacy information systems. *Proc. Conference on Semantics of Interoperable Database Systems*, Australia, eds. Hsiao, D., Neuhold, E.J., Sacks-Davis, R. Elsevier North Holland, Amsterdam.

BTS (2003) National Transportation Atlas Databases (NTAD) 2003, Federal Railroad Administration (FRA) National Rail Network 1:100,000 (line) 2003 ed., Bureau of Transportation Statistics, US Department of Transportation, Washington, DC.

BTS (2007) Amtrak Ridership Transportation Statistics Annual Report, 2007. Bureau of Transportation Statistics, US Department of Transportation, Washington, DC.

CFR (2005) 49 CFR Parts 209, 234, and 236 Standards for the Development and Use of Processor Based Signal and Train Control Systems; Final Rule. Office of the Federal Register, Washington, DC.

Chai, S. et al. (2007) Surface Transportation and Cyber-Infrastructure: An Exploratory Study. *Proceedings of the 2007 IEEE International Conference on Intelligence and Security Informatics.*

Chittester, C. and Haines, Y. (2004) Risks of Terrorism to Information Technology and to Critical Interdependent Infrastructure. *Journal of Homeland Security and Emergency Management* **1**(4), Berkley Electronic Press.

CRS (2005) Terrorist Capabilities for Cyber Attack – Overview and Policy Issues Report RL33123. Congressional Research Service of the Library of Congress, Washington, DC.

DOD (1985) Department of Defense Trusted Computer System Evaluation Criteria. US Department of Defense, Washington, DC.

Ferraiolo, D. and Kuhn, D. (1982) Role Based Access Control. 15th National Computer Security Conference.

FHWA (2006) Freight Facts and Figures 2006. Office of Freight Management and Operations, Federal Highway Administration, US Department of Transportation, Washington DC.

FRA (1994) Railroad Communications and Train Control, Report to Congress Federal Railroad Administration, US Department of Transportation, Washington, DC.

GAO (2004) GAO Testimony Before the Subcommittee on Technology Information Policy, Intergovernmental Relations and the Census, House Committee on Government Reform, Critical Infrastructure Protection Challenges and Efforts to Secure Control Systems US General Accounting Office, Washington DC.

ISST (2002) Diversification of Cyber Threats. Institute for Security Technology Studies at Dartmouth College, Investigative Research for Infrastructure Assurance Group, Dartmouth, NH.

Menezes, A., van Oorschot, P. and Vanstone, S. (2001) Handbook of Applied Cryptography 5ed, CRC Press.

NSA (2002) Information Assurance Technical Framework (IATF), Release 3.1. Information Assurance Solutions Group, US National Security Agency, Fort Meade, MD.

NSTAC (2003) Wireless Security. The Presidents National Security Telecommunications Advisory Committee Wireless Task Force Report, Washington, DC.

NTSB (2004) NTSB Most Wanted Transportation Safety Improvements 2004–2005. National Transportation Safety Board, Washington, DC.

NTSB (2005) Collision of Norfolk Southern Freight Train 192 with Standing Norfolk Southern Local Train P22 with Subsequent Hazardous Material Release at Graniteville, SC, January 6, 2005. National Transportation Safety Board, Washington, DC.

NTSB (2008) Positive Train Control Systems Symposium March 2–3, 2005 Ashburn, Virginia. http://www.ntsb.gov/events/symp_ptc/symp_ptc.htm.

RSAC (1999) Report of the Railroad Safety Advisory Committee to the Federal Railroad Administrator, Implementation of Positive Train Control Systems. Federal Railroad Administrator, US Department of Transportation, Washington, DC.

Sandhu, R. (1993) On Five Definitions of Integrity. Proceedings of the IFIP WG11.3. Workshop on Data Base Security, Lake Guntersville, AL.

Sindre, G. and Opdhal, A. (2005) Eliciting Security Requirements with Misuse Cases. *Journal of Requirements Engineering* **10**(1), Springer, New York.

STB (2003) Statistics of Class I Freight Railroads in the United States 2003. Office of Economics, Environmental Analysis and Administration, Surface Transportation Board, Washington, DC.

TRB (2003) Cyber Security of Freight Information Systems. Transportation Research Board of the National Academy of Sciences, Washington, DC.

USC (2002) Federal Information Security Management Act of 2002 (Public Law 107-347). United States Code.

USC (2008) Rail Safety Improvement Act of 2008 (Public Law 108–432). United States Code.

USS (2003) Senate Report 107–224, Department of Transportation and Related Agencies Appropriations Bill. Congressional Record, Washington, DC.

7

Automotive Standardization of Vehicle Networks

Tom Schaffnit

Schaffnit Consulting, Inc.

This chapter focuses on the various global standardization activities related to the automotive industry, where vehicles are communicating directly with other vehicles and with roadside infrastructure. It begins with a discussion of the basic requirement for interoperability within a major region, and then provides some insight into why standardization is absolutely necessary, but is not sufficient, to ensure interoperability.

The next portion of the chapter provides an overview of the Open Systems Interconnection (OSI) seven-layer protocol model, and begins the characterization of relevant standards development activities according to the related layers of this OSI model.

This chapter then introduces and describes standards development progress in the three regions with strong activity: North America, Europe and Japan. Differences in the various standardization processes are described, along with relationships and interactions among standards development organizations, with the overall goal of providing the reader with both a regional and a global perspective on relevant standardization processes.

7.1 General Concepts

In-vehicle networks have become standardized, up to a point, over the past twenty years. In particular, the Controller Area Network (CAN) protocol for fairly high-speed serial data exchange over twisted pair wiring in motor vehicles has been standardized at the lower protocol layers on an international basis (ISO 11898 2007). This standard has been adopted by most of the vehicles being produced, since it offers an economical approach toward reliable interconnection of sensors and processors within a motor vehicle. The stated

Vehicular Networking   Edited by Marc Emmelmann, Bernd Bochow, C. Christopher Kellum
© 2010 John Wiley & Sons, Ltd

qualification 'up to a point' refers to the fact that the upper layer portion of the CAN protocol implemented on a vehicle is likely to be a manufacturer-proprietary protocol. This situation has occurred as different manufacturers have innovated in their internal networks and the associated sensors and processors, and as the manufacturers attempt to protect their critical engine and safety system control from outside interference or unauthorized manipulation. The good news to be drawn from the CAN bus experience is that a technology can be standardized over the longer time-frame (spanning multiple decades) required for implementation in motor vehicles. The cautionary note from this experience is that having an international standard relating to just a portion of the protocol stack does not necessarily provide interoperability for applications.

This chapter describes the conceptual motivation behind automotive standardization of vehicle networks, and then discusses relevant standardization activities and progress from a global perspective.

7.1.1 Vehicle-to-Vehicle communications

One of the fundamental driving forces underlying the movement toward automotive standardization of vehicle networks is the belief by automotive safety engineers that the capability for Vehicle-to-Vehicle (V2V) communications offers the opportunity to enable cooperative vehicle safety applications that will be able to prevent crashes. The types of cooperative collision-avoidance applications that are envisioned for initial deployment would:

- identify other vehicles in the immediate vicinity;
- maintain a dynamic state map of other vehicles, including their location, speed, heading and acceleration;
- perform a continuous threat assessment based on this dynamic state map;
- identify potentially dangerous situations that require driver intervention;
- notify the driver at the appropriate time and in the most appropriate manner.

Of course, the long-term potential for automatic vehicle intervention to avoid or mitigate crashes is implied in these considerations. It will take some time, however, for field experience to validate the reliability of the V2V communications to a level where such automated responses are deployed. There is also a sense that the potential for public acceptance of automated intervention for crash avoidance must be carefully gauged before deployment could be realistically considered. Finally, the concept of cooperative collision-avoidance applications, as described above, has generated some intrinsic requirements for the wireless communications technology necessary to support these applications.

7.1.2 Vehicle-to-Infrastructure communications

In conjunction with the considerations regarding V2V communications as described in the previous section, automotive safety engineers have identified the potential to use the same communications approach to allow vehicles to communicate with the infrastructure. This would enable an expanded range of vehicle crash-avoidance safety applications using the same wireless technology. One of the principal areas where research has been devoted to vehicle safety enabled by Vehicle-to-Vehicle/Infrastructure (V2X) communications is intersection collision avoidance. For example, if a vehicle knows its own dynamic state (location, speed, heading, acceleration, control inputs), as well as the intersection geometry, and, in the case of a signalized intersection, the status of the signal phase and timing, with very low communications latency, then the vehicle could conceivably warn the driver if it appeared likely that a violation would occur. The communications requirements for such safety applications are very similar to those described above for V2V crash-avoidance applications.

In contrast, 'soft safety' applications (such as notification of congestion or adverse weather conditions ahead), mobility applications (vehicle probe data) and convenience applications (for example, driver information or electronic payments) are viewed as generating less stringent communications performance requirements, in terms of latency, range and time to establish communications.

7.2　Interoperability

It appears possible, or even desirable, to deploy 'soft safety', mobility and convenience applications, as mentioned in the previous section, using a diverse variety of wireless communications technologies, along with proprietary systems and applications. However, the crash-avoidance 'active' safety applications obviously require that all participating vehicles and roadside units be able to 'hear' each other, and that they 'speak the same language'. The implication of this requirement is that, within the area of normal operation of a vehicle, the same standardized communications technology should be used.

Consider the hypothetical case where different wireless technologies are deployed initially to support 'soft safety', mobility and convenience applications. Then, after such deployment, there is a desire to enhance the overall benefits of vehicle communications by including the crash-avoidance applications. If Vehicle A uses cellular communications and Vehicle B uses satellite communications to support 'soft safety', it seems difficult, if not impossible, to enhance the existing systems to support the interoperability requirement. This illustration also ignores how the existing technologies could attempt to meet the other requirements of crash-avoidance vehicle safety applications (such as latency and range). As a net result of such considerations, it appears to be most beneficial to agree (within a major region, at least) on the preferred wireless technology to support the crash-avoidance vehicle safety applications, and then make that technology the standard.

From the hypothetical case described above, it appears that there are very different forces driving the deployment of 'soft safety', mobility and convenience applications, in comparison with the rationale for crash-avoidance 'active' safety applications. However, in the longer term, the same technology used to support 'active' safety applications could also be used to support many of the 'soft safety', mobility and convenience applications. Interoperability is clearly a fundamental requirement for the crash-avoidance vehicle safety applications, making detailed standardization mandatory. This interoperability could also prove to be a long-term benefit for 'soft safety', mobility and convenience applications, by providing a unified technical deployment environment. As with the Internet or GPS systems, such a unified technical deployment environment for vehicle communications could potentially spur innovation, and create a vast market for proprietary, differentiated products and services that use the same basic, standardized communications technology.

In order to realize this opportunity, however, this environment must be stable over the longer term, with any technology enhancements providing backwards compatibility, as with the Internet and GPS experience over the past several decades.

The basis for this required interoperability is agreement on thorough, open standards that are stable over decades-long time horizons. This long-term focus is necessary not only to allow the development of innovative products and services based upon this standard technological foundation, but also to support the deployment life cycle of automobiles – many of which remain in service for over a decade.

7.2.1   Regional requirements and differences

Having made the argument for interoperability through standardization at the beginning of this section, it is necessary to qualify this somewhat to address spectral reality on a regional basis. Although the world's manufacturers of automobiles would strongly benefit from global standardization of vehicle communications systems, due to the associated economies of scale and global single-design savings, this is at present (and for the foreseeable future) highly unlikely.

Radio Frequency (RF) spectrum has traditionally been allocated on a national basis, with a few notable exceptions (like international short-wave radio bands or satellite links), where international usage of the spectrum must be coordinated. This has resulted in spectral allocations for almost all usable frequency ranges to have developed very differently in different countries and regions. Now, in order to 'open up' a particular frequency range to be used for vehicle communications, legacy users must be moved from their existing spectral allocations. This is particularly true for crash-avoidance type applications that involve safety of life issues, and which generally require dedicated spectrum, or, at least, highest priority access to the spectrum. This can be a costly and time-consuming process.

Generally, administrations seek to coordinate assignment of spectrum for such new services as vehicle safety communications with either:

- spectrum being vacated as new technology migrates particular existing services to other frequencies (for example, Digital Television, or DTV); or

- new spectrum (generally higher frequencies) becoming economically usable due to improvements in technology.

Thus, a single global frequency range for vehicle safety communications is likely to take a long time to realize. However, would potentially be major benefits to economies of scale and automotive systems design if there could be a global agreement on the general frequency range to be used (such as 'around' 5.8 GHz), channel widths (for example, 10 MHz), and the over-the-air protocol (such as IEEE 802.11p). If such basics were globally agreed, then regional differences could be readily accommodated – potentially while using the same radio chipsets but with different programming, for example.

From a user perspective such regional differences would be unlikely to create any noticeable impact, as long as the regions were large enough to include all of the geographical area within which a vehicle purchased in an area would be likely to be driven during its entire deployed life. For example, if North America had a slightly different frequency range from that of Europe, users would only rarely notice this difference, since most vehicles that are purchased and driven in North America will never be driven in Europe. This is a good situation, since such differences between regions are likely to be experienced for the foreseeable future.

7.2.2 Necessity of standards

In Section 7.2, standards are described as necessary for interoperability. In this subsection the areas of necessary standardization are discussed, along with the limits beyond which standards are not required. This type of discussion assumes a balancing point between necessary standardization and the flexibility to innovate and potentially create a market advantage through proprietary technology.

First, consider the requirement for a specific frequency to be standardized for vehicle safety communications. Imagine that a Software Defined Radio (SDR) technology were available that could:

(a) scan a gigahertz-wide spectral range in detail within one millisecond;

(b) identify the width of the transmitted signal;

(c) adjust channel-edge filters to eliminate adjacent signals from interfering with the desired one;

(d) identify the modulation scheme being used on the detected signal;

(e) employ the appropriate de-modulation scheme in real time for the detected signal;

(f) identify and employ the appropriate protocol stack in real time;

(g) present the resulting message content at a standard Applications Programming Interface (API).

Even with such an advanced receiver technology, many potential vehicle safety messages could be expected to be 'lost'. For example, at a transmission rate of six megabits per second, a 100-byte vehicle safety message would take approximately 133 microseconds of 'airtime'. The scan by the receiver would need to hit the correct frequency for such a signal at *exactly* its transmission initiation time, and then stay there throughout the packet reception. Either during this 'stay-on-frequency' time or while scanning other frequencies, additional signals could be transmitted on different frequencies from those currently being scanned or received. This illustrates the need to standardize a particular frequency to be monitored by all system nodes for safety messages.

A similar example could be used to illustrate the need for a standard modulation scheme and a standard protocol stack. It should be noted that it appears to be possible to accommodate limited protocol branching, for example allowing separate protocol pathways for unaddressed packets versus addressed packets, within a standardized approach.

At the top of a standard protocol stack there is a need to 'speak the same language', as noted in an earlier section. This language can be a standardized message set consisting of standardized data elements (basically, the words of the language) connected in a standard way.

The standard message set allows different organizations to construct very different applications from one another while still being interoperable. This is a meaningful distinction from the perspective of automobile manufacturers. For example, one manufacturer may want to apply the unique capabilities of an internal sensor unit to best advantage to initiate an action based on a received message. A proprietary application could therefore be used to combine the standard message set with the unique internal sensor data. Thus, a standard message set can allow interoperable proprietary applications.

7.2.3 Insufficiency of standards

Standards are necessary for interoperability, as described in the previous subsection. However, standards are insufficient to guarantee interoperability. Additional 'rules of use' will be necessary for that purpose in addition to the required standards. For example, the Federal Communications Commission (FCC) in the United States specifies the maximum transmitted power for a particular frequency, and provides some form of enforcement sanctions. Such rules are also necessary, for example, for:

- the spectral envelope of channel filtration;

- priorities for messages;

- creation of intentional interference;

- many other aspects of radio communications operations.

Another entire area where standards are necessary, but insufficient, is the area of security. In this case a particular security scheme can be standardized, but some legal status must be provided to a certificate authority, for example, and some operational rules must be put in place and enforced regarding distribution and management of such certificates.

Finally, unless there is some level of agreement to use particular standards, then the standards cannot provide interoperability. Such an agreement could be a voluntary industry agreement, a government mandate, or something in between. In any case there must be some form of uniform adherence to the use of the agreed standards if interoperability is to be supported.

7.3 Wireless Protocols and Standardization Activities

In this section the wireless protocols and standardization activities relevant to automotive standardization of vehicle networks for V2X communications are discussed in the context of a commonly cited layered protocol model.

7.3.1 OSI seven-layer protocol model

The OSI seven-layer model (Figure 7.1) (ITU-T X.200 1994) represents one of the earliest attempts to classify similar protocol functions into distinct layers, such that each layer would provide the functionality necessary to support the layer above it. More modern protocol stacks tend to blur the distinctions between such well-defined layers, making it difficult to directly map the different protocols onto the original OSI seven-layer model. However, this model proves useful in discussing the standards development activities relating to standardization of vehicle networks for V2X communications.

The *physical layer*, shown at the bottom of the protocol stack in Figure 7.1, represents the physical connection to the transmission and reception medium. In this role it provides a mechanical, electrical and procedural interface to the wireless communications link. This includes specific aspects of the wireless link, such as modulation scheme, frequency and other bit-level details of it.

The main function of the *data link layer* (Layer 2) is to arrange the bits present in the physical layer into frames (logical sequences) and to detect errors. Some Layer 2 protocols also provide error correction.

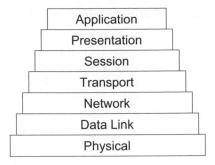

Figure 7.1 OSI seven-layer model

The third layer is called the *network layer*, and it performs the network routing function. It may also provide fragmentation and reassembly functions, when a packet from higher layers is too long to fit into the required frame size for the transmission medium being used.

The *transport layer* (Layer 4) performs error control and flow control functions to allow reliable data transfer to the layers above.

The *session layer*, immediately above the transport layer, manages the connections between local and remote applications, including setting up and terminating the necessary connections.

Layer 6, the *presentation layer*, provides encoding functions to package the data from the application layer into a form that can be understood by the layers below. The presentation layer may also include some forms of protocol conversion.

The *application layer* (Layer 7) at the top of the stack provides the overall interface to the entire protocol stack. A software application that needs to communicate to a remote location or locations, such as the vehicle safety applications described elsewhere in this chapter, would use Layer 7 to access all the functions contained in the protocol stack to support the desired communications.

7.3.2 Standards activities relative to protocol layers

The IEEE 802.11 group (IEEE 802.11 2009), which is well known for Wi-Fi wireless communications standards, is focused primarily on the first two layers of the stack. At the lower portions of the data link layer, there is a functional portion called the Medium Access Control (MAC) sublayer. The major focus for Wi-Fi standardization is at this MAC sublayer, along with the entire Physical (PHY) layer (Layer 1).

The ISO 11898 standard for the vehicle CAN bus (ISO 11898 2007), mentioned in the initial sections of this chapter, specifies the PHY layer (Layer 1 of the OSI model) and the data link layer (Layer 2).

One of the most widely used protocols at Layer 3 – the network layer – is the IP (IETF RFC 1122 1989). IP is developed and maintained by the Internet Engineering Task Force (IETF) (IETF 2009).

Mapping into both the transport layer (Layer 4) and the session layer (Layer 5) are two standard protocols that are widely used to support the Internet. These are the TCP (IETF RFC 793 1981) and the UDP (IETF RFC 768 1980), for which the standards are also developed and maintained by the IETF.

7.3.3 Cooperation required among different standards

Based upon the concept that the protocol layers below a particular layer provide the functions necessary to support it, cooperation between the layers is a fundamental requirement. In an ideal world, the entire protocol stack would be developed by the same standards development organization to ensure a consistent functional approach throughout the stack. However, specific functional expertise relating to the functionality expressed at a particular layer of the stack often resides is a different standards development organization from the one developing the standards for an adjacent layer protocol. When the standards for different layers are being developed in different organizations, it becomes imperative that strong liaisons are in place between the various development groups in order for each group to understand the functional requirements of layers above and the technical constraints of layers below the ones upon which they are focused.

The level of cooperation required can become quite complex when a sequential development approach is not being followed. If the protocols for different layers are being developed at the same time, then changes in one layer can have an impact on another layer. This can lead experts in a particular functional area to express opinions regarding protocol functioning on other layers that are somewhat outside their specific area of expertise. It is most desirable to keep the protocol experts focused upon the areas in which their expertise is strongest, and provide clear functional requirements, as well as specification of technical constraints, to experts working on other protocol layers. This includes detailed specifications of the interfaces between various layers. As mentioned above, such detailed specification can be very difficult when the detailed protocol functions on each layer have not yet been fully determined.

The coordination process required to ensure the necessary cooperation among different protocol layers that are being developed in different standards development organizations can be further complicated where the standards development organizations are functioning in different global regions, or where regional variations exist to the point of generating different protocol approaches for similar layer functionality. This situation would seem to beg for an international standardization approach. However, such an approach is only partially being followed in the case of standardization of vehicle networks for V2X communications. Different regions appear to have different priorities for the types of applications to be supported by such a vehicle communications system, and experts in different standards have different motivations and different technical preferences. In addition, resources (time from the functional experts and funding for participation in standards development activities) have to be allocated toward those activities most likely to affect the company or the nation supplying them. As well as this, related legacy systems in particular areas may be promoted by those with a vested interest in that type of technology as a global solution, without a thorough consideration of the requirements of the desired future system. Thus there are technical as well as political challenges involved in fostering cooperation in international standards development.

Even with significant challenges to major international standards development, though, there have been notable successes in this area recently. Bluetooth technology (Bluetooth 2009), for example, has now reached a stage where interoperability is generally well supported on a global basis. This success required a significant amount of coordinated work that was supported by a large consortium of companies who were motivated toward a positive outcome. Even so, it took much longer to reach the final result than most parties imagined at the initial stages.

In the case of cooperative collision-avoidance applications, it may be difficult for many of the companies likely to be involved in development of the relevant standards and deployment of the necessary technology to monetize the estimated benefits of such deployment to the extent necessary to gain major support within their respective organizations. One of the main reasons for this is that the estimated safety benefits of the contemplated system accrue mainly to areas included in public-sector accounting processes. This has led some participants to adopt a strategy toward development of a vehicle communications system that would be expected to be immediately commercially viable, based upon immediate market estimations, and let the public-sector safety benefits arise as a secondary aspect of the vehicle communications system. This may be a viable approach, as long as the technical capabilities of the system that is developed continue to meet the often more stringent requirements of the cooperative crash-avoidance vehicle safety applications.

7.4 Regional Standards Development Progress

Although many of the standards development organizations have a stated international focus, they are most often strongly influenced by the specific concerns and interests of the regions in which they are most active. In this section a number of the standards development activities toward automotive standardization of vehicle networks to support V2X communications are discussed.

7.4.1 North America

There appears to be reasonably good coordination among the United States, Canada and Mexico in terms of spectrum allocation, relevant standards development, and, arguably, plans to deploy systems that utilize these standards.

In the United States 75 MHz of spectrum, from 5.850 to 5.925 GHz, was allocated by the FCC in 1999 for usage by Intelligent Transportation Systems (ITSs) (FCC 1999). This spectral allocation has been one of the focal points for the development of the 5.9 GHz Dedicated Short Range Communications (DSRC) systems in North America.

American Society for Testing and Materials International

In 2003 the FCC issued a report and order regarding the use of the 5.9 GHZ spectrum for ITS services (FCC 2003). This order specified the use of American Society for Testing and Materials (ASTM) E2213-03 – 'Standard Specification for Telecommunications and Information Exchange Between Roadside and Vehicle Systems – 5 GHz Band DSRC Medium Access Control (MAC) and Physical Layer (PHY) Specifications'. This ASTM standard was strongly based upon the IEEE 802.11a standard, which uses Orthogonal Frequency Division Multiplexing (OFDM) in the 5.8 GHz unlicensed spectrum. However, the ASTM version made several adjustments to the IEEE 802.11a standard to accommodate the movement of vehicles at highway speeds and other special circumstances unique to the anticipated requirements of V2X communications. This ASTM standard, however, is being replaced by the work that has been undertaken in the IEEE 802.11p committee (as described in the next section).

IEEE 802.11p

After concentrated consideration of the ASTM standard to specify the PHY and MAC for 5.9 GHz DSRC systems, it was decided by the technical participants in this effort that this work should be moved back to the IEEE 802.11 group. This decision was based, to a large extent, upon the heavy use of IEEE 802.11a in the ASTM standard, and on the anticipated need to keep the 5.9 GHz PHY and MAC consistent with the ongoing evolution of IEEE 802.11.

In general, the IEEE 802.11p standardization focus is on the enhancements to IEEE 802.11 necessary to support wireless short-range communications for ITS. The intention (as of late 2008) is to petition the FCC to change its designation of the required standard for use in the 5.9 GHz DSRC spectrum from ASTM E2213-03 to IEEE 802.11p, as soon as the IEEE 802.11p standard passes its ballot approval process in the IEEE 802 committee.

As mentioned above, the IEEE 802.11p PHY has adopted the OFDM approach of IEEE 802.11a, with 52 sub-carriers, but uses 10 MHz channel widths rather than the 20 MHz channel widths in IEEE 802.11a. The OFDM approach of IEEE 802.11p uses the same Binary Phase-Shift Keying (BPSK), Quadrature Phase-Shift Keying (QPSK), and Quadrature Amplitude Modulation (QAM) (at both the 16-QAM and 64-QAM levels) modulation as IEEE 802.11a, but each modulation level results in half the data throughput since the duration of each symbol is twice as long within the 10 MHz channel width. For IEEE 802.11p, the mandatory data rates are 3, 6 and 12 Mbit/s, even though up to 27 Mbit/s operation is possible with 64-QAM modulation. The default data rate being used in the current testing for collision-avoidance type safety applications is 6 Mbit/s, since the higher data rates also have higher associated error rates (IEEE 802.11p 2009).

These modifications to IEEE 802.11a allow IEEE 802.11p to operate more reliably in the vehicular environment, where very short-duration communications are required for vehicle safety and other applications, and where the multipath situation changes very rapidly with vehicles moving in relation to each other (and the roadside) at highway speeds. In addition, due to the speed differentials involved, the normal Wireless Local Area Network (WLAN) approach of 'joining' a network may not be possible in many roadway scenarios. IEEE 802.11p allows vehicles to exchange messages, without necessarily having to 'join' a network, by using broadcast message protocols. However, IEEE 802.11p also allows for the more traditional WLAN two-way communications mode.

IEEE 1609

IEEE 1609 represents a family of standards that function in the middle layers of the protocol stack. These standards have been designed to flexibly support safety, mobility and convenience applications that use short-range V2V and vehicle-to-roadside communications. Examples of the types of applications to be supported by these standards include collision avoidance vehicle safety, traffic management, enhanced navigation and automated tolling, to mention but a few of the contemplated uses. The individual standards in this family are identified and briefly described in the following paragraphs (IEEE WAVE 2007).

- *IEEE Trial-use Standard 1609.1^{TM}-2006 – Wireless Access in Vehicular Environments (WAVE) – Resource Manager*: This standard is a resource manager in the sense that it manages the flow of command–response communications with multiple remote applications. IEEE 1609.1 also defines data flows and resources, command message formats and data storage formats.

- *IEEE Trial-use Standard 1609.2^{TM}-2006 – Wireless Access in Vehicular Environments (WAVE) – Security Services for Applications and Management Messages*: IEEE 1609.2 defines the security approach for this protocol stack; in particular, an authentication approach that allows one-way secure communications for low-latency application requirements. It further defines the circumstances for using secure message exchanges and specifies the necessary handling of those message exchanges, depending upon the purpose of the exchange.

- *IEEE Trial-use Standard 1609.3^{TM}-2007 - Wireless Access in Vehicular Environments (WAVE) - Networking Services*: IEEE 1609.3 covers the network and transport layer services that are necessary to support secure data exchanges. These services, as is expected at these protocol layers, include addressing and routing. Another important aspect of this standard is its definition of the WAVE Short Messages (WSM). WSM offers an efficient mechanism by which to transfer short messages, such as those used for collision-avoidance vehicle safety applications. In a bit of a stretch from the network and transport layers, this standard also defines the Management Information Base (MIB) for the whole IEEE 1609 protocol stack.

- *IEEE Trial-use Standard 1609.4^{TM}-2006 – Wireless Access in Vehicular Environments (WAVE) – Multi-Channel Operations*: This standard has been considered an 'upper MAC' in terms of functionality. It allows for multi-channel operation through a channel switching approach and provides an enhanced interface to the normal IEEE 802.11 MAC, which generally only supports single channel at a time operation.

- *IEEE P1609.0* is a newer standard under development that will provide an architectural perspective on the entire IEEE 1609 family of standards, along with their relationship to related standards for short-range V2V and vehicle-to-roadside communications.

- *IEEE P1609.5* is another standard within the 1609 family whose development has been initiated more recently (2008). P1609.5 is targeted toward providing communication management services for WAVE to support V2X applications. More specifically, P1609.5 is intended to utilize trial use knowledge gained in order to extract and collect in a single document various communications management services that had previously been embedded within 1609.3 and 1609.4.

SAE J2735

Automobile manufacturers have expressed a strong need for a standard message set that provides interoperability to support many of the planned vehicle safety applications, while still allowing for innovative, non-standardized applications. SAE J2735 provides this necessary functionality through the use of standardized message sets, data frames and data elements (USDOT 2006). Other potential users of DSRC systems have expressed similar needs for interoperability with flexibility in application design.

SAE J2735, for example, provides a vehicle safety message called the Basic Safety Message (BSM), or, unofficially, the 'heartbeat' message, which has been explicitly designed to support the anticipated collision-avoidance V2V communications. The draft BSM set includes speed, heading, acceleration, position (latitude, longitude and elevation), time, a temporary identification and the status of various driver control inputs in the mandatory 'Part One', which has been optimized to fit within around 40 bytes. There is also a 'Part Two' of the BSM that contains optional data elements. This approach allows for specific information that is not often required to be readily attached to the 'Part One' when it is needed.

This standard has been designed specifically for the 5.9 GHz DSRC protocol stack, comprising the IEEE 1609 series of standards and IEEE 802.11p at the lower layers. Although the message sets have been intended for use with the 5.9 GHz DSRC protocol, to the extent that is possible they have been designed so that they can be deployed over other wireless protocols as well. This is expected to assist with international harmonization, where the same spectrum might not be available in different regions.

7.4.2 Europe

In Europe progress on V2X communications for safety-related ITS purposes has accelerated, due to the allocation of 30 MHz of spectrum, from 5.875 to 5.905 GHz, in August 2008 (EC 2008).

This allocation is meant to be harmonized across all the member states of the European Union. The members were instructed to designate this band for V2V safety communications within six months, and to plan for allocation and licensing for Vehicle-to-Infrastructure (V2I) as soon as their local circumstances permit. The different spectrum allocations on a nation by nation basis had been a major impediment to progress in the European Union prior to this recent agreement. Besides a consistent spectrum across the European Union, this allocation also provides a degree of consistency with similar developments on a global basis. Although this allocation is not exactly the same as the 5.9 GHz spectrum allocation in North America, it is within its range, which should facilitate the use of the same radio frequency transceiver chipsets and antennas.

Typically regions like Europe have indigenous standards organizations that relate international standards to the particular requirements of the region. Europe therefore has a regional organization – the European Committee for Standardization (CEN) – that represents regional interests related to International Organization for Standardization (ISO) standards. Similarly, the European Committee for Electrotechnical Standardization (CENELEC) provides the regional relationship with the International Electrotechnical Commission (IEC). The third major European standards organization is the European Telecommunications Standards Institute (ETSI), which has the same kind of relationship with the International Telecommunication Union (ITU).

Relevant standards developments in the European Union have been harmonized to a major extent with similar developments in North America (please see Section 7.5 for more information on global harmonization activities). In general, current developments support an optimistic view that vehicle safety communications standards can be harmonized globally within very similar basic technologies. This, potentially, enables a global view toward planning for optimal safety benefits, and cross-regional conceptual facilitation of applications development.

European Committee for Standardization (CEN)

As mentioned above, CEN reflects European interests related to ISO's international standards. Its mandate is in technical areas other than the electrotechnical and telecommunications fields, and it draws its membership from nations within the European Union.

The standards that it develops are considered to be voluntary standards; they are called European Standards and are designated by 'EN'.

European Committee for Electrotechnical Standardization (CENELEC)

In the case of the electrotechnical field, a technical area not addressed by CEN, CENELEC is the designated regional organization. As mentioned earlier in this section, this group reflects European interests related to the IEC standards.

European Telecommunications Standards Institute (ETSI)

In its role as the European regional standards development organization for telecommunications, ETSI has established a specific Technical Committee (TC) to develop relevant standards for ITS (the ETSI TC ITS). This committee has been working on standards and specifications that will support interoperability among vehicle-mounted and roadside equipment throughout the European region, in response to policy guidance from the European Commission.

Harmonized standards are meant to be universally adopted throughout the European Union. The ETSI TC ITS is currently engaged in the development of harmonized standards that relate to various ITS applications areas involving automobiles. One of these areas is Communications Access for Land Mobiles (CALM), which, as its name suggests, is focused on long and medium range communications involving automobiles. The application areas addressed by CALM over a variety of communications media include both safety and entertainment. Harmonized standards for specific wireless communications systems are also being developed for interconnecting vehicles, using spectrum in both the 5 GHz and the 63 GHz ranges. Another major application area being addressed is Electronic Fee Collection (EFC), which includes tolling and congestion pricing considerations. This effort is focused onDSRC in the 5.8 GHz range to provide the necessary connectivity between automobiles and roadside units.

CAR-2-CAR Communication Consortium

European vehicle manufacturers have established the CAR-2-CAR Communication Consortium (C2C-CC) to enhance safety and efficiency by enabling interoperable communications between and among automobiles. Besides the vehicle manufacturers, membership also includes research organizations and suppliers, as well as other interested stakeholders.

The technical concept for C2C-CC is based on the WLAN approach and existing IEEE 802.11 standards and includes the idea of automatic ad hoc network connections between vehicles in close proximity. This connected network is envisaged as providing a mechanism for vehicles to share information with each other. This network concept includes the idea of multi-hop communications, where each vehicle might function as a packet router to connect vehicles that are not in direct radio communications range of one another (C2C-CC 2009).

Communications for eSafety

The Communications for eSafety (COMeSafety) project addresses the promotion of cooperative ITS interests enabled or enhanced by V2V and V2I communications. Basically, the COMeSafety Project provides a focal point for the development of concepts and requirements related to European interests. Inherent in this group's activities is the desire to formulate and promote a consolidated European perspective in both European and international standards.

7.4.3 Japan

As stated by the Prime Minister in 2003, the Japanese government has focused upon a goal to:

> Achieve the world's safest road transportation through cooperative vehicle safety support systems. (CAO 2004)

With a deployed base of over 23 million electronic toll collection transponders operating in the 5.8 GHz band, Japan has a large legacy system that must be considered in the development of more advanced V2X communications capabilities. The main standardization activities for these systems in Japan are conducted by the Association of Radio Industries and Businesses (ARIB).

The ARIB is a public service corporation that serves multiple roles associated with the use of radio technologies. As a public service corporation, it has a focus on using radio technologies for the good of society. Besides serving as a standards development organization, ARIB also conducts research and development related both to broadcasting and to telecommunications, and including research in fundamental radio technologies.

There is ongoing research in Japan regarding the most appropriate spectrum for future vehicle safety communications. This includes consideration of spectrum in the 5.8 GHz range as well as in the range of 700 MHz. Frequencies from 715 to 725 MHz with room for 5 MHz guard bands above and below could become available after analog to digital television conversion. This spectrum is therefore being studied in addition to 5.8 GHz for possible use to support V2V communications.

Association of Radio Industries and Businesses STD-T75

As the main standards development group for ITS in Japan, ARIB is one of the focal points for technological considerations regarding ITS systems, such as V2X communications. The ARIB STD-T75 standard covers communications operating in the range from 5.770– 5.850 GHz using different frequencies for uplink (higher end of band) and downlink (lower end of band) purposes. This standard defines standardized objects on Layers 1, 2 and 7 of the OSI Basic Reference Model. In particular, at Layer 1 (PHY) ARIB STD-T75 uses 14 separate channels. Each channel is 4.4 MHz in width. Two different modulations are used – Amplitude-Shift Keying (ASK), which supports a 1 Mbps data rate; and QPSK, which supports data rates of 1 or 4 Mbps. This standard also uses a (TDMA/FDD) scheme, which incorporates eight time slots. The maximum number of vehicles that can be supported within one zone is 56. The nominal range of this system is 30 meters, so the resulting communications zones are quite a bit smaller than North American 5.9 GHz DSRC zones (Abdalla, Abu-Rgheff and Senouci 2009). Such differences in communications capabilities may require unique adaptations of applications from region to region.

Association of Radio Industries and Businesses STD-T88

ARIB STD-T88 defines an Application Sub Layer (DSRC-ASL) that is positioned on top of Layer 7 of the protocol stack defined in ARIB STD-T75. The main purpose of this standard is to allow multiple applications to use the existing DSRC systems (ARIB STD-T75) communications capabilities. This DSRC-ASL functionality supports networking protocols such as TCP/IP as well as non-network applications, allowing for more flexible use of the ARIB STD-T75 communications systems (ARIB 2004).

7.5   Global Standardization

As discussed in Section 7.2, global standardization could potentially provide many benefits to stakeholders, in particular global automobile manufacturers (and, by extension, their suppliers). While it would be likely to be very beneficial for universal wireless interoperability for vehicle safety applications, the variations in spectrum allocations between different regions, and even between different countries, make this a difficult goal for the foreseeable future. However, some of the benefits of global standardization could potentially be realized if at least portions of the standards were harmonized on a global basis. For example, it might be possible to design and deploy V2V safety applications fairly consistently if there could be global agreement on a standardized message set and data dictionary to be used for these communications exchanges. This could conceivably be accomplished even though the underlying spectrum, physical layer modulation and intermediate communications protocols varied between regions, if those regional communications systems could meet the requirements of the V2V safety applications in terms of latency and range, for example.

Another possible scenario for global standardization is to agree worldwide standards that affect only the lower portions of the protocol stack. This approach is analogous to the current automotive CAN bus situation. The CAN bus standard fully supports proprietary message sets and data definitions, while providing standardized interoperability from the physical layer up through the data link layer. Another successful example of this lower-layer interoperability approach is the TCP/IP protocol of the Internet. In this case, a standard interface is provided to the transport layer. If an application uses this standard interface, then the protocol stack is able to format and send the message. However, at the received end of the communication, no interoperability is present above this interface in the stack, unless both parties to the communications have previously agreed on a higher-layer protocol, or they have both agreed to use the same application (and the application directly accesses the standard interface). To continue the Internet analogy, such an agreement on protocol above the TCP/IP defined interface would be, for example, to use File Transfer Protocol (FTP), or Hyper Text Transfer Protocol (HTTP). For cooperative collision-avoidance vehicle safety applications, however, it appears that standards will be required to cover the full range of protocol layers from the PHY through the data dictionary and message set.

To further this discussion toward understanding the future of automotive standardization of vehicle networks for vehicle safety, it may be beneficial to assume that global standardization of short-range wireless communications could be achieved to the point where different PHY and MAC layers might be utilized, but something akin to the WAVE Short Messages (WSM) could be universally agreed to be delivered through a standardized interface at the top of protocol Layer 7. The only apparent consideration of which PHY and MAC were being used within a major geographical region would likely be some requirements concerning range, latency, and so forth.

One cautionary note that should be emphasized is that interoperability for the collision-avoidance safety applications will necessarily require that the same frequency spectrum, PHY and MAC be used by all these applications within a major geographical region, as described in Section 7.2.1.

If the two concepts, lower-layer standardization (up to a designated point) and message-set standardization (over the lower layers), could be combined, it might be possible to support a global interoperability approach that allows, for example, different spectrum usage in different major regions and also allows for proprietary applications to support safety innovation and product differentiation. This appears to be one of the most promising directions for consideration towards the future global harmonization of the standards necessary to support collision-avoidance applications. However, there are efforts underway to work toward the achievement of a common global spectrum for vehicle communications, and, in particular, vehicle safety communications. These efforts, as well as other standards activities targeted towards global harmonization, are described in the following subsection.

7.5.1   Global standards development organizations and mechanisms

Although a number of the regional standards development organizations consider themselves to be international in scope, there are several standards-setting bodies that are widely recognized to represent the international agreement on particular standards in areas directly related to automotive standardization with respect to vehicle networks.

International Organization for Standardization

ISO represents one of the most influential international standards-setting bodies. It is made up of representatives from the different national and regional standards development organizations. This body has been in existence for over 60 years, and plays a significant role in promoting global standardization in many areas. ISO does not have any direct authority to enforce standards compliance. However, the ISO standards often become enforceable through their adoption as national standards within various countries, legally enforceable within the domains of those countries. In the discussion of regional standards development progress in Section 7.4, ISO standards related to various regional initiatives were identified. The interest arising from particular regional initiatives can often provide the motivation for the development of an ISO standard.

There are currently over 200 TCs within ISO, working on global standards for everything from nuts and bolts (literally) to nanotechnologies. Of particular relevance to automotive standardization for vehicle networks is the TC that is focused on ITS standardization – TC 204. Within ISO, activities related to ITS surface transportation standards development, including communications systems, control systems and related information, are addressed by TC 204. Applications that are relevant to these TC 204 activities include traveler information and traffic management areas, with a focus on ITS aspects of various transport systems on an intermodal and multimodal basis. However, 'in-vehicle transport information and control systems' are specifically excluded from the scope of TC 204. These areas have been allocated to TC 22.

TC 204 contains 13 Working Groups (WGs) that address a fairly wide-reaching range of topics in the ITS area. A number of these groups, as shown in the 'vehicle' category in Table 7.1, are involved in work that can relate directly to vehicle networks, while some of the others have a portion of their work that may relate less directly to them.

Table 7.1   ISO TC 204 CALM WGs

Category	Focus Area	Working Group
Vehicle	Automatic vehicle and equipment identification	WG 4
	Traveler information systems	WG 10
	Route guidance and navigation systems	WG 11
	Vehicle/roadway warning and control systems	WG 14
Infrastructure	Architecture	WG 1
	Transport Information and Control System (TICS) database technology	WG 3
	Fee and toll collection	WG 5
	General fleet management and commercial/freight	WG 7
	Public transport/emergency	WG 8
Communications	Dedicated short range communications for TICS applications	WG 15
	Wide area communications/protocols and interfaces	WG 16
Device	Nomadic devices in ITS Systems	WG 17

Although a number of these WGs are relevant to automotive standardization with respect to vehicle networks, WG 16 is one of the most active in this regard at the present time. WG 16 has focused on developing a family of international standards for CALM. CALM standards development efforts are focused primarily on the air interface for various communications technologies, operating within different spectrum ranges. The main unifying element of these technologies is that they function in the medium- and long-range communications domains. The most distinctive aspect of the CALM standards is the concept that applications should be able to switch to different communications media according to availability, capabilities and other parameters. Therefore, in addition to developing standards for V2V, V2I and infrastructure-to-infrastructure using specific technologies or media, CALM is also developing the standardized mechanisms to switch among different media according to the requirements of the applications.

An ambitious standards development program is underway within WG 16, including the standards listed in Table 7.2, which are in various stages of development with one already published.

The IS designation in the table indicates that the CALM standard for infrared media has already been published. Harmonization with fairly mature related standards is underway for CD 21215 (which is being strongly coordinated with IEEE 802.11p) and CD 29281 (which is being strongly coordinated with the IEEE 1609 group of standards).

The International Telecommunication Union – Radiocommunication Sector

The International Telecommunication Union – Radiocommunication Sector (ITU-R) establishes regulations and recommendations for radio frequency spectrum management and satellite orbits on a global basis. Since radio spectrum is a scarce resource, a fair and equitable approach needed to be developed to assign these resources to the various uses and users. The ITU-R implements 'Radio Regulations and Regional Agreements' and handles the ongoing maintenance and updating of these regulations through regularly scheduled 'World and

Table 7.2   ISO TC 204 CALM WG 16 activities

Category	Description	CALM Classification	Related Document Number
Media	Legacy and existing digital cellular networks are called second generation, or 2G. This group is developing air interface and parameter standards for 2G communications.	CALM 2G medium	CD 21212
	Newer digital cellular networks with enhanced data carrying capabilities are called third generation, or 3G. This group is developing air interface and parameter standards for 3G communications.	CALM 3G medium	CD 21213
	Although typically providing a much shorter range than cellular systems, infrared (IR) communications are used in various regions, for example Japan, for V2I communications. This group is developing air interface and parameter standards for IR communications.	CALM IR medium	IS 21214:2006
	Also providing shorter-range communications are systems operating in the 5 GHz microwave spectrum range (M5). This group is developing air interface and parameter standards for M5 communications.	CALM M5 medium	CD 21215
	Millimeter microwave (MM) refers to wavelengths in the millimeter range, implying frequencies ranging an order of magnitude higher than M5. This group is developing air interface and parameter standards for M5 communications.	CALM MM medium	WD 21216
	There are various communications protocols currently being used for satellite communications. This group is working on a standardized approach for CALM communications over satellites.	CALM applications using satellite	WD 29282
Networking	This group is working on standards that enable IP communications over the various CALM media.	CALM networking for Internet Connectivity	CD 21210
	This group is working on standards that enable protocols other than IP to be supported over the various CALM media.	CALM non-IP networking	CD 29281
Management	This standards development activity is focused on the service access points that allow services in the infrastructure to be accessed by vehicles.	CALM lower layer service access points	DIS 21218
	Since the CALM concept is to support multiple applications with the same system, there needs to be a standard method for managing the various applications and their access to the communications link.	CALM application management	CD 24101
	Each CALM station has a requirement to perform various functions. This group is developing a standardized way to manage these functions.	CALM station manager	CD 24102

WD: Working draft; CD: Committee draft; DIS: Draft International Standard; IS: International Standard

Table 7.2   Continued

Category	Description	CALM Classification	Related Document Number
Architecture	This part of the CALM standards work defines the overall architecture for the CALM system and provides the overview of how applications in various communications modes would be able to utilize different media flexibly.	CALM global architecture	CD 21217

Regional Radiocommunication Conferences' (WRC/RRC). The ITU-R also conducts studies that result in 'Recommendations' for operational performance of radio communications systems. The intention is to facilitate harmonized development and efficient operations of radio communications systems (ITU-R 2009).

The ITU-R is working on a broadband communications standard for the 5.9 GHz spectrum, but is dependent upon the completion of the ISO CALM M5 standard. The intention is to maintain compatibility with the upper layers in the ISO CALM standard. As mentioned earlier, the CALM M5 standards are being harmonized with the IEEE 802.11p and 1609 standards. ITU-R, however, will focus only on the physical and network layers of the protocol stack.

The International Telecommunication Union – Telecommunication Standardization Sector

Telecommunications standards were among the first activities of the ITU in its initial structure over 140 years ago. International Telecommunication Union – Telecommunication Standardization Sector (ITU-T) is the current portion of ITU that is focused on telecommunications standards. Many of the standards developed by ITU-T (and its predecessors) have been deployed globally, and provide the technological basis for the global telecommunications network. The standards developed by the ITU-T (and the ITU-R) are called 'Recommendations', because they do not become mandatory until they are legally adopted by nations.

One other international group that should be mentioned is the Advisory Panel for Standards Cooperation on Telecommunications related to Motor Vehicles – APSC TELEMOV. This group has been established as an advisory standards cooperation panel that addresses standards harmonization relating to motor vehicle communications. Participants in this panel include international standards development organizations and industry consortia. One of the main purposes of encouraging international cooperation is to avoid duplication of efforts, and a related purpose is to focus attention on common issues that require resolution.

7.5.2   Allowances for regional differences

As mentioned earlier in this chapter, different regions may not be able to allocate exactly the same spectrum to support short-range V2X communications, particularly in the case of collision-avoidance vehicle safety applications. The present hope is that most regions will initially be able to assign spectrum for this use in very similar spectral ranges, which could allow for the same transceiver chipset to be used on a global basis. This kind of diversity in regional deployments, if limited to such minimal differences, could be expected to be manageable through the provision of some flexibility in the appropriate standards. This could be as simple as adjusting the settings for the available channels and power levels, for example, in the 5.8–5.9 GHz ranges currently being planned for these types of applications.

This situation becomes more complex when, for example, entirely different lower-layer protocols, or very different frequency ranges, are contemplated to be used. It appears to be possible to support such diversity, however, as long as the functional communications requirements of the supported applications are met by the proposed protocols.

At (or above) the upper layers, such as the level of the SAE J2735 message set standard, it should be possible to support more differences in wireless protocols and spectrum utilization. This may require some adjustments in applications from one region to another, however, since the transmission range, latency, channel capacity and other associated parameters of the particular protocols and spectrum characteristics could significantly affect the performance of the applications in these different environments.

Finally, as discussed in Section 7.2.1, interoperability for collision-avoidance vehicle safety applications and other low latency applications must be maintained by the collective agreement to use the same frequency and wireless protocol throughout the upper layers of the protocol stack, as well as common message sets above the applications layer, within the region where a vehicle would be likely to travel throughout its lifespan.

References

Abdalla, G., Abu-Rgheff, M. and Senouci, S. (2009) Current Trends in Vehicular Ad Hoc Networks. http://www.tech.plym.ac.uk/see/research/cdma/Papers/Paper%20for%20UBIRO%ADS%202007.pdf.

ARIB (2004) DSRC Applications Sub-Layer, ARIB Standard, ARIB STD-T88, Version 1.0, May 25, 2004. Association of Radio Industries and Businesses (ARIB). English Translation.

Bluetooth (2009) Introduction to Bluetooth. http://www.gsmfavorites.com/documents/bluetooth/introduction/.

C2C-CC (2009) Car-2-Car Communication Consortium – Technical Approach. http://www.car-to-car.org/index.php?id=8&L=0.

CAO (2004) Cabinet Office White Paper on Traffic Safety in Japan 2004. International Association of Traffic and Safety Sciences.

EC (2008) 2008/671/EC: Commission Decision of 5 August 2008 on the harmonised use of radio spectrum in the 5875–5905 MHz frequency band for safety-related applications of Intelligent Transport Systems (ITS) (notified under document number C(2008) 4145) (Text with EEA relevance). http://eur-lex.europa.eu/LexUriServ/LexUriServ.do?uri=CELEX:32008D0671:EN:NOT.

FCC (1999) FCC Report and Order (99–305). Federal Communications Commission.

FCC (2003) FCC Report and Order (03–324). Federal Communications Commission.

IEEE 802.11 (2009) IEEE 802.11 WIRELESS LOCAL AREA NETWORKS The Working Group for WLAN Standards. http://www.ieee802.org/11/index.shtml.

IEEE 802.11p (2009) IEEE P802.11 – Task Group p – MEETING UPDATE. http://www.ieee802.org/11/Reports/tgp_update.htm.

IEEE WAVE (2007) IEEE COMPLETES FOURTH "WAVE" RADIO COMMUNICATION STANDARD. http://standards.ieee.org/announcements/PR_radiocomstd.html.

IETF (2009) Overview of the IETF. http://www.ietf.org/overview.html.

IETF RFC 1122 (1989) RFC 1122, Requirements for Internet Hosts – Communication Layers. Internet Engineering Task Force.

IETF RFC 768 (1980) RFC 768 User Datagram Protocol. University of Southern California Information Sciences Institute.

IETF RFC 793 (1981) RFC 793 Transmission Control Protocol. Defense Advanced Research Projects Agency.

ISO 11898 (2007) ISO 11898 Controller Area Network (CAN), Part 1:2003, Part 2:2003, Part 3:2006, Part 4:2004, and Part 5:2007. International Organization for Standardization.

ITU-R (2009) Welcome to ITU-R. http://www.itu.int/ITU-R/index.asp?category=information&rlink= itur-welcome&lang=en.

ITU-T X.200 (1994) ITU-T Rec. X.200 (1994 E), Data Networks and Open System Communications, Open Systems Interconnection – Model and Notation, International Telecommunications Union, Section 6–7, pp. 28–52. ITU-T Telecommunications Standardization Sector of ITU.

USDOT (2006) Fact Sheet for SAE J2735 – Dedicated Short Range Communications (DSRC) Message Set Dictionary. http://www.standards.its.dot.gov/fact_sheet.asp?f=71.

8

Standardization of Vehicle-to-Infrastructure Communication

Karine Gosse, Christophe Janneteau, Mohamed Kamoun, Mounir Kellil, Pierre Roux, Alexis Olivereau, Jean-Noël Patillon, Alexandru Petrescu

CEA LIST, Département des Technologies et des Systèmes Intelligents

David Bateman

EDF R&D

Sheng Yang

Supelec

Wireless communication standards useful for Vehicle-to-Infrastructure (V2I) interaction have been delivered by various standardization bodies around the world since 2002. Systems covered range from specialized vehicular communication systems such as Dedicated Short Range Communications (DSRC) up to cellular systems primarily designed without high mobility support. A variety of telecommunication protocols that have been developed and

Vehicular Networking   Edited by Marc Emmelmann, Bernd Bochow, C. Christopher Kellum
© 2010 John Wiley & Sons, Ltd

standardized for non-vehicular usage are also deployed in vehicular environments, and these have occasionally required enhancements to cope with the special needs of highly mobile users. Such enhancements are mostly developed in novel working groups evolving within the various standardization bodies.

This chapter provides a comprehensive overview of vehicular-related standardization activities. It covers the work of the IEEE, European Telecommunications Standards Institute (ETSI) and Internet Engineering Task Force (IETF) working groups focusing on vehicular communication. Also, although it is not yet included in the agenda of official standardization bodies, this chapter highlights the work of selected research projects that have developed protocol enhancements to cope with the vehicular environment.

8.1   Introduction

Wireless communication standards useful for V2I have been delivered by various standardization bodies around the world since 2002. Broadly speaking, they can be divided into two categories: specific vehicular communication systems and generic cellular systems. On one hand, DSRC systems are designed to operate in the 5.9 GHz frequency band allocated to Intelligent Transportation System (ITS) services and provide a short to medium range communication service that supports both public safety and private operations in vehicle-to-roadside and Vehicle-to-Vehicle (V2V) communication environments. DSRC systems are being standardized to provide very high data transfer rates in circumstances where minimizing latency in the communication link and isolating relatively small communication zones is important. On another hand, cellular communication systems are not specifically designed to support V2I links but represent a powerful wireless solution for high-speed vehicular communication needs. In turn, DSRC systems can be divided in two classes: Radio Frequency Identification (RFID)-type systems dedicated to specific applications such as toll collection, freight fleet management, parking management, and public transport applications; and the more generic Wireless Local Area Network (WLAN) type of systems. For the sake of concision, this chapter focuses on the second class of systems, which are currently receiving most attention from the standardization bodies, and addresses in detail the design of both DSRC and WLAN and cellular wireless communication standards in support of V2I communications.

In order to deploy ITS applications, standardized means of wireless communication have to be complemented by networking solutions supporting the needs of the vehicular applications. Fully developed architectures have been proposed within several fora and standards bodies to provide the required end-to-end services. In particular, the IETF designs protocols and architecture applicable in the Internet network of computers; communication protocols with potential use in the vehicular industry have often been mentioned in several Working Groups (WGs). This chapter will also provide details of these efforts, complementing the Physical (PHY) layer and Medium Access Control (MAC) layer designs to result in a complete protocol stack optimized for vehicular needs.

It is also worth noting that other types of wireless communication standards do exist, to support vehicles such as trains or aircraft, for instance. In particular, the ETSI has created Technical Committees (TCs) covering road transports (ETSI 2009). TC ITS has five working groups, which are referred to as WGs 1 to 5 and cover User and Application Requirements; Architecture; Cross Layer and Web Services; Transport and Network; and

Media- and Medium-related Security. The TC on Electro-Magnetic Compatibility and Radio Spectrum Matters (ERM) also deals with transportation-related issues within TC ERM Task Group (TG) 37 for ITS, but also for ultra wide band automotive radar (TC ERM TG 31B), for aeronautical applications (TC ERM TG 25), and for maritime needs (TC ERM TG 26). Finally, Railway Telecommunications (RTs) are handled within the TC RT group (ETSI RT 2009), which is in charge of standardizing the GSM-Rail (GSM-R) system, operating in a dedicated and harmonized frequency range and adding to the Global System for Mobile Communication (GSM) functionalities to support applications such as In-train GSM-R Mobile Station (CAB Radio), ticketing, dispatching, logistics, signaling, and passenger information. Indeed, the European Rail Traffic Management System (ERTMS) has been set up by the European railways and the supply industry supported by the European Commission to create unique signaling standards throughout Europe for safer and more competitive rail transport. GSM-R is one of two basic components of ERTMS; it uses the bands 876–880 MHz (uplink) and 921–925 MHz (downlink). These standards will not be covered in this chapter, and the reader is kindly requested to refer to GSM-R Industry Group (2009), for instance, for more details on this matter.

This chapter is divided into three main sections, as follows.

- Section 8.2 provides a brief overview of the main standards and fora in the area of V2I communication systems.

- Section 8.3 describes the radio access standards dedicated either to vehicular-specific usage, such as IEEE 802.11p, or to more generic connection needs, such as Worldwide Interoperability for Microwave Access (WiMAX) or Third Generation Partnership Project (3GPP) Long-Term Evolution (LTE).

- Section 8.4 describes the key networking elements standardized by the IETF to support vehicular applications.

8.2 Overview of Standards and Consortia Providing Vehicle-to-Infrastructure Communication Solutions

8.2.1 Spectrum

The development of V2I communication systems has been enabled by the worldwide allocation of spectrum to ITS. In 1999 the USA allocated the 5.850–5.925 GHz band to be used exclusively for Intelligent Transportation System Radio Services (ITS-RSs), and the US regulation allowing DSRC deployments is defined within Federal Communications Commission (FCC) 47 Code of Federal Regulation (CFR) (FCC 2006; FCC 47 CFR 2009a,b). The associated usage rules are referred as 'license by rule': though the licensing is non-exclusive for area of operation with site registration, it does not involve any financial compensation for obtaining usage rights.

In Europe, on August 5, 2008, the European Commission allocated the 5.875–5.905 GHz frequency band for ITS for road safety and mobility applications (EC 2008). The related standardization work was undertaken within the ETSI TC ERM TG 37, which delivered the EN 302 571 (ETSI ITS 2008) specifications in September 2008. They form the legal basis for deployment of WLAN-type V2I standards (IEEE 802.11p systems) within Europe. In Japan, it is foreseen that the 715–725 MHz band will be available in 2012 for the WLAN type of service.

8.2.2   Standards

Given the early availability of dedicated spectrum for ITS in the USA, standardization bodies consequently have started working on system specifications.

Wireless Access in Vehicular Environments

The main set of standards dedicated to V2I communications is denoted as Wireless Access in Vehicular Environments (WAVE), and is being established within the IEEE standardization body. WAVE is composed of five parts:

- IEEE 802.11p (PHY and MAC layer) (IEEE 802.11p 2008);

- IEEE 1609.1 (2006) (Resource Manager);

- IEEE 1609.2 (2006) (Security Services for Applications and Management Messages);

- IEEE 1609.3 (2007) (Networking Services);

- IEEE 1609.4 (2006) (Multi-Channel Operations).

The scope of the IEEE 802.11p project, as defined in the Project Authorization document, is to propose modifications of the IEEE 802.11a standard PHY and MAC layer necessary to support communication between vehicles themselves and between vehicles and the roadside. The components involved on the vehicle and roadside parts are referred to as On-Board Units, or OBUs, and Road-Side Units, or RSUs, respectively. The standard aims at supporting operations in the 5.850–5.925 GHz band, thus focusing on the North American spectrum, and at supporting all forms of surface transportation, including rail and marine. It has been identified that existing IEEE 802.11 standards are not suitable for the intended use on streets and highways. This environment, the speed of the vehicles (speeds up to 200 km/h must be supported), and the very short latencies (associate and complete multiple data exchanges within 50 ms) are the primary problems that need to be solved in order to adapt IEEE 802.11 specifications to ITS usages. More details on the IEEE 802.11p standard will be given in Section 8.3.1.

The IEEE 1609 set of standards complements IEEE 802.11p by describing wireless data exchange, security, and service advertisement between OBUs and RSUs. They describe the physical mechanism of communication as well as the command and management services, and provide two options, WAVE short messages and Internet Protocol Version 6 (IPv6), for communicating with OBUs. They give the basis for the design of applications interfacing with the WAVE environment and providing network services including data storage access mechanisms, device management, and secure message passing. The protocol stack of WAVE is depicted in Figure 8.1.

ISO TC 204 WG 16 CALM

In parallel with the WAVE specifications work, International Organization for Standardization (ISO) TC 204 WG 16 is currently drafting a series of standards under the acronym CALM

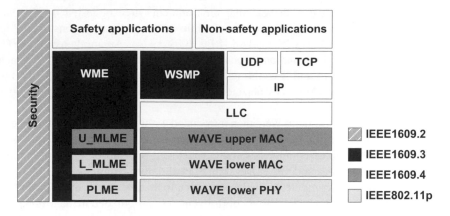

Figure 8.1 WAVE standard protocol stack

(Communications Access for Land Mobiles) that aim at providing architecture and a set of protocols to separate applications from the communication media. A wide selection of media are covered, and a basic network protocol (IPv6) is used together with a comprehensive management stack (CALM 2004, 2009).

The key principles embraced within the CALM network layer are an IPv6 kernel; Mobile IPv6 elements for handover; Network Mobility (NEMO) (see Section 8.2.2) support; header compression; Internet connectivity; mobile connectivity and routing in fast ad hoc network situations; and Common Service Access Points (SAPs) toward the lower layers (LSAP) and for management services. The radio access technologies and standards supported include WLAN (IEEE 802.11a/b/g/p), RFID-based DSRC, 2G/3G cellular systems, satellite and GPS, millimeter wave radar, and infra-red.

It has to be noted that another consortium, the CAR-2-CAR Communication Consortium (C2C-CC) (C2C-CC 2009), dedicated to increasing road traffic safety and efficiency by means of V2V communications, is also addressing V2I architecture and protocols. It was initiated by European vehicle manufacturers, but is now open for other partners. The two groupings have developed architectures and protocols that are mostly complementary, but also partly overlapping and in some cases conflicting. There is currently intensive work being done to harmonize and incorporate the C2C-CC needs into the CALM protocols.

In addition, ETSI is also contributing to CALM with respect to frequency aspects and market access requirements, as well as interface specifications from CALM applications to existing telecommunications networks.

Internet Engineering Task Force

The IETF is an organization that designs protocols and architectures for running on the Internet network of computers. In particular, it specified IP (IETF RFC 791 1981), which forms the basis of all communications on the Internet. Since then, several of its Working Groups have specified communication protocols with potential use in the vehicular industry.

Very early efforts to deal with vehicular environments are suggested by the Mobile Router (MR) concepts and the mention of 'automobile' in the earliest versions of the Mobile Internet Protocol Version 4 (IPv4) standard (RFC 2002 1996). As another example, Lach et al. (2003) described, for information purposes, some experiments with a vehicle connected to the Internet by using IPv4 and IPv6 through General Packet Radio Service (GPRS) while on the highway, and through WLAN when close to a WLAN Access Point (AP) in a hot-spot area.

Vehicular industry requirements were then introduced within the NEMO protocol between 2002 and 2005 in the NEMO Working Group. This protocol was designed for any kind of moving network (a group of computers moving together); it could be applied to vehicles, but also to ships, aeroplanes and more (Devarapalli et al. 2005). Here, the moving network, composed of all on-board computers, is to be connected to the Internet through an IPv6 stack. A parallel effort for developing IETF protocols for vehicles was pursued in the Mobile Ad hoc Network (MANET) WG, in which the topic of vehicles moving around and interconnecting with IETF routing protocols was mentioned, especially for defense applications.

Obviously, the two efforts to design IETF protocols for vehicles are related and may serve the same purpose. As a consequence, in 2007, a preparation for an IETF Birds of a Feather (BoF), the seed for an official WG, was pursued. The goal of the MANET and NEMO (MANEMO) BoF, as its name suggests, was to combine NEMO protocols with MANET protocols so as to gain advantages from both sides: on one hand to have the vehicles communicate with one another when the Internet fixed infrastructure is not available (using MANET protocols), and on the other hand to have the vehicles connected to the Internet and to one another in a manner respecting the Internet architecture and routing when the Internet fixed infrastructure is available (using NEMO). Although this effort has seen participation from members of other Standards Development Organizations (SDOs) related to the vehicular industry (for example, ISO CALM and the C2C-CC, or ETSI), it has only produced a few ephemeral documents on problem statement and requirements, and it has never become an official WG.

More recently, though, the Mobility Extensions (MEXT) WG, in charge, among others, for maintaining and extending the NEMO protocol (Devarapalli et al. 2005), has shown increasing interest in the vehicular industry. As part of this activity, it has been identified that one of the most important problems in the use of the NEMO protocol is the artificially long paths of communication. This problem of 'triangular routing' is inherited from the Mobile IPv6 protocol (Johnson et al. 2004): instead of communicating on direct paths between two entities, a third segment is added, to or from the Home Agent (HA). For NEMO, the problem is exacerbated (it is sometimes called the 'multi-angular' paths problem) since the path includes not only the HA of one mobile entity, but all the HAs of all the MRs in an aggregation of moving networks ('nested' moving networks). To address the problem of multi-angular paths in NEMO deployment (or in Route Optimization, or RO, mechanisms), the WG approaches it in three domains of activity: consumer electronics, the aeronautics industry and the automotive industry. The requirements from the automotive industry are formulated in a draft document, 'Automotive Industry Requirements for NEMO Route Optimization' (Baldessari et al. 2009). This is probably the single most important IETF document today that addresses requirements from the vehicular industry. It contains requirements as formulated in three SDOs specifically related to the vehicular industry: ISO CALM, C2C-CC and IEEE 802.11p. It will be described in Section 8.4.2.

Main European projects

Significant technology development and validation efforts have also been undertaken within collaborative European projects, and the three main ones, in terms of number of partners and relevance of the project scope with respect to standardization work, running from 2006 to 2010, are:

- CVIS (Cooperative Vehicle–Infrastructure Systems) (CVIS 2009);

- SAFESPOT (Cooperative Vehicles and Road Infrastructure for Road Safety) (SAVESPOT 2009);

- COOPERS (Cooperative Systems for Intelligent Road Safety) (COOPERS 2009).

The *Cooperative Vehicle–Infrastructure Systems (CVIS)* consortium, 61 partners strong, aims at delivering: (a) a multi-channel terminal capable of maintaining a continuous Internet connection over a wide range of carriers employing cellular, WLAN, infra-red or DSRC; (b) an open architecture connecting in-vehicle and traffic management systems and telematics services at the roadside; (c) techniques for enhanced vehicle positioning and the creation of local dynamic maps; (d) extended protocols for vehicle, road, and environment monitoring to allow vehicles to share data with other vehicles, infrastructure, or roadside service centers; (e) application designs for a wide range of cooperative V2I services; (f) a deployment enabling toolkit. One of the goals of CVIS is also to validate, improve and qualify the CALM standards for European use.

The main objective of the *Cooperative Vehicles and Road Infrastructure for Road Safety (SAFESPOT)* project, which also brings together more than 50 partners, is to understand how intelligent vehicles and intelligent roads can cooperate to produce a breakthrough for road safety, within an open, flexible, and modular architecture and communication platform. The aim is to prevent road accidents by developing a Safety Margin Assistant that detects in advance potentially dangerous situations and that extends, in space and time, drivers' awareness of the surrounding environment. The Safety Margin Assistant will be an Intelligent Cooperative System based on V2V and V2I communication, exploiting a new generation of infrastructure-based sensing techniques, ad hoc dynamic networking, accurate relative localization, and dynamic local traffic maps. Test scenario-based applications as well as practical system implementation strategies are also being considered. The SAFESPOT approach is aligned with CALM and C2C-CC work.

The *Cooperative Systems for Intelligent Road Safety (COOPERS)* project provides vehicles and drivers with real-time local situation-based, safety-related infrastructure status information, distributed via a dedicated V2I link. This approach extends the concepts of in-vehicle autonomous systems and V2V communication by adding tactical and strategic traffic information that can only be provided by the infrastructure operator in real time. V2I communication significantly improves traffic control and safety via effective and reliable transmission of data fully adapted to the local situation of the vehicle (or ensemble of vehicles) (COOPERS 2009). The greatest effect from such communications will be achieved in areas of dense traffic where risk of accidents and traffic jams is extremely high. The real-time communication link between infrastructure and vehicle also allows vehicles to be utilized as floating sensors to verify infrastructure sensor data as the primary source for traffic control measures.

8.3 Radio Access Standards for Vehicle-to-Infrastructure Communications

8.3.1 IEEE 802.11p

As explained in Section 8.2.2, IEEE 802.11p (2008) is not a standalone standard. It is intended to amend the overall IEEE 802.11 standard (IEEE 802.11 2007) and thus relies on the IEEE 802.11a Orthogonal Frequency Division Multiplexing (OFDM) PHY, and the IEEE 802.11 MAC amended by IEEE 802.11e Quality of Service (QoS) mechanisms. In summary, the IEEE 802.11p standard is meant to:

- operate in the 5.85–5.925 GHz frequency band;

- support high vehicular speeds, while the IEEE 802.11a standard addresses low mobility indoor WLAN systems;

- describe the functions and services required by WAVE conforming stations to operate in a rapidly varying environment and exchange messages without having to join a Basic Service Set (BSS), as in the traditional IEEE 802.11 use case (ad hoc operation);

- define the WAVE signaling technique and interface functions that are controlled by the IEEE 802.11 MAC.

In consequence, Radio Frequency (RF), PHY and MAC layer modifications are to be adapted for IEEE 802.11p, and will be described in this section.

Radio Frequency layer design for IEEE 802.11p

Adaptation of IEEE 802.11a devices' hardware in order to cope with vehicular operation requires some constraints on the RF front-end design in order to address the traditional limited range of operation. This range is intrinsically linked to the maximum transmission power in the band considered.

At baseband level, the same PHY is used for both IEEE 802.11p and IEEE 802.11a, leading to the same performance in terms of Packet Error Rate (PER). The extended range of operation for IEEE 802.11p is thus achieved by increasing the amount of power that is transmitted per bit, which in turn can be done by:

- increasing the maximum transmission power, which is limited to 23 dBm in IEEE 802.11a in the 5.2 GHz bands, and 30 dBm in the 5.725–5.85 GHz band (in the USA) with 6 dBi additional antenna gain;

- reducing the bandwidth of the system, effectively reducing the data rate, but increasing the power spectral density;

- using directional antennas on either the receive or the transmit side of the system.

Although this is not explicitly mentioned in the IEEE 802.11p standard, the roadside equipment might be fitted with directive antennas (Zaggoulos and Nix 2008) or even multiple antennas, allowing single stream transparent eigen-beamforming to be perceived as an increase of the antenna gain for all users; this concept has been introduced in IEEE 802.11n

(IEEE 802.11n 2009). The topic goes beyond the scope of this chapter, and the interested reader can find more information from the references quoted above.

It has to be noted that the first performance measurement results presented in the framework of the CVIS collaborative project indicate that the proposed specification of IEEE 802.11p enables transmissions up to 800 m, with Line-of-Sight (LoS) limitation. This nominal range limits the capability of the radio system to address vehicular application needs (IEEE 802.11p 2008; Kaul et al. 2007).

Physical layer modifications for IEEE 802.11p

Two basic properties of a vehicular radio system impact the deployment of an IEEE 802.11a-like PHY for use in the 5.9 GHz band: the length and characteristics of the outdoor vehicular channel impulse response, as well as the resistance to the Doppler effect at high velocity. User velocity is currently limited to 3 m/s in IEEE 802.11a usage scenarios, but has to be extended at least by a factor of 10 for ITS systems. These are well addressed in the recent article by Cheng et al. (2008), and the following summarizes some of the details of this paper.

Characteristics of the vehicular channel A typical IEEE 802.11a propagation channel is indoors and has a relatively short impulse response. In an outdoor vehicular channel, the reflections from distant objects increase significantly the channel impulse response, which is 0.6 μs and 1.5 μs long (Cheng et al. 2008) for suburban and rural environments respectively.

For low complexity equalization of the broadband channel, IEEE 802.11a uses OFDM modulation, splitting the band into 64 sub-carriers. The OFDM symbol contains a Guard Interval (GI) (Nee and Prasad 2000) that is designed to be larger than the channel impulse response length so that inter-symbol interference is avoided and a simple frequency domain scalar equalization scheme is implemented. The GI of 0.8 μs in IEEE 802.11a is insufficient to cope with the measured rural channel response. However, reducing the bandwidth of the system from 20 MHz to 10 and 5 MHz enables this guard interval to be enlarged, to 1.6 μs and 3.2 μs respectively. Therefore, operating on a smaller channel bandwidth for IEEE 802.11p is critical in order to obtain the expected resistance to inter-symbol interference present in outdoor vehicular environments.

In addition, the increased length of the vehicular channel will also result in smaller coherence bandwidth. However, in any case, and for operation at 5, 10 or 20 MHz, this coherence bandwidth remains significantly larger than the IEEE 802.11ap carrier spacing, and the correct functioning of the system is preserved. A measured worst case coherence bandwidth is of 400 kHz (Cheng et al. 2008), which is larger than the 156.25 kHz and 78.125 kHz carrier spacings for the 10 and 5 MHz IEEE 802.11p channels respectively.

Resistance to Doppler The IEEE 802.11a standard was initially designed to be able to support user velocity of at most 3 m/s or 10.8 km/h, which is significantly less than the 100 km/h needed for a vehicular system. The IEEE 802.11a system's resistance to the Doppler effect is indeed limited by the fact that the channel is estimated using a known preamble transmitted at the start of the frame. It is then assumed that the radio channel does not change during the entire length of the frame. In order to improve the Doppler resistance, the only option is thus to limit the physical layer frame length, but, in turn, the preamble and MAC overhead becomes larger and there is less useful payload transmitted relative to the control signaling. The measured minimum coherence time (Cheng et al. 2008) for a rural vehicular environment is 300 μs, whereas the OFDM symbol lengths for a 5 and 10 MHz channel are 16 and 8 μs respectively, including the scaled GI. The IEEE 802.11a PHY includes a 16 μs Physical Layer Convergence Protocol (PLCP) preamble and a single OFDM signalization symbol to define the packet length, rate, etc. The equivalent PLCP preamble length for a 10 and 5 MHz bandwidth would be 32 and 64 μs respectively. The length of the Physical Layer Service Data Unit (PSDU) in the frame can therefore be derived as

$$\text{(Max. PSDU length)} = 300\ \mu s - \text{(PLCP length)} - \text{(OFDM symbol length)}, \qquad (8.1)$$

and is 260 μs for a 10 MHz channel and 220 μs for 5 MHz. Therefore a total of 32 OFDM symbols can be transmitted in the PSDU for a 10 MHz channel and 13 OFDM symbols for a 5 MHz channel, while respecting the coherence time of 300 μs. An estimate of the percentage of useful data in the OFDM frames can then be given by the length of the PSDU divided by the total frame length, giving 72.2% and 86.5% for 5 and 10 MHz channel bandwidths, respectively. Therefore, even assuming a difficult rural channel and a user velocity of 30 m/s, it is verified that a reasonable percentage of the air time can still be dedicated to user data. Finally, advanced Doppler tracking techniques, e.g. employing decision-directed feedback equalization, might also be used in this context. However, this would require a new dedicated chip-set design, preventing the appealingly easy silicon reuse, and would introduce a significant additional complexity to the implementation. Thus, a simpler solution is to replace the OFDM preamble by introducing a 'mid-amble', used for updating the channel estimation at the middle of the frame. Such a scheme has been proposed for use within IEEE 802.11p (Oh et al. 2008).

IEEE 802.11p MAC layer modifications

When considering the IEEE 802.11p MAC performance, a distinction has to be made between safety- and non-safety-related traffic. Safety-related traffic uses a broadcast communication mode, since its content is generally intended for all other users in the vicinity. Non-safety-related traffic relies typically on unicast transmissions to access a remote service; these are treated very similarly to regular IEEE 802.11a packets. The IEEE 802.11p MAC thus provides a number of unique features related to safety packet management, which are discussed below.

The major change to the IEEE 802.11a MAC is the fact that a Terminal Station (STA) is allowed to transmit without being associated to a network or to a BSS (O'Hara and Petrick 2005). The reason for relaxing this constraint is that the latency needed to associate to the BSS is unacceptably long for safety applications. However, operating outside the context of a BSS imposes certain requirements, as follows.

- A traditional BSS uses a beacon for synchronization between the STAs, and therefore IEEE 802.11p terminals operating outside the context of a BSS have to derive their clock from another source, such as a GPS.

- The transmitter cannot determine with certitude what frequency band is in use through scanning, and thus the frequency planning has to be known in advance, e.g. being a fixed allocation set by regulation.

The use of a broadcast transmission mode for safety-related transmissions in a mobile environment poses particular questions (Chen et al. 2007). Due to the fact that the safety-related messages are transmitted in a broadcast mode to more than one receiver, the so-called 'hidden terminal' problem is aggravated, and the zone where a hidden terminal might be located is the union of the coverage zones of all of the receivers. Therefore, the modeling of the hidden terminal problem is particularly important for evaluating IEEE 802.11p performance in the field. Chen et al. (2007) analyzes this phenomenon and derives the estimated delay and packet reception probability against the traffic density. For the sake of simplification of the analysis, road positions of the vehicles can be assumed to be limited to a straight line due to the geometry of the road itself, and the transmission medium can be considered as being unsaturated, since the safety-related messages are transmitted in a dedicated channel and the messages themselves are short.

The packet delay prediction is under 1 ms in all cases, and easily meets the 100 ms requirements of DSRC systems. However, the packet reception probability drops significantly for longer packets and higher traffic densities. It is therefore critical to constrain safety-related messages to be short, but even this might be insufficient to meet the requirements of a safety-related application.

One suggestion for addressing the packet loss in this context is to take account of this issue at the infrastructure level, where coordination between terminals could take place within the WAVE protocol itself (Ferreira et al. 2008). Alternatively, as the estimated packet delay is largely superior to the requirements, a scheme with multicast acknowledgments might be implemented to ensure transmissions to a known set of terminals (Si and Li 2004).

8.3.2 Applicability of generic wide area radio access standards to V2I communications

Apart from dedicated standardization efforts for vehicular radio access solutions, it appears that, for a number of use cases and applications, wide area communication technologies can prove to be very relevant for addressing V2I communications links, since they are, intrinsically, designed to cope with wide area coverage deployments and high terminal speeds.

3GPP Long-Term Evolution

2G cellular systems have often been used for data communications-based home, entertainment, industrial, and public safety applications. In particular, the GSM/GPRS network has been extensively explored in the ITS context. Catling and de Beek (1991), for instance, propose the use of cellular systems such as GSM instead of short range communication networks, thus providing an inherent mobility support and a large coverage area at the cost of a reduced data rate with respect to WLAN-based solutions. In such a network, there is no need for a new infrastructure deployment on the road-side: the concept of virtual beacons is defined in order to emulate a set of regularly spaced short range devices. A virtual beacon consists of a piece of software (located both in the vehicle and in the infrastructure) that relies on specific data structures, which allow the existence of road-side beacons for vehicles to be faked.

In Europe, many collaborative projects have also worked on the use of GSM networks for road transportation, including SOCRATE, DELTA, CORDIS, PATH, and CARTALK2000, to name a few (Catling and de Beek 1991; Reichardt et al. 2007). More recently, Third Generation (3G) Universal Mobile Telecommunications System (UMTS) has also been considered for vehicular communications in European projects such as DRIVER, developing a digital repository infrastructure vision for European research, and OVERDRIVE, dealing with spectrum efficient uni- and multicast services over dynamic multi-radio networks in vehicular environments (Toenjes et al. 2002).

The 3GPP Long-Term Evolution (LTE) standard (3GPP LTE 2009) as such has not been considered yet for V2V or V2I communication since it is still undergoing finalization. However, it offers a set of communication features that make it suitable for such a context. The LTE standard has been developed by 3GPP in parallel with the High Speed Downlink Packet Access (HSDPA) specification as an evolution of the UMTS standard in order to support the higher data rates suitable for upcoming multimedia applications. While HSDPA deployments have already emerged in Europe, the first commercial launches of LTE are currently expected by the end of 2010. Holma and Toskala (2007) provide an extensive overview of the PHY and MAC layers of LTE.

LTE quick overview   The 3GPP LTE standard, frozen in February 2009 in its first stable version, is based on the following characteristics:

- peak downlink rates up to 316.6 Mbit/s for a 4×4 Multiple-Input Multiple-Output (MIMO) system (20 MHz bandwidth, rate 1 and 4 data flows, 1 symbol PDCCH) and 167.8 Mbit/s for a 2×2 MIMO, with peak uplink rates of 86.4 Mbit/s (20 MHz bandwidth, rate 1);

- 10 ms Terminal–Base Station–Terminal round trip delay;

- increased spectrum footprint flexibility, with frequency bandwidth support ranging from 1.4, 3, 5, 10, 15 to 20 MHz, and sub-carrier spacing of 15 kHz;

- user velocity supported up to 500 km/h;

- advanced antenna techniques based on single data stream methods (such as Space-Time Coding or transmit and receive array processing) and spatial multiplexing (MIMO and SDMA/multi-user MIMO) – using multiple antennas at both sides of the channel can linearly enhance (with the number of antennas) the throughput (Telatar 1999; Wolniansky et al. 1998) or alternatively increase the robustness of the link (i.e. the range of coverage for achieving at least a given data rate);

- connection-oriented MAC and opportunistic scheduling in frequency and time, benefiting from multi-user diversity (Knopp and Humblet 1995) (i.e. users with better channel conditions are chosen and served, subject to certain fairness constraints – see Tse 1999), and allowing control of QoS parameters of the user applications;

- interference coordination and/or avoidance.

Impacts of mobility on PHY performance   Generally speaking, mobility brings both challenges and opportunities to cellular systems. The main challenge with respect to mobility comes from the Doppler effect. The latter has impact mainly on the sub-carriers' orthogonality and the channel estimation quality. Accurate estimation of channel state information is essential both for the receiver to coherently detect the signal and for the transmitter to perform efficient closed-loop transmit array processing, based on channel knowledge at the transmitter end.

For the decoding process, the receiver needs to track the channel in order to decode the signal correctly. In the Downlink (DL) (Base Station, or BS, to Mobile Station, or MS, channel) case, a training sequence in the preamble is transmitted at the beginning of the DL frame to perform the channel estimation at the receiver. The MS derives initial channel coefficients based on the received preamble. Then pilot symbols are embedded in each DL data burst according to a known pattern, for the MS to track the channel. In the Uplink (UL) (MS to BS channel) case, the so-called uplink sounding can be applied. Specifically, the MS transmits a dedicated training sequence for the BS to estimate the channel. Note that the OFDM channel can be computed in both the time and the frequency domains, depending on the system implementation. In both the UL and the DL cases, the channel estimator should work fine if the channel coherence time is five times larger than the frame length. In this case, the receivers can estimate the Channel State Information (CSI) with both the training sequence and the pilot symbols. Even if this condition cannot be met, the channel can be estimated coarsely with the embedded pilot symbols in the DL and with the sounding sequence in the UL. Experiments show that the performance loss caused by high mobility is within 5 dB.

For the closed-loop pre-coding, a transmitter CSI is required, enabling transmit array processing; this is also referred to as eigen-beamforming. One way for the BS to obtain the DL CSI is to have the MS quantize its own estimation and feed back this information to the BS. Both scalar quantization and vector quantization can be implemented by using a constant, finite set of quantization values predefined as a 'codebook': 'vector quantization (codebook based).' However, such a scheme suffers from CSI delay, and the impact of mobility can be severe. For instance, the channel coherence time at 2 GHz and 300 km/h is of 0.9 ms, i.e. statistically the channel changes completely every 0.9 ms. In LTE, since the sub-frame length is 1 ms, the transmitter CSI is almost outdated and barely usable at 300 km/h. Closed-loop pre-coding is therefore not recommended at high velocity.

On the other hand, mobility creates the following opportunities to explore: temporal diversity and convergence to ergodic capacity, as well as multi-user diversity. With respect to the former, with high mobility the channel changes rapidly. An additional temporal diversity gain can be achieved. In this case, multiple antennas can be used to improve multiplexing gain in a fairly simple way, and neither space-time coding nor closed-loop pre-coding is needed. It is also known that in a fast fading channel ergodic capacity is achievable, and that outage events can be avoided (Glisic 2007). For the latter, it is known that multi-user diversity schemes provide higher gain with faster fading. In such a case, the overall throughput can be significantly improved (de Courville et al. 2007; Gross et al. 2007, 2009).

LTE resistance to Doppler and applicability to V2I communications   Overall, it appears that LTE is very well suited for wide area V2I communications. In addition to the possibility of making vertical handovers with existing 3GPP standards such as GSM, UMTS and HSDPA, and also non-3GPP ones such as WiMAX, LTE subscriber units are able to offer connectivity at a user speed of up to 500 km/h, which is higher than typical vehicle speeds. It has to be noted, however, that the system is optimized for slower speeds (15–30 km/h), and slight performance degradation is expected for higher velocities, since the Doppler effect has an impact on inter-carrier interference and channel estimation accuracy (Dahlman et al. 2007).

Unlike UMTS, LTE systems are only capable of a hard handover procedure. The handoff can be performed either between two neighboring LTE cells or between a GSM or UMTS base station and a LTE Base Station. This allows for a smooth transition path between old GSM-based vehicular systems and the new ones powered by LTE.

In all cases, the maximum service interruption duration is constrained by design not to exceed 500 ms for non-real-time traffic and 300 ms (Dahlman et al. 2007) for real-time services. Critical vehicle communications should thus be categorized as real-time services in order to be able to benefit from the shortest handover duration.

WiMAX

Worldwide Interoperability for Microwave Access (WiMAX) is a wireless transmission technology based on the IEEE 802.16 standard (IEEE 802.16 2004; WiMAX Forum 2009). Initially, it was designed to enable the delivery of last mile wireless broadband access as an alternative to cable and Digital Subscriber Line (DSL). With the recent IEEE 802.16e-2005 amendment, support for mobility has been added to WiMAX. Mobile WiMAX thus becomes yet another possible candidate for the fourth generation of cellular technologies. As such, it should be possible to build a V2I communication system on top of a WiMAX network, as advocated by Aguado et al. (2008).

Overview of WiMAX   WiMAX could function, theoretically, at any frequency below 66 GHz. Orthogonal Frequency Division Multiple Access (OFDMA) is used over various channel bandwidths (typically 1.25, 5, 10, or 20 MHz). The typical peak bit-rates, split between downlink and uplink, are respectively 31.7 Mbit/s and 5.0 Mbit/s, for a spectral efficiency of 6.3 bits/s/Hz/sector and 1 bit/s/Hz/sector (WiMAX 802.16e, Wave 2 with MIMO, 10 MHz bandwidth, 29/18 DL/UL symbol ratio, TDD). Some key features of mobile WiMAX, mostly shared with LTE in the physical and medium access layer, are as follows:

- advanced antenna techniques (single data stream or, alternatively, spatial multiplexing methods);

- mobility support – in mobile WiMAX, on most of the channel bandwidths, the sub-carrier spacing is set to 10.94 kHz, compatible with vehicular mobility up to 120 km/h when operating in 3.5 GHz (it has to be noted that a new standard, IEEE 802.16m, is under definition within IEEE with a charter to support IMT-advanced requirements with speeds up to 350 km/h, thus being closer to expectations in the vehicular communications arena; handover between cells is also supported);

- QoS support at the MAC layer, with defined service flows that can map to Differentiated Services (DiffServ) code points, enabling an end-to-end IP-based QoS.

WiMAX MAC protocols and network architecture   Few initial studies have been conducted to evaluate the suitability of WiMAX for vehicular applications, as described by Aguado et al. (2008) and in Costa et al. (2008). Costa et al. (2008) map the traffic classes supported in WiMAX to vehicular applications, as follows.

- Unsolicited Grant Services for fixed-size data packets of constant bit rate, and Real-Time Polling Service for periodic variable bit-rate traffic can be used for safety warning and assisted driving.

- Non-real-time polling services and best effort services can be used for traffic management, for instance.

It is shown that a 100 ms maximum latency for safety messages can be met with mobile WiMAX; however, the network entry process is time-bounded to up to 3 s.

Aguado et al. (2008) present and evaluate a mobile WiMAX network architecture using Opnet in two scenarios, assessing that WiMAX is a valid access technology for V2I applications. It is highlighted, though, that some handover situations across different Access Service Network (ASN) entities require more than the 50 ms for handover delay.

8.4 Networking Standards for V2I Communications

8.4.1 Non-IP networking technologies for critical messaging

In order to be able to introduce specific services for V2I communications, classical networking paradigms are currently being reviewed and adapted to devise new standards. More specifically, safety-related services that have very stringent requirements in terms of latency and reliability are driving such efforts, although other objectives such as infotainment services are also being considered.

One important paradigm which is addressed in this context is 'geographical routing' (Stojmenovic 2007). The purpose of traditional routing is to bring data packets towards their final destination address, which identifies a specific node in the network (or a set of nodes, in case of broadcast or multicast). With geographical routing the address concept is extended in order to cover not only node identification but also node geographical location. It is then possible to address packets either to nodes or to locations, or to combinations of both. Of course, this principle is particularly pertinent if the nodes are mobile ones, which is the case for vehicles.

Geographical routing should also be considered in the context of ad hoc networks. Even though fixed infrastructures on the road-side will play a significant role in future vehicular communication, it is considered that vehicles will also form ad hoc networks between themselves in order to propose new safety services to drivers and vehicle users. In the context of ad hoc networks, vehicles will typically play the roles of packet consumer, packet provider, and also packet forwarder and packet router. In this context, a typical use case of geographical routing occurs when a vehicle detects a potential danger, such as the presence of ice, in a given road area. This vehicle will share its knowledge with neighboring vehicles even though it is not connected to any fixed road infrastructure at that time. It may create a warning message and then rely on geographical routing in order to let this message reach all vehicles in the vicinity through the Vehicular Ad hoc Network (VANET).

Geographical areas will be defined by combining several geographical shapes. Once the address concept has been extended to include geographical position information, the node address validity is time limited for a mobile node.

Routing algorithms are being reviewed in order to address geographical routing. Each vehicle should maintain routing tables involving the neighboring vehicles under radio coverage, because of its router behavior. Since the positions of neighboring vehicles change very quickly, the routing tables must be updated in a highly dynamic way. Movement characteristics should be exchanged between vehicles, in order to let a given vehicle anticipate the expected movement of each neighbor. More frequent information exchanges are needed when vehicles are accelerating or decelerating.

New criteria are also being elaborated in order to develop routing strategies. As an example, if a packet is being routed in a vehicle in order to reach a given position, a good strategy may consist of minimizing the number of hops by selecting the neighbor vehicle with the minimum distance to the target, among those neighboring vehicles that are under direct radio coverage.

It is also anticipated that, besides geographical packet forwarding, there will also be a need for geographical message forwarding. A vehicle that has received a message will interpret it and enrich it with its own view of the environment, in order to refine the pertinence of the information carried, before forwarding the reviewed message to other vehicles. Message forwarding does not allow for separating layers, such as, for example, the networking and application layers. However, it will improve service efficiency in many cases. It is therefore anticipated that both packet and message forwarding will be used, and that both should be considered in future standards.

As mentioned in Section 8.2.2, WAVE includes the WAVE Short Message Protocol (WSMP), which provides the ability to exchange short messages between mobile nodes without going through the traditional IP stack. WSMP provides upper layers with the means to control physical layer characteristics, such as the WAVE channel number or the transmitted power. Such a capability is useful because network reliability and low latencies must be guaranteed when safety services are involved. There is a requirement, for example, not to share bandwidth with non-safety applications, and to provide very strict priority mechanisms among safety applications. It is even needed to assign radio channels for exclusive usage by a specific safety application, and to minimize the interference affecting such channels through appropriate control of the transmitted powers of all WAVE channels.

8.4.2   IP-based vehicular networking

IPv6 in ITS

The IETF standardizes protocols that are used as essential components for architectures built by the ITS standardization organizations. Among these components, the sixth version of the Internet Protocol (IPv6) is being given a strong emphasis due to its upcoming deployment as a complement for, and eventually the successor of, IPv4 (Deering and Hinden 1998).

The IPv4 address shortage is a well-known problem: it has been determined that the IPv4 address space will be exhausted at the beginning of 2011 (Huston 2009). This essential limitation of IPv4 prevents the assignment of a unique IPv4 address per vehicle, in view of the 3 billion vehicles that it is expected will be IP-connected by 2050. In contrast, IPv6 solves this issue by providing 2128 available addresses as compared with 232 provided by IPv4. IPv6 allows not only the assignment of an address to a vehicle but the provision of an entire IPv6 prefix, giving a plurality of IPv6 addresses that can be used by multiple IP devices in the vehicle. The addressing space that is allocated to a single vehicle is typically large enough to allow for multiple hierarchical levels of sub-networks (e.g. one for sensors and one for passengers). It must also be noted that IPv6 addresses in the vehicle will be public, global ones as opposed to the private IPv4 addresses that are usually offered to citizens by their Internet Service Providers (ISPs) today whenever more than one IP address is needed. As a consequence, and unlike IPv4 devices located behind a Network Address Translation (NAT) gateway, the IPv6 devices in the vehicle will be easily reached by their remote correspondents without having to initiate the connection.

IPv6 mobility protocols (Mobile IPv6 – Johnson et al. 2004; fast Mobile IPv6 – Koodli 2005; NEMOv6 – Devarapalli et al. 2005) do take advantage of the large addressing space, the simple address auto-configuration mechanism, and the ability to reach a node within the vehicle from the outside. Their IPv4 counterparts have not proven to be scalable enough to accommodate large fleets of vehicles. Therefore, their use has been restricted to specific areas, such as military/emergency vehicles or premium public transport services.

Various other enhancements offered by the IPv6 protocol, such as end-to-end pervasive Internet Protocol Security (IPsec) or simplified support for multicast, are also highly relevant to ITS networks and the specific services that are being designed. Preventing the forgery of security-critical messages while allowing their delivery to a plurality of vehicles is a very realistic scenario where specific IPv6 features may be used to enable new services.

IP mobility: NEMO and vehicular networks

The NEMO protocol is an extension to the Mobile IP protocol for managing network mobility, which is an important protocol in the ITS domain where several devices move simultaneously on board a vehicle, a train or a plane. For detailed reference, the NEMO version for IPv6 described by Devarapalli et al. (2005) is an extension to Mobile IPv6 (described by Johnson et al. 2004), whereas NEMOv4 (Leung et al. 2008) is an extension to Mobile IPv4 (Perkins 2002). In the following, we use the generic term NEMO to mean either NEMOv6 or NEMOv4; when clarification is needed, one of the specific terms NEMOv6 or NEMOv4 is used instead of the generic 'NEMO'. NEMO is a key baseline protocol for vehicular applications and the CALM Handbook (CALM 2004) describes the use of the NEMO protocol extensions in the context of the ISO standard.

The salient characteristics of the moving networks can be paraphrased as follows. First, all computers in the moving network move together, as a unit, and have the need to change their IP address. Indeed, 'movement' is defined not only physically, when for example a computer A travels from point B to point C, but rather as a modification of the IP address assigned to that computer. A computer needs to change its IP address when it moves too far away from its current point of attachment, such as a wireless access point or an Ethernet RJ-45 socket. Second, the mobility of the entire set of computers in the moving network is typically managed by a single computer, designated the Mobile Router (MR). The MR has, typically, an egress interface connecting to the IP fixed infrastructure and an ingress interface connecting to the set of computers within the moving network. It is possible and often desirable to have several egress and ingress interfaces as well as several MRs in a single moving network, for reliability purposes, since such a multiplicity of interfaces may provide access to several different access technologies on the egress, and may even separate the moving network's internal subnets. Finally, the applications running on the MR(s) and on the computers within the moving network should not be interrupted by the mobility events: for example, when a video streaming application runs between a user laptop deployed in a train and a server on the Internet, and the MR deployed in the train changes its IP address because it enters a new coverage area, the user's video stream should not be interrupted.

Figure 8.2 depicts a typical network topology to accommodate moving networks. An IP fixed infrastructure, depicted at the top of the figure, is connected to the Internet and links together the Correspondent Node (CN), the HA and the Access Routers (ARs). The HA is connected on the home link of the moving network, and serves the purpose of redirecting the packets addressed to the moving network when this latter is not at home. The CN is the entity with which a Local Fixed Node (see below) communicates, for example a web or video server. The IP infrastructure offers wireless access to the moving network by using two ARs.

The moving network connects to the Internet via these ARs. An AR is akin to a UMTS Base Station or to a WLAN Access Point.

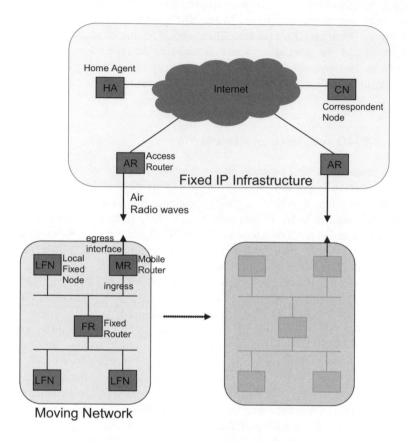

Figure 8.2 Topology of a moving network

The moving network is depicted at the bottom left of Figure 8.2. It contains a set of computers deployed and moving together, as, for example, in a vehicle. As per NEMO terminology, they are called the Mobile Router (MR), Fixed Routers (FRs), and Local Fixed Nodes (LFNs). The latter are the end user systems. The MR is the main entity connecting the moving network to the Internet; it typically has a wireless egress interface and a wired ingress interface; the details can vary, but for the purposes of this explanation this description is sufficient. All the entities in the moving network, the LFNs, the MR, and the FRs, stay linked together. When the moving network attaches to another AR (shown in the bottom right of Figure 8.2), only the MR is exposed to the mobility events (e.g. changing its Care-of Address, or CoA), whereas the entities in the moving network keep their IP addresses and other IP data structures unmodified.

From a high-level perspective, the NEMO protocol is actually a simple exchange of request/acknowledgment messages between the MR and the HA, as depicted in Figure 8.3. In short, after the MR acquires a new Care-of Address it sends a Binding Update (BU) with

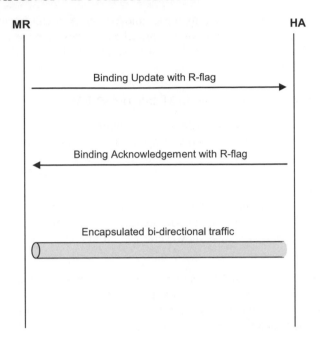

Figure 8.3 Conceptual overview of message exchange for the Mobile IPv6-based NEMO protocol

the flag R set ('R' stands for Mobile Router Flag). This flag is specific to NEMOv6, and is not set for a simple Mobile Host. NEMOv6 can use two alternative modes of registering: implicit and explicit. In implicit mode only the R-flag is set, whereas in explicit mode the MR communicates its Mobile Network Prefix (MNP) by inserting it in the BU, in addition to setting the R-flag. Following reception of an R-flag BU, the HA processes the request according to rules specified in RFC 3963 (Devarapalli et al. 2005), and subsequently replies with a Binding Acknowledgment (BA), also containing the R-flag.

NEMO also embeds Dynamic Home Agent Address Discovery (DHAAD) extensions for Mobile Routers, described by Devarapalli et al. (2005), which help in discovering a HA that supports MRs (some HAs may only support Mobile Hosts).

The NEMOv4 extension protocol for IPv4 (RFC 5177) (Leung et al. 2008) is very similar to NEMOv6, except that it substitutes Binding Updates and BAs with Registration Request and Registration Reply messages, in compliance with standard Mobile IPv4. The protocol extensions for NEMO have been prototyped and demonstrated on several occasions, for instance in the ANEMONE IST project (ANEMONE 2009), and the E-Bike demonstration (E-Bike 2009). The ANEMONE demonstration relied on several vehicles on a university campus that were equipped with a Mobile Router and a web-cam as well as GPS devices, whereas the E-Bike demonstration used sets of bicycles, with one bicycle within each set designated as the MR. It ran NEMOv6 protocol extensions.

For sake of concision, we do not address here the description of the traffic paths from the Correspondent Node towards a Local Fixed Node in the moving network, the behavior of the HA intercepting these packets by using proxy Neighbor Discovery, and the behavior of the MR auto-configuring a new CoA upon each movement. Ambient Network Consortium (2001) summarizes existing literature and approaches on these topics; the reader is also referred to RFC 3775 (Johnson et al. 2004), for instance, which is the basic description of the Mobile IPv6 protocol.

Simultaneous use of multiple accesses in Mobile IPv6/NEMO

One important characteristic of the NEMO basic support, as originally designed, is that a mobile router can register only one CoA at a time to its home agent. This is called the primary CoA. This characteristic becomes a limitation for mobile routers equipped with multiple egress network interfaces and capable of connecting to different types of access networks, since the NEMO basic support protocol would actually allow only one of these interfaces to be used at a time to convey IPv6 flows between nodes in the moving network and correspondent nodes in the infrastructure. The simultaneous use of multiple access networks by a multi-interface mobile router is, however, of great interest, as it offers the following capabilities.

- Allow different flows with different needs in term of QoS to be mapped onto different radio access networks. For example, a Voice over IP (VoIP) session could be routed via a low delay and low jitter access network, while a simultaneous file download session would be routed via a high bandwidth access.

- Increase overall available bandwidth by using multiple interfaces (possibly correspond-ing to different technologies) at the same time. This is typically useful for access networks (e.g. GPRS) where no QoS reservation is available and maximum bandwidth per terminal (i.e. per interface) is limited.

An extension of the NEMO basic support protocol is therefore needed, allowing a mobile router to simultaneously register multiple CoAs with its Home Address. A solution is currently being standardized within the IETF MEXT WG. It is based on two complementary enhancements of the base Mobile IPv6 protocol.

- The first enhancement consists in extending Mobile IPv6 (respectively NEMO) protocol to allow a Mobile Node (MN) (respectively a MR) to register multiple CoAs for a given Home Address to its home agent (Wakikawa 2008). The solution relies on a new 'Binding Identifier Option' that is sent by the MN/MR to its Home Agent along with a (Home Address/Care-of Address) binding update. This allows discrimination between different bindings of the MN/MR Home Address (HoA) to different CoAs, since distinct CoAs will be associated to distinct Binding Identifiers (BIDs). Hence this enables multiple bi-directional tunnels to be established between the MN/MR and its HA, each tunnel being bound to a different CoA (typically corresponding to a different access network).

- The second enhancement complements the above-mentioned mechanism in that the MN/MR can explicitly tell its HA how to map a specific flow to a certain binding identifier, and thus to the corresponding CoA and bi-directional tunnel (Soliman 2008). The Mobile IPv6/NEMO binding update is further extended to include a flow identification option (e.g. protocol type, port number, etc.) so that corresponding traffic for the MN/MR can be sent towards the associated CoA, within the associated bi-directional tunnel. Obviously, several flows can be bound to the same CoA, identified by a unique binding identifier; and those flows can be moved simultaneously from one CoA at a first access network to another CoA at a second access network just by updating the CoA associated to the binding identifier (using the sole 'BID Option'). This provides a simple and efficient way to handle handover of multiple flows simultaneously between two interfaces.

It is worth noting that the solution described by Wakikawa (2008) and Soliman (2008) is not restricted to MN/MR–HA bi-directional tunnels but can also be adapted to systems using a route optimization solution where MN/MR and CN communicate directly, as described in the next section. In this situation, the CN (instead of the HA) receives extended BUs from the MN/MR and is responsible for directing flows to the requested MN/MR CoA.

While this solution is applicable to both Mobile IPv6 and NEMO, the latter protocol is very likely to put additional complexity onto the mobile router. The MR would indeed have to manage not only its own flows but also the flows corresponding to the moving network nodes it is serving. Figure 8.4 represents two such Moving Network Nodes (MNNs), MNN #1 and MNN #2, which are respectively communicating with CN #1 and CN #2. The MR is sending their respective flows over the bi-directional tunnels corresponding to CoA #1 and CoA #2 respectively, so as to best fit the QoS needs of each flow. The MR can take the decision to route a particular flow over a given access network (via the associated CoA) based on policies locally configured. Alternatively, the MR can take the decision for a flow based on preferences or needs expressed by the MNN bound to the flow. While the latter approach would provide more flexibility, it introduces the need for a new signaling protocol enabling the communication of these requirements from MNNs to the MR. Such a protocol has not been standardized yet.

Route Optimization for NEMOv6

General problem The NEMO Route Optimization (RO) problem (Ng et al. 2007a) stems from the way IP routing is done according to the NEMO protocol, when a moving network is roaming outside its home network. The bi-directional tunnel that is established between the roaming mobile router and its home agent in this situation leads to packets being sent/received over a non-optimal route.

NEMO RO solutions aim at solving or mitigating this problem by establishing an almost optimal routing path between the MR and its CN, thus greatly improving performance. First, the shorter the path the shorter the transmission delay between the CN and the moving network. Second, using an optimized route between the mobile router and the correspondent node instead of a MR/HA tunnel means that the tunneling and de-tunneling operations, which had previously to be performed at both tunnel ends, are no longer required (see Figure 8.5). As a consequence, packet processing time at the MR is much decreased and less bandwidth is required to send the same amount of data. As well as just improving performance, route optimization solutions also involve fewer routing nodes, and therefore reduce susceptibility to link failure.

Route optimization is highly relevant to vehicular networks, in particular for an infrastructure node trying to send urgent data to a nearby vehicle, or when two vehicles in the same area are to communicate with each other. The non-optimized route would mandate all exchanged traffic to pass through the respective home agents of the two communicating mobile routers, leading to altered performance.

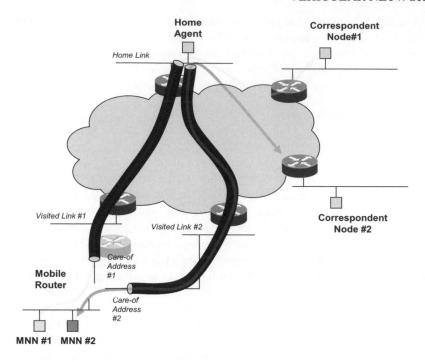

Figure 8.4 Simultaneous use of multiple accesses by a mobile router

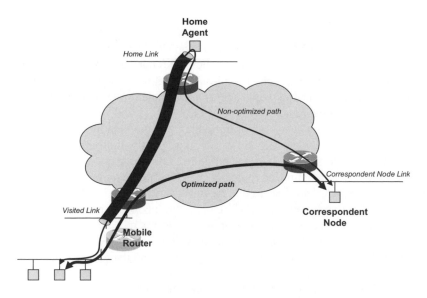

Figure 8.5 NEMO optimized vs. non-optimized path

The NEMO RO requirements from two standards development organizations (namely ISO CALM and C2C-CC) are described by Baldessari et al. (2009) in a document that is a work in progress at the date of writing. The following use cases are cited: notification services, peer-to-peer applications, upload and download services, vehicle monitoring, infotainment applications, and finally navigation services. The scenarios are described in detail, together with potential protocol stacks that could be used for CALM and C2C-CC. In addition, a current set of requirements for RO are described: addressing RO security, binding privacy protection, multihoming, minimum signaling, and switching HAs. As RO is currently the greatest concern and the main focus of the ongoing work, the following sections will further elaborate optimization approaches.

Before considering the solution space for the NEMO route optimization problem, it is worth considering briefly how RO is achieved in the Mobile IPv6 protocol (Johnson et al. 2004). An IPv6 mobile node enables the use of an optimized route with a CN by sending a BU (very similar to the one sent to the HA). This BU establishes the association between its long-term address (HoA) and its current address, routable at its current location (CoA). It has to be secured in the sense that the actual ownership of both Home Address and CoA by the mobile node has to be validated by the CN. This is ensured through a dedicated return routability procedure, during which the correspondent node verifies that the MN indeed receives traffic sent to either address.

This Mobile IPv6 RO solution cannot straightforwardly be applied to the NEMO RO problem. Its closest NEMO counterpart is based on a prefix-scope BUs sent to the CN. The CN is notified that the MR claims ownership of its entire prefix and requests that all traffic sent to that prefix be redirected towards its current CoA. This requirement, however, raises serious security issues since there is no simple way for the CN to check the MR claim. Some other solutions (Bernardos et al. 2007; Olivereau et al. 2005), logically close to Mobile IPv6, propose that the MR sends BUs to the correspondent node only on behalf of the moving network node for which RO is to be performed. That is, only traffic destined for that node will be redirected to the MR CoA, and hence only the ownership (or rather, the routing capability) of this address by the MR will have to be checked by the CN.

A question that has to be considered is the support for RO at the CN side. While it is guaranteed that a direct route from MR to CN is optimal, one cannot assume that every CN supports a NEMO RO solution. Hence, some proposals, such as that of Wakikawa and Watari (2004), recommend using a dedicated enabler (named Correspondent Router, or CR) for RO that would be located close to the correspondent node. The route between MR and CR would be optimized, while the CN would remain unaware of the optimization process taking place. This solution, however, raises a few problems. First, the location of the CR must be carefully chosen: on one hand, it will provide a more optimized route if it is closer to the CN; on the other hand, locating the CR further from the CN would allow it to serve more CNs. Second, the CR must be securely identified as having the right to handle route optimization on behalf of the CN.

Yet another family of solutions that could specifically fulfill the needs of the vehicular environment is based on mobility management infrastructure entities – HA proxies (Thubert et al. 2005) or Hierarchical Mobile IPv6 (Thubert et al. 2005) mobility anchor points – distributed along the MR path. These entities manage the MR's mobility topologically close to its current location, hence achieving the entire RO task without requiring any change on the part of the MR or the CN. Of course, the efficiency of this family of solutions is based on the availability of such intelligent entities close to the MR. Also, the MR must be given a means of locating the entity it should use, and a means to discard non-trusted malicious ones.

More details about the various solutions that have been proposed for enabling NEMO RO are given by Ng et al. (2007b). This IETF document briefly discusses their respective advantages and limitations. It also emphasizes the fact that using a RO solution may not be recommended in certain conditions. RO solutions, indeed, come with a few issues such as signaling overhead, increased protocol complexity and even possibly increased handover delay. This means that an intelligent module at MR and/or CN should detect when (and possibly how, if multiple solutions are available) to trigger the RO procedure.

Group communications for vehicular systems

Vehicular systems can benefit from group communication technologies for applications such as software download/upgrade, large-scale device monitoring, etc. IP Multicast technology (Deering 1989) as an efficient means to ensure IP-based group communications can be used in the vehicular communications scenario, and specifically in the infrastructure-to-vehicle settings.[1] A typical example could be a software upgrade process in which a vehicle manufacturer uses an IP-based vehicular infrastructure to deliver relevant information from an application server to a group of vehicles over a large area such as a city, a country, etc. Of course, each vehicle concerned is assumed to have a device (e.g. an OBU) with IP multicast capabilities (IGMP/MLD protocols – see Cain et al. 2002; Vida 2004) to receive IP multicast data from the application server. Figure 8.6 illustrates an example where the IP-based vehicular infrastructure uses a set of IP multicast routers to deliver data to different vehicular devices.

IP multicast routers are used to deliver application data from the application server to the receivers. These routers run some multicast routing protocol such as PIM-SM (Fenner et al. 2006). A multicast routing protocol is necessary to build and maintain the logical paths between multicast routers (i.e. the multicast tree).

In addition, a vehicle can continue to receive natively multicast data while moving, as long as it is under the coverage of the same IP multicast router (depending upon the network architecture, an IP multicast router may be connected to many IEEE 802.11p wireless access points in the vicinity of the vehicle). If the vehicle moves into the coverage area of a new IP multicast router, it can either reconnect (resubscribe through IGMP/MLD reports) to the delivery tree via the new IP multicast router (remote subscription) or initiate a Mobile IP bidirectional tunneling (Johnson et al. 2004; Perkins 2002) with its HA to continue to receive multicast data from the old IP multicast router (Figure 8.7). Of course, on the vehicle side, both options apply, whether the IP end point of the vehicle (i.e. the multicast receiver) is a mobile terminal or a MR.

The remote subscription option optimizes the data path length from the vehicle to the IP multicast router. The main problem with this option, though, is that it introduces overheads (update of the multicast routing state and construction of new paths), resulting in additional delays for the vehicle to rejoin the delivery tree.

---

[1]The multicast source sends a single copy of a multicast message to all the receivers of the group. Source data are replicated by the multicast router according to the number of forwarding interfaces per multicast group, regardless of the number of receivers. This results in reduced overhead on both the source and the routers, and saves network bandwidth.

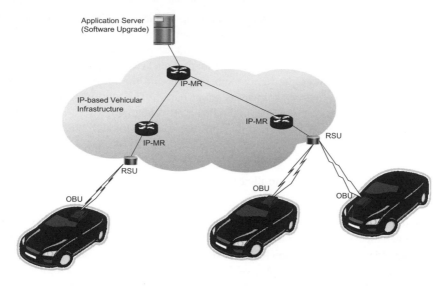

Figure 8.6 Topology for group communications in vehicular networks

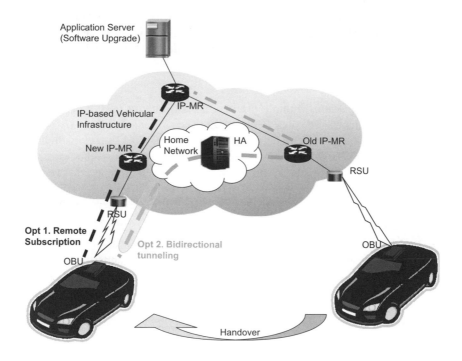

Figure 8.7 Group communications in vehicular networks: IP handover scenario

Compared with the remote subscription option, the home subscription avoids impacting the delivery tree with additional overheads. However, this is achieved at the expense of introducing non-optimal paths between the vehicle and the delivery tree, as well as potential overheads on the HA (in multiplying and encapsulating packets for many receivers).

In order to balance between the remote subscription and the bi-directional tunneling, a third option exits. It is commonly known as agent based. This option consists in placing static agents in the IP infrastructure (e.g. in the IP-based vehicular network infrastructure) to allow inter-agent handover when the mobile receiver moves. Although this option has been well discussed in the IETF (Schmidt et al. 2008), no standard exists so far.

Sensor networking for vehicular applications

While sensor networks have been deployed within vehicles for many years, new trends in communication systems allow their use to enable new applications to be considered. In particular, a vehicle connected to a fixed infrastructure will offer a wide range of new services, such as sensor-triggered malfunction identification, remote diagnostics and repair, or remote software update.

The paradigm for the communication inside a vehicular sensor network is currently evolving away from a static set of wire-connected, non-IP sensors interconnected through a Controller Area Network (CAN), possibly connected to an IPv4 backbone through a gateway. This evolution is especially driven by the following trends.

- *Radio appearance and diversification:* The emergence of dedicated radio technologies such as IEEE 802.15.4 (IEEE 802.15 2009) or WiBree is a key enabler for wireless sensor networks. Yet even older radio technologies such as IEEE 802.11 (WLAN) or IEEE 802.15.1 (Bluetooth) can be used as a basis for these when the power consumption is not a critical requirement.

- *Protocol diversification:* Multiple protocol stacks running on top of IEEE 802.15.4 are being developed, such as ZigBee (ZigBee Alliance 2009), Wireless HART (HART Communication Foundation 2009) or ISA-SP100.11a (International Society of Automation 2009). Z-Wave (Z-WAVE 2009) and proprietary technologies like SimpliciTI, MicrelNet, Cirronet, and EnOcean represent even more alternative protocols.

- *Entity diversification:* Sensor networks are evolving into more heterogeneous networks composed of sensors and actuators.

- *Application diversification:* The involvement of many complex new-generation devices (with more computing power and more memory space) means that sensor networks are expected to run more and more complex applications.

While the transition from wired to wireless sensor networks in vehicles is desirable (especially in terms of mechanical resistance, complexity and cost of installation), the multiplicity of radio protocol stacks for wireless sensor networks makes their interconnection to the fixed infrastructure quite complex, and strongly mitigates against their interoperability. Figure 8.8 shows how each type of wireless sensor network protocol stack requires its specific gateway, acting as protocol translator or proxy, to allow interconnection to an IP infrastructure. Alternatively, allowing any wireless sensor node to become a real IP node (i.e. running an IP protocol stack just like any computer today) would solve this problem. IPv6 is definitively a key enabler for this design, since its address space is large enough to provide all vehicular sensor nodes with a globally unique address.

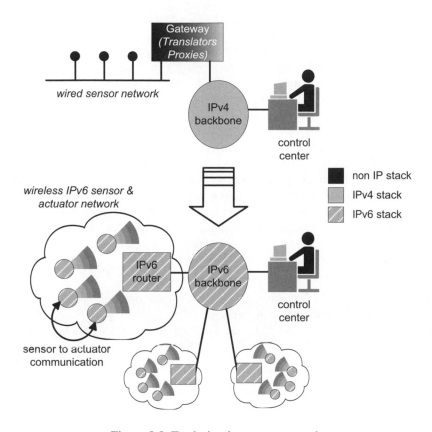

Figure 8.8 Evolution in sensor networks

Global IPv6 deployment would thus allow for flat IP connectivity from the sensors to the infrastructure, hence avoiding the current need for dedicated protocol translation gateways and proxies. It would therefore enable data collection, remote management and maintenance, identified as key applications in the vehicular environment. In addition, advanced IPv6 features like mobility, security, and group communications could be used for more complex scenarios.

The deployment of IPv6 on small wireless devices such as sensors requires the adaptation of the protocol stack to the limitations of this environment. 6lowpan (IETF 6lowpan 2009) is an example of such adaptation, done by defining a convergence layer that allows IPv6 to be run over IEEE 802.15.4 links (Montenegro 2007). It also requires that optimized IPv6 stacks be developed as part of the operating systems running on processing-power-limited and memory-limited devices. The Contiki OS (Contiki Operating System 2009), for example, has included the world's smallest IPv6 stack (uIPv6) since November 2008. Finally, IPv6 routing should be adapted to low power and lossy networks. The IETF Routing Over Low Power and Lossy Networks (ROLL) WG (IETF ROLL 2009) aims at providing routing protocols directly targeted at this kind of networking environment.

8.5   Summary

Several standardization bodies are focusing on specific aspects of vehicular communication, and hence addressing the growing commercial interest in seamlessly integrating the latter into existing terrestrial wired and wireless infrastructure. This chapter has provided a brief overview of the main 'key players' among the standardization fora dealing with this topic. A deeper view and qualitative evaluation on the current standardized, vehicular-dedicated technological approaches is given for techniques for Third Generation Partnership Project (3GPP), Worldwide Interoperability for Microwave Access (WiMAX), Long-Term Evolution (LTE), and Wireless Local Area Network (WLAN). As all of the technologies standardized by the latter fora are almost always deployed with an IP-based infrastructure, this chapter has also discussed the work of the Internet Engineering Task Force (IETF) addressing vehicular mobility. It has shown how the original approaches to supporting mobility in general have been either directly applied or enhanced to address the needs of vehicular communication.

References

3GPP LTE (2009) Overview of 3GPP Release 8 V0.0.3 (2008-11). http://www.3gpp.org/ftp/ Information/WORK_PLAN/Description_Releases.

Aguado, M., Matias, J., Jacob, E. and Berbineau, M. (2008) The WIMAX ASN Network in the V2I scenario. *IEEE VTC 2008-Fall*.

Ambient Network Consortium (2001) Deliverable d9-b.1 mobility support: Design and specification. Technical report.

ANEMONE (2009) ANEMONE Project Testbed. http://www.ist-anemone.eu/index.php/Testbed.

Baldessari, R., Ernst, T., Festag, A. and Lenardi, M. (2009) Automotive Industry Requirements for NEMO route optimization.
    http://www.ietf.org/internet-drafts/draft-ietf-mext-nemo-ro-automotive-req-02.txt.

Bernardos, C., Bagnulo, M. and Calderon, M. (2007) Mobile IPv6 Route Optimisation for Network Mobility (MIRON). IETF Internet Draft, work in progress.

C2C-CC (2009) Car-2-Car Communication Consortium – Homepage. http://www.car-to-car.org.

Cain, B., Deering, S., Kouvelas, I., Fenner, B. and Thyagarajan, A. (2002) RFC 3376 – Internet Group Management Protocol, Version 3. http://tools.ietf.org/html/rfc3376.

CALM (2004) The CALM handbook.
    http://www.tc204wg16.de/Public/TheCALMHandbookv2-060215.pdf.

CALM (2009) CALM Homepage. www.calm.hu.

Catling, I. and de Beek, F.O. (1991) Socrates: System of cellular radio for traffic efficiency and safety. Technical Report vol. 2.

Chen, X., Refai, H. and Ma, X. (2007) A Quantitative Approach to Evaluate DSRC Highway Inter-Vehicle Safety Communication. *GLOBECOM07*, pp. 151–155.

Cheng, L., Henty, B., Cooper, R., Stancil, D. and Bai, F. (2008) A measurement study of time-scaled 802.11a waveforms over the mobile-to-mobile vehicular channel at 5.9 GHz. *IEEE Comm Mag* 84–91.

Contiki Operating System, The (2009) Contiki: The Operating System for Embedded Smart Objects – the Internet of Things. http://www.sics.se/contiki/.

COOPERS (2009) Homepage of the COOPERS Project. http://www.coopers-ip.eu.

Costa, A., Pedreiras, P., Fonseca, J., Matos, J., Proenca, H., Gomes, A. and Gomes, J. (2008) Evaluating WiMax for vehicular Communication Applications. *IEEE Conference on Emerging Technologies and Factory Automation* 1185–1188.

CVIS (2009) Homepage of the CVIS Project. http://www.cvisproject.org.

Dahlman, E., Parkvall, S. and Skold, J. (2007) *3G Evolution: HSPA and LTE for Mobile Broadband.* Academic Press.

de Courville, M., Kamoun, M., Robert, A., Gosteau, J., Fracchia, R. and Gault, S. (2007) Another resource to exploit: multi-user diversity. doc. 11-07/2187, IEEE 802.11 VHT SG (Very High Speed Study Group).

Deering, S. (1989) RFC 1112 – Host Extension for IP Multicasting. http://tools.ietf.org/html/rfc1112.

Deering, S. and Hinden, R. (1998) RFC 2460, Internet Protocol, Version 6 (IPv6) Specification. http://tools.ietf.org/html/rfc2460.

Devarapalli, V., Wakikawa, R., Petrescu, A. and Thubert, P. (2005) RFC 3963 – Network Mobility (NEMO) Basic Support Protocol. http://tools.ietf.org/html/rfc3963.

E-Bike (2009) E-Bike Demonstration within the ANEMONE Project. http://www.ist-anemone.eu/images/a/ad/Ebike_demonstration_07.pdf.

EC (2008) Commission Decision of 5 August 2008 on the harmonized use of radio spectrum in the 5 875–5905 MHz frequency band for safety-related applications of Intelligent Transport Systems (ITS) (2008/671/EC). *Official Journal of the European Union.* http://ec.europa.eu/information_society/policy/radio_spectrum/ref_document.

ETSI (2009) ETSI Structure – Organization Chart. www.etsi.org/WebSite/AboutETSI/EtsiOrganizationchart.aspx.

ETSI ITS (2008) ETSI EN 302 571 v1.1.1 (2008–09) Intelligent Transport Systems (ITS); Radio communications equipment operating in the 5.855 GHz to 5.925 GHz frequency band; Harmonized EN covering the essential requirements of article 3.2 on the R&TTE Directive.

ETSI RT (2009) ETSI Technical Committee on Railway Transportation (TC RT). http://portal.etsi.org/rt/summary_06.asp.

FCC (2006) Amendment of the Commission's Rules Regarding Dedicated Short-Range Communication Services in the 5.850–5.925 GHz Band. Technical Report 06-110, FCC.

FCC 47 CFR (2009a) Regulations governing the licensing and use of frequencies in the 5850–5925 MHz band for dedicated short-range communications services (DSRCS): Part 90 Private Land Mobile Radio Services. http://www.access.gpo.gov/nara/cfr/waisidx_07/47cfr90_07.html. paragraphs 90.371, 90.373, 90.375, 90.377, and 90.379.

FCC 47 CFR (2009b) Regulations governing the licensing and use of frequencies in the 5850–5925 MHz band for dedicated short-range communications services (DSRCS): Part 95 Personal Radio Services. http://www.access.gpo.gov/nara/cfr/waisidx_07/47cfr95_07.html paragraphs 95.639 and 95.1511.

Fenner, B., Handley, M., Holbrook, H. and Kouvelas, I. (2006) RFC 4601 – Protocol Independent Multicast-Sparse Mode (PIM-SM): Protocol Specification (Revised). http://tools.ietf.org/html/rfc4601.

Ferreira, N., Fonseca, J. and Gomes, J.S. (2008) On the adequacy of 802.11p MAC protocols to support safety service in ITS. *ETFA 2008.*

Glisic, S.G. (2007) *Advanced Wireless Communications: 4G Cognitive and Cooperative Broadband Technology.* John Wiley & Sons, Ltd.

Gross, J., Emmelmann, M., Punal, O. and Wolisz, A. (2007) Dynamic multi-user OFDM for 802.11 systems. doc. 07/2062, IEEE 802.11 VHT SG Very High Throughput Study Group, San Francisco, CA, USA.

Gross, J., Punal, O. and Emmelmann, M. (2009) Multi-user OFDMA frame aggregation for future wireless local area networking. *IFIP Networking 2009,* pp. 220–233, Aachen, Germany.

GSM-R Industry Group (2009) GSM-R Industry Group Homepage. http://www.gsm-rail.com/.

HART Communication Foundation (2009) HART Communication Protocol. http://www.hartcomm.org/protocol/wihart/wireless_technology.html.

Holma, H. and Toskala, A. (2007) *WCDMA for UMTS: HSPA Evolution and LTE*. John Wiley & Sons, Ltd.

Huston, G. (2009) IPv4 Address Report. http://www.potaroo.net/tools/ipv4/index.html.

IEEE 1609.1 (2006) Trial-Use Standard for WAVE – Resource Manager.

IEEE 1609.2 (2006) Trial-Use Standard for WAVE – Security Services for Applications and Management Messages.

IEEE 1609.3 (2007) Trial-Use Standard for WAVE – Networking Services.

IEEE 1609.4 (2006) Trial-Use Standard for WAVE – Multi-Channel Operation.

IEEE 802.11 (2007) Wireless LAN Medium Access Control (MAC) and Physical Layer (PHY) Specifications.

IEEE 802.11n (2009) Enhancements for Higher Throughput, Draft Amendment to Standard for Information Technology – Telecommunications and Information Exchange Between Systems – LAN/MAN Specific Requirements – Part 11: Wireless Medium Access Control (MAC) and physical layer (PHY) specifications.

IEEE 802.11p (2008) Wireless Access in Vehicular Environment, Draft Amendment to Standard for Information Technology – Telecommunications and Information Exchange Between Systems – LAN/MAN Specific Requirements – Part 11: Wireless Medium Access Control (MAC) and physical layer (PHY) specifications.

IEEE 802.15 (2009) Homepage of IEEE 802.15 WPAN Task Group 4 (TG4). http://www.ieee802.org/15/pub/TG4.html.

IEEE 802.16 (2004) IEEE 802.16: Air Interface for Fixed Broadband Wireless Access Systems.

IETF 6lowpan (2009) IPv6 over Low power WPAN (6lowpan). http://www.ietf.org/html.charters/6lowpan-charter.html.

IETF RFC 791 (1981) RFC 791 – Internet Protocol. http://tools.ietf.org/html/rfc791.

IETF ROLL (2009) Routing Over Low power and Lossy networks (ROLL). http://www.ietf.org/html.charters/roll-charter.html.

International Society of Automation (2009) Homepage of International Society of Automation. http://www.isa.org.

Johnson, D., Perkins, C. and Arkko, J. (2004) RFC 3775 – Mobility Support in IPv6. http://tools.ietf.org/html/rfc3775.

Kaul, S., Ramachandran, K., Shankar, P., Oh, S., Gruteser, M., Seskar, I. and Nadeem, T. (2007) Effect of antenna placement and diversity on vehicular network communications. *4th Annual IEEE Communications Society Conference on Sensor, Mesh and Ad Hoc Communications and Networks. SECON '07.*, pp. 112–121.

Knopp, R. and Humblet, P. (1995) Information capacity and power control in single-cell multiuser communications. *IEEE International Conference on Communications*, pp. 331–335.

Koodli, R. (2005) RFC 4068 – Fast Handovers for Mobile IPv6. http://tools.ietf.org/html/rfc4068.

Lach, H.Y., Janneteau, C., Oliverau, A., Petrescu, A., Leinmueller, T., Wolf, M. and Pilz, M. (2003) Laboratory and Field experiments with IPv6 Mobile Networks in Vehicular Environments. http://tools.ietf.org/html/draft-lach-nemo-experiments-overdrive-01.txt.

Leung, K., Dommety, G., Narayanan, V. and Petrescu, A. (2008) RFC 5177 – Network Mobility (NEMO) Extensions for Mobile IPv4. http://tools.ietf.org/html/rfc5177.

Montenegro, G. (2007) RFC 4944 – Transmission of IPv6 Packets over IEEE 802.15.4 Networks. http://tools.ietf.org/html/rfc4944.

Nee, R.V. and Prasad, R. (2000) *OFDM for Wireless Multimedia Communications*. Artech House Inc.

Ng, C., Thubert, P., Watari, M. and Zhao, F. (2007a) RFC 4888, Network Mobility Route Optimization Problem Statement. http://tools.ietf.org/html/rfc4888.

Ng, C., Zhao, F., Watari, M. and Thubert, P. (2007b) RFC 4889 – Network Mobility Route Optimization Solution Space Analysis. http://tools.ietf.org/html/rfc4889.

Oh, H., Kim, S.I., Cho, H. and Kwak, D.Y. (2008) Midamble aided OFDM performance analysis in high mobility vehicular channel. doc. 11-08/112, IEEE 802.11p Wireless Access for the Vehicular Environment WG.

O'Hara, B. and Petrick, A. (2005) *IEEE 802.11 Handbook* 2nd edn. IEEE Press.

Olivereau, A., Janneteau, C. and Petrescu, A. (2005) A Method of Validated Communication. Patent Publication No WO/2005/015853.

Perkins, C. (2002) RFC 3344 – IP Mobility Support for IPv. http://tools.ietf.org/html/rfc3344.

Reichardt, D., Miglietta, M., Moretti, L., Morsink, P. and Schulz, W. (2007) Cartalk 2000: safe and comfortable driving based upon inter-vehicle-communication. Technical Report vol 2.

RFC 2002 (1996) RFC 791 – IP Mobility Support. http://tools.ietf.org/html/rfc791.

SAVESPOT (2009) Homepage of the SAVESPOT Project. http://www.safespot-eu.org.

Schmidt, T.C., Waehlisch, M. and Fairhurst, G. (2008) Multicast Mobility in MIPv6: Problem Statement and Brief Survey. IETF Internet Draft, work in progress.

Si, W. and Li, C. (2004) RMAC : a reliable multicast MAC protocol for wireless ad hoc networks. *International Conference on Parallel Processing*.

Soliman, H. (2008) Flow Bindings in Mobile IPv6 and Nemo Basic Support http://tools.ietf.org/html/rfc3963. draft-ietf-mext-flow-binding-00.txt, IETF Internet Draft, work in progress.

Stojmenovic, I. (2007) Position-based routing in ad-hoc networks. *IEEE Communications Magazine* **40**(7), 128–134.

Telatar, E. (1999) Capacity of multi-antenna Gaussian channels. *Europ. Trans. Telecommun., ETT* **10**(6), 585–596.

Thubert, P., Wakikawa, R. and Devarapalli, V. (2005) RFC 4140 – Hierarchical Mobile IPv6 Mobility Management (HMIPv6. http://tools.ietf.org/html/rfc4140.

Toenjes, R., Moessner, K., Lohmar, T. and Wolf, M. (2002) Overdrive spectrum efficient multicast services to vehicles. *Mobile IST Summit*.

Tse, D. (1999) Forward link multiuser diversity through proportional fair scheduling. *Bell Laboratories Journal*.

Vida, R. (2004) RFC 3810 – Multicast Listener Discovery Version 2 (MLDv2) for IPv6. http://tools.ietf.org/html/rfc3810.

Wakikawa, R. (2008) Multiple Care-of Addresses Registration. draft-ietf-monami6-multiplecoa-10.txt, IETF Draft work in progress.

Wakikawa, R. and Watari, M. (2004) Optimized Route Cache Protocol (ORC). IETF Draft work in progress.

WiMAX Forum (2009) WiMAX Forum Homepage. http://www.wimaxforum.org.

Wolniansky, P., Foschini, G., Golden, G. and Valenzuela, R. (1998) V-BLAST: An architecture for realizing very high data rates over the rich-scattering wireless channel. *URSI International Symposium on Signal, Systems, and Electronics Conference*, pp. 295–300.

Z-WAVE (2009) Homepage of Z-WAVE. http://www.z-wave.com.

Zaggoulos, G. and Nix, A. (2008) WLAN/WDS performance using directive antennas in highly mobile scenarios: Experimental results. *IWCMC Conference*.

ZigBee Alliance (2009) Homepage of ZigBee Alliance. http://www.zigbee.org/en/index.asp.

9

Simulating Cooperative Vehicle-to-Infrastructure Systems: A Multi-Aspect Assessment Tool Suite

Gerdien Klunder and Isabel Wilmink

TNO Built Environment and Geosciences

Bart van Arem

Delft University of Technology

There is a great interest in cooperative vehicle-infrastructure systems: vehicles communicating with each other and with infrastructure-based components in order to achieve coordinated behavior. This asks for a well-structured and validated development process, and supporting tools to enable effective and efficient development and performance evaluation of such systems. This chapter describes a tool suite that covers many of the important aspects of cooperative vehicle-infrastructure systems design, and assesses issues regarding technical functioning, human factors and traffic flow in a consistent way.

Vehicular Networking   Edited by Marc Emmelmann, Bernd Bochow, C. Christopher Kellum
© 2010 John Wiley & Sons, Ltd

9.1 Introduction on Design and Evaluation of Cooperative Systems

Due to the merits and potential of Information and Communications Technology (ICT) and the significant problems with growing traffic volumes, there is a great interest in cooperative Vehicle-to-Infrastructure (V2I) systems: vehicles communicating with each other and with infrastructure-based components in order to achieve coordinated behavior. Cooperative road-vehicle systems are slowly becoming available; as the potential of ICT increases, and authorities and the automotive industry realize how they can support innovation in the traffic and transport system. Singular systems such as navigation systems, traffic and travel information, traffic lights, congestion warning, local danger warning and dynamic speed limits will become much more effective by exchanging information and delivering tailored information and advice to the driver.

Many research programmes were and are carried out to test cooperative systems and analyze the impacts that stand-alone and cooperative systems may have on traffic efficiency, safety and the environment, as part of the EU 6th Framework Programme (Kerry Malone et al. 2008; Schulze et al. 2008). The complexities of cooperation between vehicles and infrastructure ask for a well-structured and validated development process and supporting tools to enable effective and efficient development and performance evaluation of such systems. This chapter describes the application of a tool suite that covers many of the important aspects of cooperative vehicle-infrastructure systems design, and assesses issues regarding technical functioning, human factors and traffic flow in a consistent way. An Integrated Full-Range Speed Assistant (IRSA) was selected as a case to guide and test the development of the tool suite. The IRSA system is a collection of functions to support a driver in maintaining an appropriate speed in a number of selected traffic conditions. IRSA-equipped vehicles communicate with other vehicles as well as with the infrastructure. The development of both the tool suite and the IRSA concept were carried out in the Netherlands Organization for Applied Scientific Research (TNO) research programme Sustainable Mobility Methodologies for Intelligent Transport Systems (SUMMITS) (Driessen et al. 2007; van Arem 2007; Wilmink et al. 2007).

This chapter will first analyze the design problems for cooperative systems. Then the Multi-Aspect Assessment Approach design methodology is presented, followed by a brief description of the various modules of the SUMMITS Tool Suite. Finally, the approach and use of the tool suite are illustrated for the IRSA application. The aim of the system and the specifications of the controllers are given in Section 9.4. Two types of simulation are discussed: detailed simulations on system robustness with the Multi-Agent Real-time Simulator (MARS), and higher-level simulations on the traffic flow impacts with the ITS modeller.

9.2 Design Problems for Cooperative Systems

Cooperative road vehicle systems are expected to lead to more efficient, safer and environmentally friendlier driving. This is the result of cooperation between the vehicle and the infrastructure (concerning the desired action of the vehicle, in relation to the road it is on and the surrounding traffic) and between the vehicle and the driver (concerning what the driver is expected to do – or not do). Communication is an important aspect linking the driver, the vehicle, the road and the other traffic on it.

In designing, testing and evaluating the performance of cooperative systems, all the elements mentioned above must be taken into account. The systems must be capable of constantly acquiring information about the surroundings of the vehicle (in the immediate vicinity of the vehicle and further away), and processing the information into an action by the vehicle, or a piece of advice to the driver. The necessary information comes from the vehicle's sensors, and messages relayed by other vehicles or the infrastructure. This means many questions have to be answered, such as: what improvements in traffic flow do we want to achieve with the system? What behavior is needed from drivers and vehicles to support this? What information do drivers and vehicles need for this? How is this information obtained, and how accurate is it? How likely are drivers to comply in various situations? In the end, to develop successful cooperative applications in traffic and transport, different performance criteria have to be considered more or less at the same time. Important aspects are:

(a) *The effects on traffic flow, safety and the environment:* Does the cooperative system improve throughput and traffic safety? Does it lower exhaust and noise emissions? To what degree, and how, can it be designed in order to maximize these effects?

(b) *The driving behavior and user acceptance in the short and long term:* Does the user accept the cooperative system? Does it support human behavior? Is it without adverse effects (for example, encouraging unsafe behavior)?

(c) *The dependability of the cooperative system:* Can one trust the system? Is it robust enough? Can it cope with real-life artifacts such as noisy sensors, failing communication channels and inconsistent information?

(d) *The technical feasibility and implementation:* Can the system be economically implemented? Is the real-life performance for the individual driver as expected?

This chapter discusses aspects (a) (the impacts on throughput, safety and the environment) and (c) (robustness of the system; capability to cope with real-life artifacts such as noisy sensors, failing communication channels and inconsistent information). Simulations were carried out to make sure that the system design was proven to be effective, comfortable and safe before it was actually introduced on the road. Aspect (b) was evaluated with tests of the system in a driving simulator; for aspect (d) several cooperating vehicles performed tests on a track (for these aspects see van Arem et al. 2007).

9.3   SUMMITS Tool Suite and Multi-Aspect Assessment

9.3.1   Multi-aspect assessment

As communication between vehicles and the infrastructure is not very common yet, very few tools were available that could be used in the design, testing and evaluation process of cooperative road vehicle systems. In the SUMMITS programme, several tools were brought together and enhanced to deal specifically with questions about cooperative systems. The resulting set of tools is called the SUMMITS Tool Suite. Typically, the tools associated with the different assessment aspects are not using the same models. For example, for assessing the robustness of an Adaptive Cruise Control (ACC) controller, a high level of detail in the vehicle and controller models is required. In contrast, for assessment of effects on the traffic flow level, simpler vehicle models are sufficient, and in fact they are needed for reasons of computational efficiency. In order to be able to use the results of the one assessment in the other tool and thus to optimize different aspects, the so-called multi-aspect assessment methodology was introduced. The methodology is based on the commonly used V-model, but adds an iterative aspect to the conceptual phase. It consists of the formulation of a common mathematical model of the system to be developed. This 'meta-model' is a high level description of the functional behavior and the parameters that can be controlled. It serves as a common basis for the formulation of specific models to assess different aspects using dedicated scenarios (Figure 9.1). A more detailed description of the methodology is given by Driessen et al. (2007).

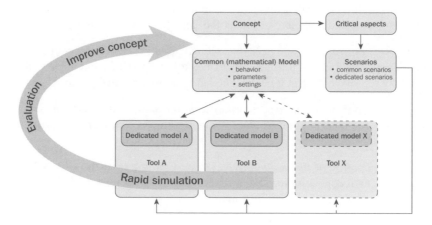

Figure 9.1 Multi-aspect assessment

The meta-model can be translated into functional behavior (conceptual algorithms) and the parameters that can be influenced and/or controlled. From this meta-model, specific models with varying levels of detail are generated to be used in the different tools.

9.3.2   The SUMMITS Tool Suite

The SUMMITS Tool Suite supports the development and evaluation of all kinds of cooperative vehicle-infrastructure systems. It consists of different modules that cover specific aspects of those systems, varying from traffic flow analysis to assessment of human factors, and from dependable cooperative control architectures to fault tolerant hardware implementation. In most cases there is no physical link between the tools (in order to keep the models as flexible as possible), although a few models have been coupled for specific experiments. However, it is always possible to use one model's output as input for another, by specifying a common format of the data. A brief description of the tools in the SUMMITS Tool Suite for cooperative systems is given below.

- *MARS* is a programmable simulator for cooperating complex dynamical systems in general, and for intelligent vehicles equipped with sensors and communication in particular. In MARS, the cooperative system implementation is embedded in a vehicle control/management environment, which provides simulated (but close to real) sensory inputs, implemented exception handling and added functionalities to enable the evaluation in realistic traffic situations. The robustness of the system can then be evaluated under various modes of degraded functionality, as well as under certain non-regular circumstances, such as platooning.

- The *ITS modeller* is used to simulate traffic flows with different shares of equipped vehicles. The ITS modeller is a microscopic traffic simulation environment developed by TNO, which allows a flexible implementation of cooperative vehicle and roadside systems. It uses less detailed models of vehicle and driver behavior than does MARS, but is capable of modeling large numbers of vehicles in a road network. The implementation of the communication system includes message generation both from vehicles and from beacons along the roadside, message receiving within a specified range, and message handling by the vehicles.

- *Vehicle Hardware In the Loop (VeHIL)* is TNO's laboratory for the development and testing of intelligent vehicle systems, including cooperative systems. The system to be tested is implemented in a real vehicle optionally running on a chassis dynamometer. The local traffic environment of the vehicle is emulated by so-called moving bases, and/or simulated as their models describe. The setup permits systems to be tested under strictly determined and reproducible laboratory conditions and in a safe way. Thus, VeHIL is an important step in the development process between fully virtual simulation and on-the-road testing. VeHIL's real-time simulator is based on MARS technology.

- *Driving simulators* can be used to study the Human–Machine Interface (HMI), driving behavior and user acceptance of cooperative systems. Important questions are whether the intended driving behavior is achieved, and whether driving behavioral adaptation effects occur. Other issues are the acceptance of a system, and, for example, the additional workload such a system requires. TNO has various driving simulators operational, including a sophisticated moving-base one as well as low-cost fixed-base simulators. These can be applied either stand-alone or coupled with each other or with other tools (e.g. MARS). For on-the-road assessment of driver behavior, an *instrumented vehicle* is available.

- For *on-the-road testing* (on test tracks as well as public roads) test vehicles equipped with prototypes of intelligent vehicle systems are used. TNO uses two passenger cars (Smart) and the above-mentioned instrumented vehicle for testing cooperative systems.

9.3.3   Some practical aspects of the approach

The meta-model was set up by a team of people with different backgrounds and expertise, among them traffic and automotive engineers, psychologists and mathematicians. Considerable time was spent on 'learning to speak the same language' – different backgrounds were needed, but all members of the team had to be able and willing to work outside their own area of expertise. The team discussed what should be part of the meta-model, for example vehicle and driver behavior, intelligence in the infrastructure, communication and so on, and considered the vehicle, vehicle cluster and traffic flow levels. The IRSA system was defined in such a way that all 'levels' could work with the same meta-model. In other words, the same

IRSA system is assessed in the ITS modeller, in the driving simulator or in an experiment on the road (albeit with different levels of detail in the algorithms).

Ideally, the different tools in the SUMMITS Tool Suite should be used in the order prescribed by the V-model, i.e. according to the stage of development. In reality, because of time limitations, several processes were carried out in parallel in SUMMITS. There was, however, one feedback loop as described in Figure 9.1: some shortcomings in the first version of the IRSA system were identified and the meta-model was improved.

9.4 Integrated Full-Range Speed Assistant

Speed is one of the key factors in road traffic. It is positively associated with the quality of travel: a high speed implies a short travel time. However, a high speed can also lead to high accident risk or high emission of exhaust gas and noise. The speed of a vehicle is traditionally controlled by the driver, who takes into account local traffic conditions as well as applicable speed limits. However, decisions by the driver are sensitive to judgment and operational errors. Many accidents are speed related and partly due to human error. In cases of congestion, human drivers are typically poor controllers. An example of an Advanced Driver Assistance (ADA) system that is commercially available is the ACC system: by extending a 'regular' cruise control system with a radar sensor, the vehicle can maintain a preset speed, but also adapt the speed to a slower predecessor. In addition to sensors on the vehicle, ADA systems can also use wireless communication to receive information from roadside systems and other vehicles. That way, information from more vehicles than just the preceding vehicle can be taken into account. The Integrated Full-Range Speed Assistant (IRSA) is such a system – a cooperative ACC. It supports a driver in maintaining an appropriate speed in a number of selected traffic conditions, such as approaching a traffic jam or a reduced speed limit zone, cut-in situations, sharp curves or leaving the head of the queue at a traffic light. The primary aim of cruise control functionalities is to increase comfort. However, because speed advice and warnings are also taken into account by the cruise control, the system is also expected to contribute to improvements in traffic throughput, safety and emissions to the environment.

9.4.1 Modes and functions

IRSA can be used in different ways: as a purely advisory system, as a system that partly intervenes in the vehicle controls (e.g. by a haptic throttle), or as a controlling system that fully manages the longitudinal speed of the vehicle. Drivers determine in which way they will use IRSA by selecting a mode of operation of the system. The possible modes of operation are IRSA off, IRSA advisory mode, IRSA intervening mode and IRSA controlling mode. From an abstract or systems engineering point of view, the major difference between the controlling mode and the advisory or intervening modes is that there is no driver 'distorting' the acceleration computed by the IRSA system in the controlling mode.

In the advisory mode, information about the desired acceleration by the IRSA system is presented to the driver in the form of audible or visual information. In the intervening mode, the information is passed on to the driver in a more active way, e.g. by a haptic accelerator pedal. The driver will react to these signals, and give a new desired acceleration to the vehicle by pushing the accelerator, brake or clutch pedals. In the controlling mode, the desired acceleration by the IRSA system is directly given to the vehicle.

9.4.2   Scenarios

The IRSA system was tested in several scenarios. These were (a) approaching a traffic jam, (b) approaching a reduced speed limit zone, (c) a cut-in situation, and (d) leaving the head of a queue (at a traffic light). In the first three scenarios, the aim of IRSA was to help the driver slow down in a safe and comfortable way, and in the fourth scenario the aim was to help drivers accelerate in an efficient way, to improve the safety and throughput at traffic lights. In this chapter we describe the 'approaching a traffic jam' scenario in more detail. The results of the other scenarios are described briefly.

9.4.3   IRSA controllers

A main functionality of IRSA is the cruise-control-like functionality. This is activated in the IRSA controlling mode. Depending on the situation the vehicle is in (for example, with or without a predecessor, i.e. Vehicle-to-Vehicle, or V2V, communication possible or not) a specific cruise control functionality will be activated. The cruise control modes are given in Table 9.1.

Table 9.1   Cruise control modes

Cruise Control Mode	Detection of Predecessor
Conventional Cruise Control	No predecessor taken into account
Adaptive Cruise Control (ACC)	Predecessor detected by radar, no V2V
Cooperative Adaptive Cruise Control (CACC)	Predecessor(s) detected by V2V/radar

The cooperative IRSA controller was developed by adding communication to a basic ACC controller and fine-tuning the controller based on the experiments carried out. This section briefly describes the three control algorithms that have been developed: one concerning common ACC and two as variants of CACC. Gieteling (2005) provides a more detailed description. Figure 9.2 shows a schematic representation of a vehicle string consisting of n vehicles. The vehicles are all equipped with V2V and an object sensor that measures the relative motion of the predecessor. The figure also indicates the relative distance $x_{r,i}$ (the headway) between vehicle i and vehicle $i + 1$. The longitudinal velocity of vehicle i is indicated as $v_{x,i}$. The relative velocity between these vehicles is $v_{x_{r,i}} = v_{x,i+1} - v_{x,i}$. Further defined are the desired distance $x_{d,i}$, the headway separation error $e_{x,i} = x_{r,i} - x_{d,i}$, the desired velocity $v_{d,i}$, the relative speed error $e_{v,i} = v_{r,i} - v_{d,i}$ and the vehicle length $l_{v,i}$. These variables are used to explain the principle of the different control strategies, given below.

ACC controller

This is a basic ACC controller, which tries to maintain a reference distance to the predecessor and to minimize the speed difference with it.

The ACC longitudinal controller of vehicle i only takes into account the relative motion of the direct predecessor $i - 1$, which is measured by an environmental sensor, such as

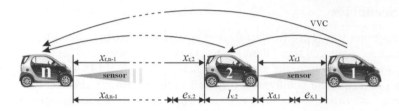

Figure 9.2 Schematic representation of a vehicle string with V2V and object sensors

laser, radar, vision or a combination of those. The controller tries to maintain a predefined velocity set-point v_{cc}, unless a slower predecessor is detected ahead. In that case, vehicle i is controlled to follow vehicle $i-1$ with equal velocity at a desired distance $x_{d,i-1}$. In velocity control mode, the ACC operates as a conventional cruise control, where the desired acceleration a_d is given by a proportional controller:

$$a_d = k_{cc}(v_{cc} - v_x) \tag{9.1}$$

in which k_{cc} is a constant gain. In distance control mode, a_d is generally given by proportional feedback control of the distance separation error and the relative speed error. Since the desired relative velocity is obviously equal to zero, the relative speed error is equal to the relative speed between the vehicles. The desired acceleration is calculated with the following equation:

$$a_d = k_2 e_v + k_1 e_x, \tag{9.2}$$

where the controller gains k_1 and k_2 are (non-linear) functions of ego-velocity v_x and distance error e_x. Equation (9.1) can be regarded as non-linear Proportional-Differential (PD) control of the distance separation error.

Because CACC uses inter-vehicle communication, the acceleration of the predecessor (which is difficult to estimate with an environmental sensor only) can be communicated to the following vehicle. With information on the acceleration a, as well as more reliable estimates for the range and range rate, the ACC control law equation (9.2) can be modified to:

$$a_d = k_3 a + k_2 e_v + k_1 e_x \tag{9.3}$$

where k_3 is a constant feed forward gain. The availability of an acceleration signal provides the opportunity to react faster to emergency braking.

The IRSA controller also takes into account vehicles that are outside the field of view of the object sensors. Because no direct relative motion measurement of these objects is available, this data has to be determined using absolute world positions (determined by fusion of GPS and on board sensors) that is communicated by the vehicles.

The CACC controller equation (9.3) only considers a single predecessor. Two methods have been developed to consider more vehicles in front. They are referred to as CACC1 and CACC2. The controllers are based on the ACC controller equation (9.2).

CACC1

This is a CACC controller in which the resulting acceleration is computed by determining the individual ACC acceleration with the basic ACC controller in relation to a number of preceding vehicles equipped with the system, and by taking the minimum of these individual ACC accelerations. The number of predecessors equipped with the system taken into account is three. The distance and speed of the first predecessor vehicle (IRSA equipped or not) are measured with the vehicle's own sensors (radar). In addition to the characteristics of that vehicle, the characteristics of the first two IRSA-equipped vehicles downstream are taken into account (Figure 9.3).

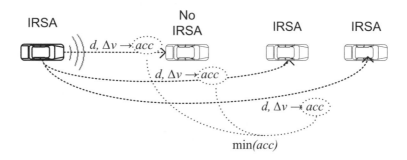

Figure 9.3 Basic CACC controller of IRSA

The controller of vehicle n calculates the desired acceleration $a_{d,n,i}$ for each preceding vehicle i, according to the ACC control law equation (9.2). The desired distance of vehicle n to vehicle i is calculated using:

$$x_{d,n,i} = (l_{v,n-1} + x_{d,n-1}) + \cdots + (l_{v,i+1} + x_{d,i+1}) + x_{d,i}. \tag{9.4}$$

The desired acceleration $a_{d,n}$ to be sent to the lower-level acceleration controller of vehicle n is then calculated by taking the minimum value of all $a_{d,n,i}$ for $n-1$ preceding vehicles:

$$a_{d,n} = \min(a_d, n, n-1, \ldots, a_{d,n,1}). \tag{9.5}$$

This controller will only function in the case where all vehicles in the platoon are equipped with V2V. This is because of the use of the desired distance in the controller. For determining the desired distance to a predecessor (9.4), the motion data of all predecessors has to be known.

CACC2

This is the improved version of CACC1, which works with a control algorithm based on the speed difference with the direct predecessor, added to a term based on the average speed difference with a number of slower predecessors (equipped with the system). The number of predecessors (equipped with the system) taken into account is three. The advantage of this method compared with CACC1 is that no distance headway needs to be determined, which is hard in cases where the penetration rate of the system is less than 100% (Figure 9.4).

The CACC2 controller consists of two parts that are added. The first part of the controller is the ACC control law (9.2) on the directly preceding vehicle. The second term consists of an error feedback of the average velocity of all remaining vehicles in front. This causes damping

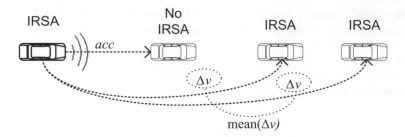

Figure 9.4 CACC2 controller of IRSA

in the string. As a result, the desired acceleration $a_{d,n}$ of vehicle n becomes:

$$a_{d,n} = (k_2 e_{v,n-1} + k_1 e_{x,n-1}) + \left(\frac{k_2}{n-2} \sum_{i=1}^{n-2} e_{v,i} \right). \tag{9.6}$$

Even when not all vehicles in the platoon are equipped with V2V, this controller will function properly because the distance is not taken into account.

Both the CACC1 and CACC2 controllers were implemented and tested in the simulations.

9.5 System Robustness – Simulations with a Multi-Agent Real-Time Simulator

9.5.1 Aims of the simulation

The IRSA CACC system consists of highly autonomous, nonlinear, interacting dynamical components. No common methods exist for the design and validation of this class of cooperative systems. However, it is clear that sophisticated evaluation, test and validation experiments have to be carried out in fully controlled environments in order to 'close the gap' between simulation of the individual components and real-world testing of interacting systems.

The IRSA project did not define application requirements on robustness or performance, necessary for verification and validation. However, robustness was evaluated quantitatively for the controlling mode of IRSA. In order to carry out these evaluations, the CACC implementations had to be embedded into a vehicle control/management environment, which provides simulated (but close to real) sensory inputs, implements exception handling (i.e. how to behave in abnormal situations) and adds new functionalities to enable the evaluation in close to real traffic situations.

This section – after introducing the evaluation environment used and describing the extended control schemes (i.e. the embedding control/management environment) implemented – summarizes the results of the robustness evaluation of the CACC component under various modes of degraded functionality. Simulations were carried out with the MARS model. The scenario used was 'approaching a traffic jam': in this case, a string of five consecutive IRSA vehicles driving at 50 km/h into a traffic jam of stationary vehicles.

All vehicles were equipped with IRSA (controlling mode CACC1). To analyze robustness, disturbances were introduced: noisy or failing sensors and communication errors.

9.5.2   Implementation of IRSA in MARS

The Multi-Agent Real-time Simulator (MARS)

MARS is a continuous time/discrete event simulator for highly autonomous multi-agent systems. In MARS terminology, the autonomous dynamical components (e.g. intelligent vehicles, roadside controllers, etc.) of the simulation are called entities, which are generalizations of agents. MARS has a model-based architecture: all experiment-specific information is stored in models and MARS loads the models when the simulation is initialized. Various models are used to describe the environment the entities 'live' in: the behavior (i.e. the dynamics) of entities themselves, the visualization etc. MARS introduces a novel and powerful way of modeling agents and their interactions using abstract sensors and actuators. Real sensors and actuators can also be modeled in the framework as an extension of the abstract functionalities. MARS serves as a foundation for a new breed of tools, which are capable of very accurate, real-time simulation and evaluation of cooperative vehicle systems and thus can effectively bridge the gap between 'desktop' simulations and field operational tests. For details about the MARS modeling concept, the computation model and a few typical applications readers are referred to Papp et al. (2006, 2003a,b).

The results from the simulations with the MARS model were used to answer questions such as 'Is the IRSA system concept robust enough and dependable?' The focus was on the added value of communication in relation to radar and on the influence of radar noise on the robustness of the CACC1 controller.

A control scheme for cooperative intelligent vehicles

In order to carry out IRSA simulations in MARS, a highly structured approach to modeling cooperative systems was adopted. The increasing complexity of in-vehicle systems and the new generation of cooperating applications combined with stringent dependability requirements pose extreme design, implementation, testing and validation problems. One of the key challenges was to find a decomposition scheme that allows for flexibility, extendibility and manageability on one hand, and safe, efficient implementation on the other. System decomposition is a way to cope with complexity, and a good decomposition gives a natural support for system implementation. A 'classic' and popular architecture is a layered scheme: the controlled process (vehicle, in our case) is at the bottom and layers of control components are built on it. Each control component covers a well-defined control function, and this control function becomes more and more complex ('intelligent') as we climb up in the hierarchy (e.g. actuating, motion, tracking, etc.). A control component has its own (sufficient) knowledge about the process and environment to generate control commands based on observations. The layered scheme has a clear signal flow (observations bottom up, and commands top down) and it is relatively easy to make the implementation modular (component based). Unfortunately the classic scheme does not integrate exception handling (i.e. deviation from nominal operation).

Figure 9.5a shows an enhanced layered scheme, as used in the IRSA experiments, which addresses these problems. In this scheme the observations (i.e. sensory readings and derived quantities or signals) are 'freely' available for control components via the vehicle's internal communication bus (i.e. the observation bus). The control components are responsible also for handling exceptions (see the upcoming signal stream). Figure 9.5b details the internals of the components. Besides the control functionality (Controller Kernel) each component executes a Monitor–Evaluate–Act (MEA) loop to track its own and related subsystems' health state. If a component receives an exception from the layer below or the exception is generated locally (by the monitor) the evaluator decides how it should be handled. If the component cannot handle the exception fully (or at all) the exception is propagated to higher control layers.

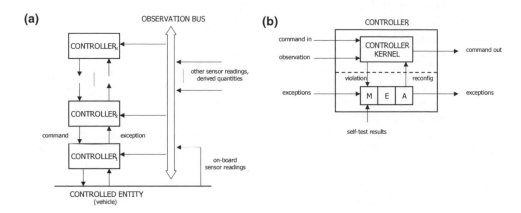

Figure 9.5 The proposed functionally layered control model with exception propagation

The figure also shows a link to the safety-related aspect. A control layer may use dedicated hardware components (e.g. sensors, communication links, etc.) to carry out its functionality. Should such a component fail, the controller should adjust its operation to this new situation, or – if not capable of doing this – should emit exception(s). Safety-critical dedicated components have to incorporate diagnostics (self-test) functionalities, and in case of a component fault the monitoring functionality has to be informed.

The architecture of the intelligent vehicles (including controllers) used in the experiments – which may serve as a basis for real-life implementation – follows the conceptual scheme described above. The implementation incorporates the following components:

- *Detectors and devices:* Each vehicle, as a controlled entity, is equipped with the following devices and detector units to provide the on-board sensor readings via the internal world model (see below) and observation bus to the controllers:

  - radar sensors to detect other vehicles ahead with a maximum range of 150 m – noise models result in inaccurate detections (expressed as a percentage deviation in distance and orientation) or missed detections (expressed as small periods of fall out), and occlusion;
  - GPS to obtain the vehicle's global position in the simulation world – noise models result in inaccurate detections, and lower sampling rates;
  - radio transmitter and receiver sets – noise models include delays in transmission (e.g. due to bandwidth restrictions), and small periods of fall out of transmission;

The detectors and devices can be used with ideal (i.e. noise free) or noisy sensor models.

- *Internal world model:* Each vehicle has an internal world model that represents the vehicle's current 'understanding' of its environment based on the available sensory and communication readings. The world model provides momentary monitoring information, such as position and speed of leaders in the CACC platoon and traffic jam obstacles, via the observation bus to the controllers.

- *Hierarchical control structure:* Each vehicle has a hierarchical control structure as in Figure 9.5a. Each hierarchical (vertical) layer of control has a specific role and responsibility: i.e. regulation, maneuvering and coordination (Netten and Papp 2005). Controllers handle both longitudinal and lateral movements and obstacle avoidance. The CACC is decomposed at three levels: a cooperation controller (coordination layer), car following (maneuvering layer) and throttle and brake regulation.

- *Vehicle:* Both the longitudinal and the lateral dynamic behaviors of the vehicle are implemented. Nonlinear effects (e.g. engine, brake, tire model, drive-chain characteristics etc.) are modeled to ensure close to real responses in the virtual environment.

More details are given by Netten et al. (2006).

9.5.3 Evaluation of robustness of IRSA CACC controllers

IRSA uses a CACC system that consists of a cruise controller, extended with a radar (or lidar) sensor and detector, to maintain a safe headway (ACC), and a short range radio for V2V. The evaluation with the MARS model addresses the robustness of the CACC1 controller, as an IRSA system component, for disturbances in the traffic environment and in sensor input.

The CACC1 algorithm requires input information about leading vehicles, including relative or absolute distances, speeds and accelerations. The radar provides this information, at high frequency sampling rates, about the direct leader of a vehicle. The short range V2V is used to exchange information that is similar to the radar information (i.e. position and speed or acceleration) for all IRSA-equipped vehicles in the leading string of n vehicles. Radio communication is also used at high frequency sampling rates (i.e. 25 or 50 Hz) here. The Most Important Object (MIO) in the CACC1 algorithm is the one vehicle in the leading platoon that requires the momentary absolute minimum acceleration or maximum deceleration. Usually either the direct predecessor, or the tail of the traffic jam, or a vehicle cutting in is identified as the MIO and is most critical for robustness. However, a vehicle further downstream may also be the MIO, and this can only be observed by communication.

Disturbances in the observation input to the CACC algorithm result from failures and processing errors in the sensors and detectors. The input disturbances, which were kept quite simple for this experiment, were:

- a complete loss of communication;

- noisy (or inaccurate) radar data, which was modeled as white noise added to the (perfect) distance to the predecessor vehicle.

Communication failure

The first set of MARS simulations provided information about the working of the IRSA controller when a string of vehicles was driving into a traffic jam of stationary vehicles, with communication errors. When working perfectly, radar and communication provide redundant data with respect to the observed information about the direct leader. Communication, however, also enables a vehicle to 'monitor' leaders beyond its own radar visibility; this information can be used to anticipate disturbances (perturbations) downstream. Figure 9.6 shows the contribution that communication makes. The extreme (minimum) decelerations and headways observed in the traffic jam scenario were similar for the cases of communication only and of communication plus radar (since the same information about a preceding vehicle is available from communication or radar, when they are functioning perfectly). However, when communication failed, vehicles anticipated the traffic jam slightly later and more strongly. This proves the added value of communication (beyond the radar visibility) for traffic stability and robustness.

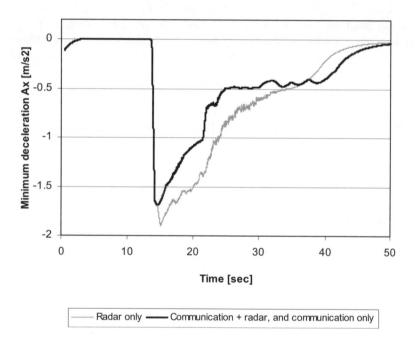

Figure 9.6 Extreme values of decelerations for the traffic jam scenario with either radar or communication devices enabled

Sensor robustness

The second experiment carried out with MARS concerned the robustness of the IRSA controller for radar noise. The CACC1 can handle both scenarios well in noise-free conditions; i.e. vehicles anticipate the traffic jam downstream and preserve stable and safe headways, damping decelerations. Figure 9.7 shows the effects of noisy CACC1 input, expressed as the average percentage deviation in distance, for the traffic jam scenario on the minimum Time-To-Collision (TTC) measured over the trajectory of each vehicle in the string. Low TTCs indicate a higher risk of collisions. As can be seen, the minimum TTC decreases with increased radar noise, but this only happens for extremely high noise values, and the TTCs do not reach critical values. For very large noise levels, the minimum TTC value of the last vehicle is more critical than that of the vehicles in front, which may indicate instability. For smaller noise levels (up to 25%, which seems much higher than expected for normal radar systems), the decrease of the TTC is very small, which indicates that, for this system, a less accurate (and cheaper) radar system will be practically as good and as safe as more accurate (and expensive) ones.

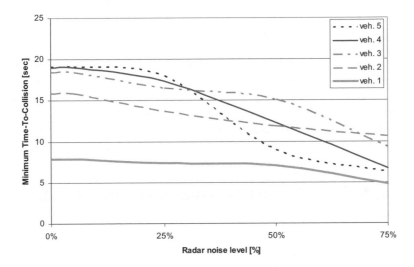

Figure 9.7 Minimum Time-To-Collision for the 'approaching a traffic jam' scenario with radar noise (the vehicle number indicates the string position – vehicle 5 is the last one)

9.5.4 Conclusions on the simulations with MARS

The advance in intelligent vehicle and transportation systems puts emphasis to the local (on-board) intelligence, vehicle–vehicle and vehicle–environment interactions. The dependability of ADA systems, like the IRSA system with sensors, communication and a CACC1 controller, becomes even more important when the mode of operation and authority increases from advisory to full controlling mode. Sophisticated evaluation, test and validation environments are required to 'close the gap' between simulation of the individual components and real-world testing of interacting systems. These environments should provide full control of the circumstances, reproducibility and – when approaching the final rounds of high fidelity testing – flexibility to mix real and virtual components.

Some observations about robustness were fed back into the development process of the IRSA concept, as follows.

- Control algorithms will be implemented differently for different purposes. Small implementation differences can result in significant variations in controller output and ultimately in system behavior. Domain application requirements (e.g. for traffic and vehicle safety and stability) should be developed for acceptable variations in input and output. Alternative implementations should be evaluated, and ultimately tested and validated, on these requirements.

- The combination of radar and communication provided some redundancy in CACC1 input that could improve system robustness for disturbances like occlusion, communication bandwidth issues, noisy detections and device failures. However, the CACC1 algorithm itself was vulnerable to variations caused by these disturbances. In some cases (if no predecessor was detected), the CACC1 algorithm switched back to Cruise Control (CC) mode and started accelerating instead of decelerating for a traffic jam. The improved controller, CACC2, was much less vulnerable to these variations.

The experiments with MARS showed the importance of high fidelity simulation-based evaluation and testing. The logical next step in the evaluation of cooperative systems is to gradually bring in real hardware components (sensors, DSP boxes, vehicles etc.) and move the experimenting into the mixed 'virtual–real world'. TNO's VeHIL facility covers exactly these needs. The fact that VeHIL relies on MARS technology makes the shift to the 'mixed world' relatively easy.

9.6 Traffic Flow Impacts – Simulations in the ITS Modeller

9.6.1 Aims of the simulations

The aim of the experiments with the ITS modeller was to assess the impacts of IRSA on the traffic flow level: how does the IRSA system, under different conditions, help to achieve safe, clean and efficient driving? In other words, how does IRSA influence the interactions of vehicles on a network? This requires a different model from that used for assessing the robustness of the system, since it needs to be able to simulate large numbers of vehicles on a network, with complete infrastructural attributes such as motorways, merging sections, intersections, right of way rules, traffic light control and communication between vehicles and the infrastructure. Simulation with a large number of vehicles in an extensive road network would require too much initialization effort and computational time in the MARS model, while for simulations on the traffic flow level it is not necessary to model the sensors in a high level of detail. IRSA influences traffic flows in several ways. Depending on the mode and the settings, following distances, accelerations and decelerations are influenced. This has effects on throughput (characterized by variables such as travel times, speeds, volumes etc.), safety (characterized by variables such as speeds, speed distributions, headways, times-to-collision etc.), and the environment (characterized by variables such as speeds, speed and acceleration distributions etc.). All these effects can be simulated and analyzed with the ITS modeller.

In the experiments with the ITS modeller, one of the aims was to analyze the impact of a gradual deployment of the system. Do the effects keep increasing with increased penetration rates, or is there an optimum at penetration rates below 100%? The following penetration rates were selected:

- 0% – no IRSA system; default traffic behavior;

- 20% IRSA: low penetration rate – some effects expected;

- 50% IRSA: medium penetration rate – IRSA vehicles likely to influence non-equipped vehicles;

- 100% IRSA: all vehicles equipped.

The possible modes of operation are advisory, intervening or controlling. In designing the IRSA system, the focus was on the controlling mode. The advisory and intervening modes were derived from the controlling mode by making assumptions about driver behavior.

In the ITS modeller, three scenarios were modeled to assess the impact of IRSA on traffic flows, namely 'approaching a reduced speed limit zone', 'approaching a traffic jam' and 'leaving the head of a queue' (at a traffic light). For the three scenarios assessed, the objectives to be achieved by IRSA were different.

- For the 'approaching a traffic jam' scenario, the aim was to reduce congestion and thus to improve throughput and safety, and consequently to reduce the emissions.

- For the 'approaching a reduced speed limit zone', the aim was to help drivers to slow down for a lower speed limit in a safe and efficient way, which could, in some situations, help to prevent congestion.

- For the 'leaving the head of a queue' scenario, the aim was to improve the efficiency of traffic accelerating at a traffic light, without compromising traffic safety.

The results discussed in this section helped to answer the question of whether and how IRSA can contribute to achieving these objectives.

9.6.2   Implementation of IRSA in the ITS modeller

The implementation of IRSA in the ITS modeller started with the elaboration of the meta-model: the IRSA controller, and the communication needed to make the IRSA system functional. In order to ensure a comparable implementation in MARS and the ITS modeller of IRSA, the implementations were not only based on the same meta-model but also made use of exactly the same Java code as did the IRSA controllers.

The following aspects are specific to simulation in the ITS modeller (and are not elaborated in other models from the tool suite):

- large-scale communication with other vehicles and infrastructure;

- in-vehicle message storing and handling;

- warning messages;

- penetration rates;

- controlling mode vs. the advisory/intervening modes.

As an extension to the cruise-controllers defined in Section 9.4.3, the IRSA system in the ITS modeller is equipped with additional V2V and V2I communication. The CACC controllers normally receive continuous information about location and speed of up to three vehicles downstream, which is used for normal car-following driving. In addition to that, in the ITS modeller, the system also receives information (from any vehicle within a range of 200 meters) about whether or not vehicles are driving at less than 70% of the speed limit (indicating a traffic jam), or about reduced speed limits further downstream (from roadside beacons). The system may use this information to control the speed differently from the normal CACC controller, e.g. by braking earlier. In the following subsections these IRSA variants are indicated with a '+'. The goal of both messages is to increase traffic safety and to homogenize the traffic flow by having the vehicles slowing down earlier than normal, and braking less hard.

In the ITS modeller the communication was assumed to be noise free and perfect (within a certain range of the receiver). Nevertheless, the communication system requires sophisticated message handling algorithms, in order to keep it manageable. Due to the high number of vehicles in the network, the number of received and stored messages grows exponentially. Therefore, criteria were implemented for sending, storing and removing of messages from vehicles or beacons. For instance, messages from roadside beacons need to be 'remembered' until the reduced speed limit zone has been reached. Speed warning messages from other vehicles need to be remembered for some time, but should be removed after this time (the traffic may have sped up in the meantime). A suitable duration was determined by trial and error, using the simulations in the ITS modeller. In the end, messages from roadside beacons were stored for 30 seconds. This is because there are only a few beacons (located upstream of the speed limit zone), broadcasting 'static' messages at 3-second time intervals, always with the same content. The assumed broadcast range is 300 meters.

Messages from other vehicles were only stored for 0.1 seconds. Vehicles equipped with the IRSA system broadcast messages to warn upstream traffic of low speeds (up to 200 meters away). Each IRSA-equipped vehicle could send a warning message every 0.1 seconds. In the ITS modeller, test runs with different settings for the speed boundary were performed. The chosen boundary, which performed best in the test simulations, was 70% of the local speed limit.

If a vehicle received a message, it first checked if this message was relevant. For example, a message from another vehicle was only relevant if this vehicle was traveling downstream on the same road and in the same lane. If this was the case, it copied and stored it. However, if the speed of the receiving vehicle was below 70% of the speed limit as well, the messages were considered not to be urgent. The receiving vehicle then removed all stored warning messages.

Then, each IRSA equipped vehicle decided whether it needed to brake, based on the received warning messages. It only considered those messages where its own speed was at least 10% above the speed of the sending vehicle. For these messages, it determined the necessary deceleration, based on its own speed and the distance to, and speed of, the sending vehicle, with a linear deceleration.

For warning messages from vehicles, the desired deceleration was the minimum of all necessary linear decelerations of the warning messages. For warning messages from beacons, the most urgent message was selected, based on the required linear deceleration, but the final desired deceleration was computed according to a braking profile based on measurements with an instrumented vehicle. The way an average driver reacts to a changed speed limit or a sharp curve was investigated. The aim was that the IRSA system braked with approximately the same braking profile, such that the braking manoeuvre would be comfortable and similar to the way drivers would brake themselves, only starting earlier and braking less strong. This braking profile was used to model braking for the reduced speed limit zone in the controlling mode. The shape of the measured braking curve was approximated and incorporated in the braking algorithm of IRSA. The measured braking profiles were approximated with a standard curve that consists of three segments: linearly increasing deceleration, constant deceleration and linearly decreasing deceleration. The vehicle braked autonomously according to this curve, unless the car-following acceleration was more restrictive.

9.6.3   Results for the 'approaching a traffic jam' scenario

In the 'approaching a traffic jam' scenario, the reference situation was a three-lane motorway with a lane drop halfway along. The traffic was near capacity, so congestion occurred near the lane drop. Vehicles equipped with IRSA, driving at speeds below 70% of the speed limit, sent out warning messages. The hypothesis was that IRSA can help to improve safety and/or reduce congestion by warning the drivers and helping them (or their vehicles, in controlling mode) to slow down in a safe and comfortable way, so that they would be driving at an appropriate speed when they arrived at the congested section.

All versions of the ACC and CACC controllers were tested for penetration rates of 20, 50 and 100%.

In the advisory mode, the system advised the drivers of these vehicles when to start slowing down, and how hard to brake. The system advised drivers to start braking when their speed was at least 10% higher than the speed of the sending vehicle, as in the controlling mode. Ideally, the deceleration was achieved by just releasing the accelerator pedal. However, we assumed that the driver brakes with the same constant deceleration as in the controlling mode (the difference was that, in the advisory mode, the vehicle was not equipped with an ACC or a CACC system).

The simulations with IRSA in a situation with congestion showed that the system has a positive impact on traffic flow. Vehicles slowed down earlier, having to brake less hard. The lane changing process at the lane drop appeared to benefit from this (although the IRSA system does not directly influence the lane changing behavior) and the congestion was reduced, with safety indicators staying at the same level or improving slightly. Figure 9.8 shows the changes in total travel times for the different variants of IRSA experimented with, as compared with a reference case with no IRSA. Both the travel times and the variation in speed decreased (with the average speed increasing slightly), when the whole section was looked at. Just before the congested area, the variation in speed actually increased (as the equipped vehicles start braking at different times, depending on the difference in speed with the vehicles in the queue and how far away they were). On the whole, however, traffic appeared to be more homogeneous, which was confirmed by a decrease in the variation in accelerations (by up to 40%). From this it can be concluded that, in this scenario, IRSA contributes to improved traffic safety and lower exhaust emissions. Delays (causing about 10% longer travel times in the reference case) were reduced by more than 30% for all CACC versions and by more than 20% for advisory IRSA.

The CACC2 version performed best. Incorporating speed information from preceding *equipped* vehicles (at all times) helped to smooth the traffic. The added value of messages

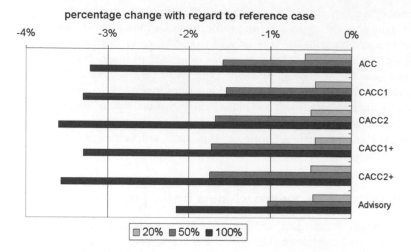

Figure 9.8 Changes in total travel time, for different versions and penetration rates of IRSA in the 'approaching a traffic jam' scenario

from vehicles driving at speeds below 70% of the speed limit was clear for the (less efficient) CACC1 controller and for lower penetrations rates of the CACC2 controller. At a penetration rate of 100% of CACC2, it appeared that the information from three predecessors alone was enough to reach the maximum impact.

The speed distribution changed with higher average speeds at higher penetration rates. The distance and time headways between vehicles did not change much, because the IRSA system setting for headway was quite short (1 second). However, the TTC decreased, which means that the differences in speed between vehicles following each other are smaller than they are without the IRSA system. Traffic thus became more homogeneous.

9.6.4   Results for the 'approaching a reduced speed limit zone' scenario

Reduced speed limits can be applied for several reasons, e.g. to improve air quality or safety, or to smooth traffic and avoid congestion. In this scenario, inspired by the implementation of a reduced speed limit on the A13 motorway in Overschie (Rotterdam) to improve air quality and reduce noise annoyance (Riemersma et al. 2004), there was a reduced speed limit (at the time of implementation from 120 km/h to 80 km/h) with strong enforcement on the mean section speed by camera observations with license plate recognition. This large difference in speed limits may cause shock waves, since drivers may brake hard when they enter the section. The hypothesis is that IRSA can support the drivers to slow down in a safe and comfortable way (earlier than they normally would), by giving speed or deceleration advice. In some situations, this may prevent congestion. The reduced speed limit and the location of the start of the reduced speed limit zone were communicated to the equipped vehicles by two roadside beacons: one located 1200 m before the start of the reduced speed limit zone and one located at the start of the reduced speed limit zone. Vehicles within 300 m of a beacon could receive the messages.

In the controlling mode the vehicle slowed down automatically due to the IRSA system, with the braking profile as measured from the instrumented vehicle. In the advisory mode the driver received a warning to start braking for the reduced speed limit. In our simulations we assumed that drivers with advisory IRSA start braking later than does the controlling mode version of IRSA, but earlier than they would without the system. Two versions were compared: IRSA 'far' (with drivers starting to brake quite early) and IRSA 'close' (with drivers starting to brake quite late). Some variation in the moment when drivers start braking was introduced in both versions, as the drivers' reaction time to the advice will vary. Penetration rates of 20, 50 and 100% were tested.

The simulations clearly showed that the equipped vehicles slowed down much earlier than did non-equipped vehicles. When looking at the whole road network, no significant effects on throughput or safety were found. However, an important effect of IRSA in this scenario was that, on the level of a cluster of vehicles, the differences in speed became smaller, especially for higher penetration rates. At a 100% penetration rate, there are practically no small TTCs. The performance (with respect to safety indicators) improved with increasing penetration rates, and the controlling mode was, as expected, more effective than the advisory modes. At lower penetration rates, especially for the controlling mode, the variation in speed in the area just before the reduced speed limit zone could be quite large.

9.6.5   Results for the 'leaving the head of a queue' scenario

The 'leaving the head of a queue' scenario differs from the two previous ones in that it focuses on acceleration, not deceleration. The scenario was elaborated for a traffic light. The hypothesis was that IRSA can help vehicles or their drivers to accelerate in a safe and efficient way. In practice, this could be an advisory or intervening version, which encourages drivers to accelerate quickly while alerting them only when they accelerate too fast and their headway with their predecessors becomes too small.

Three different settings were tested (for penetration rates 20, 50 and 100%), which were tuned to improve throughput or safety (as there appears to be a trade-off between these two aspects – faster acceleration may increase throughput but may decrease safety). The first setting was the basic CACC controller of IRSA. These settings resulted in a rather slow acceleration from standstill. Therefore, a second version of IRSA was simulated, with optimized parameter settings for accelerating from standstill, referred to as 'IRSA turbo'. Furthermore, a third ACC controller was implemented that was expected to perform well in this scenario. This controller tried to keep the headway to a fixed value, and was therefore referred to as 'fixed headway'. This was based on the assumption that keeping a minimum fixed time headway is necessary for safety reasons.

The reference scenario was calibrated such that the average acceleration from 0 to 20 km/h was $1.9 \, \text{m/s}^2$, which is reported by Bennett and Dunn (1995) and Brouwer et al. (2003) as the average (measured) real-world value.

The results of this scenario very clearly showed how different approaches and/or settings affect throughput and safety. The initial IRSA controller (CACC1) was somewhat 'cautious' compared with the reference case (Figure 9.9). This means that the throughput (measured as the number of vehicles passing the intersection during the green time of a single cycle) was smaller than for the reference case, which has been calibrated using values for accelerations found in practice. With the 'turbo' version of the IRSA controller, the number of vehicles passing the intersection during green time was larger, which resulted in lower average travel times. The simple fixed headway controller had an improved throughput compared with the reference case, but less than that of 'IRSA turbo'. This means that only keeping a fixed time headway does not optimize throughput when leaving the head of a queue. Since both controllers have the same headway setting of 1 second, it appears that a more complex, cooperative controller is needed to improve throughput.

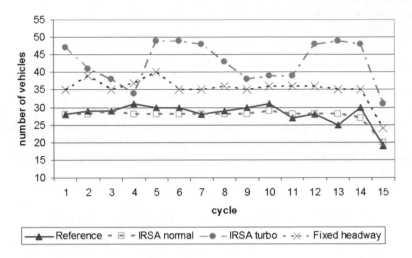

Figure 9.9 Number of vehicles passing the intersection during one cycle in the 'leaving the head of a queue' scenario (100% penetration rate)

9.6.6   Conclusions on the ITS modeller simulation results

The ITS modeller is a valuable tool in the assessment of a cooperative system like IRSA. It enables users to adapt vehicle and driver behavior, such that different algorithms and settings can be tried out easily. The changes in traffic patterns can be seen immediately in the traffic model's user interface, an animation of vehicles moving on a road. The output of completed runs can subsequently be used to assess the effects on traffic flows. Depending on the scenario and the hypotheses tested, different indicators are used to assess the effects: from aggregated variables such as the average travel time and speed to disaggregated results, such as speed profiles. All of these indicators together provide the full picture needed to assess cooperative systems, especially in dense traffic with many interactions between vehicles.

9.7   Conclusions

Cooperative road-vehicle systems are expected to have major benefits for throughput, safety and the environment. But they are complex systems, and need to be tested extensively before they can safely and efficiently be part of everyday traffic. Simulation plays a large role in this. The models used need vehicle and driver models that can take into account how the systems change their behavior, and communication protocols have to be incorporated.

The various research questions that need to be answered, ranging from how noisy sensor data affects the performance of the system to what traffic and safety impacts can be expected, require models with different levels of detail. In order to ensure consistency between evaluations at different levels of detail, a meta-model of the system should be developed that can be implemented in all the tools used in the development, testing and evaluation of a system. When analysis with one tool leads to improvements in the controller algorithms, the changes need to be implemented in all tools. This requires pro-active cooperation between the teams working with the different tools.

In the SUMMITS project, the multi-aspect assessment approach described in this chapter was applied successfully. The SUMMITS Tool Suite offered several models that are all geared towards the assessment of cooperative road vehicle systems. People in the SUMMITS project team worked well together – after getting used to each other's visions on the project. It took time to learn to 'speak each other's language', but in the end it was very rewarding to see promising and consistent results from the analyses at different levels of detail – and to see the test vehicles perform the 'approaching a traffic jam' scenario on the road.

The tool suite served its purpose well but there is always room for improvement. Some examples of areas in which further improvement is needed are human factors aspects and the modeling of communication. Human factors information, e.g., for IRSA, how people react to the speed advice (Do they comply? How hard will they brake?) is not widely available. It is quite costly to obtain; behavior varies between drivers and is stochastic and unpredictable in nature, but needs to be modeled accurately to get realistic results for systems that have the driver in the loop. As for modeling communication: the challenge is to design a system in which the number of messages is limited (but to the point and reaching the right vehicles), so that simulation running times are kept low.

Acknowledgments

The authors would like to thank the members of the SUMMITS team who provided the results described in this chapter.

References

Bennett, C. and Dunn, R.C.M. (1995) Driver deceleration behavior on a freeway in New Zealand. Transportation Research Record 1510, pp. 70–75.

Brouwer, R., Hogema, J. and Janssen, W. (2003) System evaluation of a stop-and-go system with subjects. Soesterberg, TNO report TM-03-D003.

Driessen, B., Hogema, J., Wilmink, I., Ploeg, J., Papp, Z. and Feenstra, P. (2007) The SUMMITS Tool Suite: supporting the development and evaluation of cooperative vehicle-infrastructure systems in a Multi-Aspect Assessment approach. Delft, TNO Memo 073401-N017, available at www.tno.nl.

Gieteling, O. (2005) Co-operative vehicle controllers for the SUMMITS 4 IRSA Pilot. Helmond, TNO report 05.OR.AC.031.1/JPL.

Malone, K., Wilmink, I., Noecker, G., Roßrucker, K., Galbas, R. and Alkim, T. (2008) Final Report and Integration of Results and Perspectives for Market Introduction of IVSS. Deliverable D9 & D10 of the eIMPACT project, available at www.eimpact.eu.

Netten, B. and Papp, Z. (2005) Conceptual Design of intelligent vehicles in MARS. Delft, TNO Memorandum IS-MEM-050016.

Netten, B., Papp, Z., den Ouden, F. and Zoutendijk, A. (2006) High-fidelity Evaluation of Cooperative Vehicle Systems in MARS Environment – SUMMITS/IRSA Experiments. Delft, TNO report MON-RPT-033-DTS-2007-00413.

Papp, Z., den Ouden, F., Netten, B. and Zoutendijk, A. (2006) Scalable HIL Simulator for Multi-Agent Systems Interacting in Physical Environments. In: *Proceedings of the 2006 IEEE Workshop on Distributed Intelligent Systems*, Prague, pp. 177–182.

Papp, Z., Dorrepaal, M. and Verburg, D. (2003a) Distributed Hardware-In-the-Loop Simulator for Autonomous Continuous Dynamical Systems with Spatially Constrained Interactions. In: *Proceeding of the Eleventh IEEE/ACM International Workshop on Parallel and Distributed Real-Time Systems*, Nicc.

Papp, Z., Dorrepaal, M., Thean, A. and Labibes, K. (2003b) Multi-Agent Based HIL Simulator with High Fidelity Virtual Sensors. In: *Proceedings of the IEEE Instrumentation and Measurement Technology Conference 2003*, Vail.

Riemersma, I., Gense, N., Wilmink, I., Versteegt, H., Hogema, J., van der Horst, A., de Roo, F., Noordhoek, I., Teeuwisse, S., de Kluizenaar, Y. and Passchier, W. (2004) Quickscan optimale snelheidslimiet op Nederlandse snelwegen. Delft, TNO report 04.OR.VM.016.1/IJR (in Dutch: Quick scan Optimal speed limit on Dutch motorways).

Schulze, M., Mäkinen, T., Irion, J., Flament, M. and Kessel, T. (2008) IP_D15: Final report. Final report of the PReVENT project, available at http://www.prevent-ip.eu/.

van Arem, B. (2007) Cooperative vehicle-infrastructure systems: an intelligent way forward? Delft, TNO Report D-R0158/B, available at www.tno.nl.

van Arem, B., Driever, H., Feenstra, P., Ploeg, J., Klunder, G., Wilmink, I., Zoutendijk, A. and Papp, Z. (2007) Design and evaluation of an Integrated Full-Range Speed Assistant. Delft, TNO Report 2007-D-R0280/B, available at www.tno.nl.

Wilmink, I.R., Klunder, G. and Mak, J. (2007) The impact of Integrated full-Range Speed Assistance on traffic flow – Technical report of the SUMMITS-IRSA project. Delft, TNO, TNO Report 2007-D-R0286/B.

10

System Design and Proof-of-Concept Implementation of Seamless Handover Support for Communication-Based Train Control[1]

Marc Emmelmann

Technical University Berlin

Wireless systems for commercial and public use applications, especially if they are used for mission critical communication, were in the past dominated by proprietary equipment. This equipment is usually not built in large quantities and is, hence, rather expensive. The use of commercial off-the-shelf equipment found in the consumer market is an appealing option. For example, IEEE 802.11 is considered the most mature, commercially relevant wireless LAN technology offering system components at low cost. This chapter focuses on system design and proof-of-concept implementation of seamless handover support for a real-time mission critical application, communication-based train control, employing 802.11 chipsets. The entire design and implementation methodology – starting from the system requirements, covering the design and implementation phase, and ending with the final performance evaluation – is covered here. As the design phase includes an detailed analysis of IEEE 802.11's capability to support fast handover using a system independent

---

[1]This work is based on an earlier work: 'System design and implementation of seamless handover support enabling real-time telemetry applications for highly mobile users' in Proc. ACM International Symposium on Mobility Management and Wireless Access (MobiWac 2008) http://doi.acm.org/10.1145/1454659.1454661

Vehicular Networking   Edited by Marc Emmelmann, Bernd Bochow, C. Christopher Kellum
© 2010 John Wiley & Sons, Ltd

taxonomy for handover support, the latter findings are also applicable to all application scenarios requiring fast handover support and even go beyond the scope of mere vehicular communication.

10.1 Introduction

Over the last few decades, robust and reliable wireless communication has become one key component for mission critical communication. Vehicular application scenarios range from its origins in the military sector (Mitola 2000) through homeland/public safety and emergency response scenarios (Chittester and Haines 2004; Lewis 2006; Miller 2009) to commercial industry applications (Nijkamp 1995; Virtual Automation Networks Consortium 2006). A feature that all of the employed systems have in common is that they depend heavily on the reliability and availability of a wireless communication path; specialized systems have been developed and deployed in the past. These systems can guarantee high availability at the price of extremely high system costs. In contrast, commercial off-the-shelf devices such as IEEE 802.11 Wireless Local Area Networks (WLANs) are available at a low price and have tremendously matured in reliability over recent years. Hence, the use of those low-cost devices in an environment and for application scenarios for which they were not originally designed is becoming prevalent.

The focus of this chapter is the provision of seamless, fast handover support using IEEE 802.11 commercial off-the-shelf devices in a vehicular environment. Communications-Based Train Control (CBTC) is used as an exemplary application scenario for which the chapter describes the entire design and implementation process – starting from the requirement phase and proceeding via the system design up to a proof-of-concept demonstrator-based performance evaluation. One important step within this design phase is the presentation of a system-independent handover taxonomy, and the evaluation of whether IEEE 802.11 devices can theoretically meet the application requirements. As this step analyzes the capabilities of 802.11 *in general*, it is easily applicable to *all* system designs providing fast handover support, even beyond the scope of vehicular application scenarios. Also, it shows whether 802.11 can be used in an entirely standard-compliant way while fulfilling fast handover requirements, or whether 802.11 components have to be used within a closed environment in a non-standard-compliant way.

The overall structure of this chapter is as follows: Section 10.2 provides a systematic approach to evaluating whether IEEE 802.11 can be used to provide fast handover support for CBTC. It starts by summarizing the performance requirements of CBTC as regards fast handover support. The resulting taxonomy of handover phases is then used to analyze the theoretically achievable handover performance of systems compliant to the IEEE 802.11 standard. This analysis allows the reader to assess, for any application scenario, whether 802.11 can theoretically meet the Quality of Service (QoS) requirements with respect to limiting the service interruption time during a handover to a given threshold. Having knowledge of which parts of today's standard can be left unmodified for CBTC applications and which need modification, we categorize existing work using the introduced taxonomy, evaluating whether existing handover schemes do already achieve the theoretical performance limit of IEEE 802.11 handover. Section 10.2 ends with a summary of these findings and shifts the focus towards challenges that come along with a complete system design enabling fast handover support for CBTC.

Section 10.3 presents such a complete system concept and design using 802.11 Network Interface Cards (NICs). Starting from thoughts on the required system architecture, the reader is introduced to a deterministic access scheme imposed on top of the IEEE 802.11 Carrier Sense Multiple Access Collision Avoidance (CSMA/CA) Medium Access Control (MAC). This scheme guarantees collision-free transmission, even under high system load, in a closed CBTC system. The section closes with a description of how fast handover support is realized.

Finally, Section 10.4 focuses on the implementation of a proof-of-concept demonstration and the evaluation of the latter's performance. It starts with an overview of the design and implementation methodology employed, assuring functional correctness of the code from the initial specification up to the run-time code on the Linux target system. Afterwards, a short description of the Access Point Transition Time (APTT) metric is given. This is the standardized measurement for characterizing handover performance of WLANs, and is therefore used within the subsequent empirical evaluation of the handover process. Although the implementation of the demonstrator uses automatically generated code that is not optimized for best possible performance, measurements already show that handover delays of less than 500 μs are achievable. Such short delays are in full compliance with those acceptable for real-time telemetry applications including CBTC.

10.2 Fast Handover for CBTC using Wi-Fi

10.2.1 Requirements of Communications-Based Train Control for fast handover support

Wireless deployments providing communication between a train and its wayside (and vice versa) are generally referred to as Positive Train Control (PTC) systems. Communications-Based Train Control (CBTC) is one application field for them. In general, PTC systems are classified in five groups according to the reliability and hence the functionality they can provide. At the lowest level they provide 'best effort' communication, while only the highest two levels ensure the reliability required for CBTC enabling remote, real-time control of all mainline switches and engines as well as retrieving position information about the train (RSAC 1999).

The service to retrieve the train's location information via the wireless control channel implies an important requirement for the communication system used for CBTC. The proper functionality of the deployed wireless system must not depend on any knowledge carried via the wireless link. For example, a system enabling CBTC may employ position information for optimized performance, but it needs to obtain this autonomously. Because of this, a coarser precision of location information can be acceptable, since it is not being used to provide PTC functionality such as calculation of speed and subsequent braking distances or control of the train's engine.

The second set of requirements includes guaranteed upper limits for the Inter-Arrival Time (IAT), jitter, and Round Trip Time (RTT) of datagrams. For real-time applications including CBTC, the guaranteed upper limit should be below 10 ms for an acceptable IAT/jitter and well below 100 ms for the roundtrip delay (Virtual Automation Networks Consortium 2006). Those values directly imply a guaranteed medium access delay of less than 10 ms. Note that this value is actually an upper bound, and does not require access to the media *precisely* every 10 ms.   Hence, for a controlled maximum system load/low number of users in the

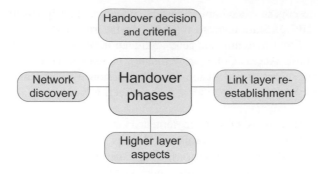

Figure 10.1 Handover taxonomy

system, Carvalho and Garcia-Luna-Aceves (2003) have shown that such an upper bound may be achieved also by a stochastic MAC scheme (IEEE 802.11 2007b). If precise media access times while operating at maximal system load have to be guaranteed, Fettweis (2007b) requires a hybrid (IEEE 802.11e 2005) or deterministic MAC scheme.

Finally, a high reliability of successful packet delivery is required. Values larger than 99.9% are stated, and companies 'compete' for the number of nines after the decimal point. Such performance is mostly provided by using most robust modulation and coding schemes, which leave a challenge to the design and implementation of the handover scheme to guarantee only occasional losses of at least one packet (Fettweis 2007a). Especially for supporting very high speed vehicles traveling at several hundreds of kilometers per hour (e.g. 'bullet trains' targeting 450 km/h), this could raise an issue: handovers are no longer a 'rare event'; the dwell time of a train in a radio cell is easily less than 10 seconds, resulting in a high handover frequency and making the handover performance and its costs an important factor in system design.

10.2.2 Taxonomy of handover phases

Regardless of the underlying technology, a handover comprises three main phases: *network discovery*, *handover decision* and *criteria*, and *link layer reestablishment* (see Figure 10.1). For the discussion of handover schemes at lower layers, for example at the MAC layer, it is occasionally useful to add a forth taxonomic group, namely *higher layer aspects*, to deal explicitly with the interaction between lower and higher layers. These handover phases do not necessarily occur in sequence. In fact, functionalities associated with different phases may be performed in parallel or may by executed well before the mobile user's transition from one radio cell to another (Dutta et al. 2006; Emmelmann et al. 2007a, 2006, 2007b).

Network discovery

During the network discovery phase, the mobile user obtains information on radio cells/ access technologies to which a handover can potentially be conducted. Two approaches exist: the mobile can detect available handover destinations itself or it may request this information via signaling, using its current communication link. As the wireless channel is usually divided, employing space-, time-, frequency, or code division multiplexing, the former approach requires the mobile to interrupt its ongoing communication, leave the current sub-channel and start detection procedures in a new one. Please note that the approach utilizing a Software Defined Radio (SDR) with an extremely wide-band Radio Frequency (RF) front-end would theoretically allow communication on one sub-channel while running network detection schemes on another. However, this technique, as of today, is too expensive in terms of computational and monetary costs and hence is not discussed further here. It is already apparent that signaling information on available handover destinations via the current link has a clear advantage: the communication is not interrupted. However, this simply shifts the problem of 'how to obtain the information to be signaled' away from the mobile, either to the operator, who would provide this statically configured information known from network planning, or to other stationary or mobile nodes that would interrupt their ongoing communication to actively search for alternative access methods within a given vicinity.

Handover decision

While the network discovery phase creates a set of potential destinations to which to conduct a handover, the handover decision phase uses this list to determine when and to which radio cell the handover occurs. Common criteria for such decisions include lower layer parameters, – for example, Receive Signal Strength Indication (RSSI), packet loss and jitter, or delay – as well as higher layer aspects – for example, total load on a given link. Either of the parameters may be measured locally at the mobile proceeding with the handover, but also remotely at the network level or at the Access Point (AP) holding the mobile's current position (and even at all APs in the set of potential handover destinations). Note that remote measurements impose an overhead on the system in terms of management messages signaling the outcome of the information acquisition to the handover decision entity. Again, this decision entity can either be co-located with the mobile or placed in the (backbone) network.

Link layer re-establishment

The link layer re-establishment phase initializes between the mobile user and the (new) AP a communication link that can be used to exchange user data, that is MAC Service Data Units (SDUs). This step usually involves synchronization on the Physical (PHY) layer and may include a renegotiation of resources at the destination AP. Two derivatives of link layer re-establishment are commonly distinguished: the 'break-before-make' (hard handover) and the 'make-before-break' (soft handover) approaches. In the former, the communication between the mobile and the originating AP is interrupted before a new link to the destination AP is established, whereas in the latter a link to both the originating and the destination AP is held for a certain time. Packet loss and reordering may occur in both cases, though it can be reduced for soft handovers as packets may be transmitted via either or even both APs involved in the handover process while the mobile user is in the critical phase of leaving the coverage of the originating AP and entering the destination AP.

Other aspects such as security-related credential exchange or admission control schemes are also part of the link layer re-establishment phase. They may increase the dwell time of the mobile in this phase, hence adding to the overall handover delay.

10.2.3   IEEE 802.11 fast handover support

The handover performance of IEEE 802.11 devices depends heavily on the manufacturer. This shows the tremendous influence of the vendor-specific algorithms controlling the handover on the delay experienced. Mishra et al. (2003) provided an empirical analysis of the handover delay, showing a high variance in measurement results depending on the device's vendor. Typical handover delays range in the order of seconds where network discovery is the dominating factor. Optimized active network discovery can reduce the latter to approximately 150–200 ms (Velayos and Karlsson 2004) while link layer reassociation can be conducted within 3–4.5 ms.

In view of these findings, IEEE 802.11 seems to be a rather unlikely candidate for fast and seamless handover support for a vehicular environment. However, this assumption could be wrong. Until today, a thorough analysis of the theoretical performance limits strictly imposed by the IEEE 802.11 standard has not existed. Such an investigation is briefly presented next, and it may be used to judge how close most recent approaches to reducing the handover delay of IEEE 802.11 devices come to this performance boundary.

Theoretical performance limits of IEEE 802.11 handover

Handover decision   IEEE 802.11 (2007a) does not specify any algorithms to decide when to initiate a handover. Hence, in full compliance with the standard, the lowest possible handover delay associated with the handover decision process is *zero*.

As example – even though not a practical one in terms of deployment cost – such a system performance could be achieved with an extremely high density of APs, resulting in very large overlaps of the coverage area such that APs always reside within the coverage areas of their direct neighbors. Having a preconfigured knowledge of the position of each AP, a mobile that has a GPS receiver could simply decide to trigger a handover to a new AP if the latter is geographically closer.

Network discovery   Terminal Stations (STAs) have two methods for discovering alternative APs: passive and active scanning (IEEE 802.11 2007a, Cls. 11.1.3.1–2). As active scanning is prohibited in some frequency bands and regulatory domains, STAs are required to implement at least passive scanning. The scanning schemes respectively allow the STA to wait for a beacon frame or actively to trigger a message exchange with all APs in its surrounding area. Receiving a frame from an AP indicates the latter to be within the proximity of the STA. Interestingly, the motive for making passive/active scanning a mandatory feature is not the provision of network discovery schemes but synchronization issues! In fact, IEEE 802.11 (2007a) does not impose any specific constraints on how to discover other APs/ networks. Hence, the minimal achievable handover delay associated with this phase is *zero*.

Link layer re-establishment   IEEE 802.11 devices have to go through three steps before user data can be exchanged via the wireless link: joining the Basic Service Set (BSS) of the destination AP, authentication, and (re-)association. The handover delay associated with joining and with authentication/(re-)association will be analyzed next.

The request of the MAC sublayer management entity (MLME) to join a BSS does not necessarily result in a message exchange via the wireless media; it merely requires the STA to synchronize its local Timing Synchronization Function (TSF) timer with that of the AP. Passive or active scanning (IEEE 802.11 2007a, Cls. 11.1) are the only means permissible to achieve such synchronization. If the STA previously conducted network discovery using either scanning scheme, it may use the TSF information retrieved for each detected AP during the scanning process to immediately synchronize or join the BSS. As conducting network discovery is not mandatory, the STA might not have used passive or active scanning before, but instead could have obtained information about available APs by other means, for example requesting IEEE 802.11k-2008 (2009, Cls. 10.3.30) neighborhood reports. In this case, the STA has to scan for a specific AP on a specific channel (known from network discovery) for synchronization.

For passive scanning, the STA simply returns its RF front-end and waits to receive a beacon from the AP. The associated delay $\delta_{\text{PassiveScan}}$ is a random variable having a uniform distribution over [0, TBTT]:

$$P[\delta_{\text{PassiveScan}} \leq \xi] = \begin{cases} \dfrac{\xi}{\text{TBTT}} & \xi \leq \text{TBTT} \\ 1 & \xi > \text{TBTT} \end{cases} \tag{10.1}$$

where the Target Beacon Transmission Time (TBTT) is the interval between the transmission of consecutive beacons. Values of 100 or 200 ms for the TBTT are typical in deployments.

Active scanning in turn involves an exchange of ProbeRequest and ProbeResponse frames as defined in IEEE 802.11 (2007a, Cls. 11.1.3.2.2). Figure 10.2 illustrates the corresponding Message Sequence Chart (MSC) including timing parameters. As the STA just switched the communication channel, it does not have any knowledge of the channel state and has to perform a 'basic access procedure' (defined in IEEE 802.11 2007a, Cls. 9.2.5.1), which involves waiting for a Distributed Interframe Space (DIFS) assuming an idle channel. The latter is followed by the regular message exchange. The end of the active scanning phase is determined either by having received the Acknowledge (ACK) confirming the reception of the ProbeResponse frame plus the mandatory DIFS afterwards, or when the Post Transmission Backoff (PTB) has started after the transmission sequence of the ProbeRequest frame expires. The event that occurs last determines the duration of the active scanning phase, since it represents the first possible time when the STA can initiate any further message exchange with the AP. Figure 10.3 shows the Cumulative Distribution Function (CDF) of the duration of the active scanning phase $\delta_{\text{ActiveScan}}$. For the most robust Modulation and Coding Scheme (MCS), a constant delay of 2050 μs occurs. For higher MCSs, the effect of the PTB starts to dominate. It is also noteworthy that increasing the data rate does not reduce the duration of the active scanning phase beyond a certain limit. The reason is that the header of IEEE 802.11 management frames is always transmitted at a lower, fixed base-rate. The graph of the CDF also includes the theoretical performance limit for an infinitely high MCS that allows the coding of an infinite number of data bits within a single Orthogonal Frequency Division Multiplexing (OFDM) symbol. The smallest possible duration varies accordingly between 290 μs and 740 μs. In order to achieving this lower limit, a STA would need to be close to the AP being queried, as only this would assure a correct transmission of data at such a high data rate. (Note that this indirectly implies having knowledge of being so close.) Hence, in a vehicular environment where handover usually occur at the borders of cells, this lower limit is not realistically achievable.

To conclude link layer (re-)establishment, IEEE 802.11 STAs have to exchange credentials and information on their capabilities (the authentication and (re-)association phase). In its currently ratified version, IEEE 802.11 (2007a, Cls. 11.3) makes this process mandatory. Figure 10.4 illustrates the message exchange involved, including timing parameters. In this 'four-way handshake', the STA and the AP each have to transmit a message twice in a relatively small timeframe. Hence, the effects of the PTB timers on the total duration of this phase have to be closely considered. PTB2 in Figure 10.4 is one example when a STA is

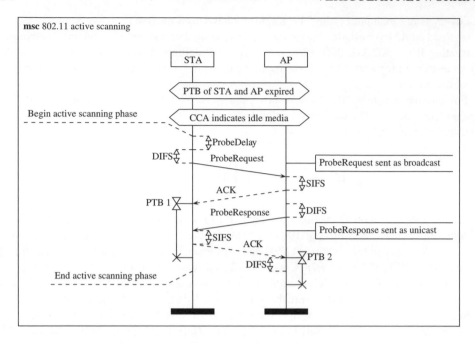

Figure 10.2 Message Sequence Chart (MSC) of IEEE 802.11 active scanning

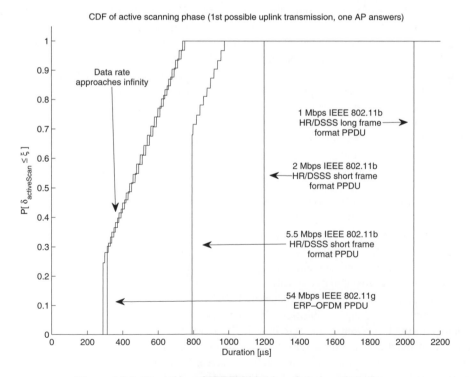

Figure 10.3 Duration of IEEE 802.11 active scanning phase

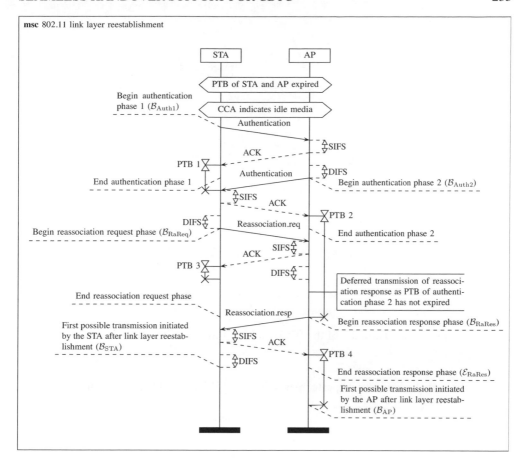

Figure 10.4 Message Sequence Chart (MSC) of IEEE 802.11 link layer reassociation

deferred from continuing with the message exchange even though the medium is idle. These deferrals dominate the duration of the entire phase for high data rates, and this means that one cannot merely add the simple transmission times of the messages involved in order to determine the total duration of the authentication/(re-)association phase. Figure 10.5 shows the CDF of the total duration for selected data rates. Again, the theoretical performance limit for an infinitely high data rate was considered. The best achievable latency is within the 480–1480 μs range. The most robust data rate (1 Mbit/s High Rate Direct Sequence Spread Spectrum (HR/DSSS) long frame format Physical Layer Protocol Data Unit (PPDU)) would lead to an increase, to values within the range 2465–3397 μs.

Future revisions of the standard   Table 10.1 summarizes the delays that come along with each handover phase. Clearly, the 4.8 ms delay for the most robust rate is already of the same order as the delay acceptable for CBTC or other telemetry applications in a vehicular network. This drawback was also seen by the IEEE Working Group (WG) Wireless Access in Vehicular Environments (WAVE), which defines a so-called 'outside the context of a

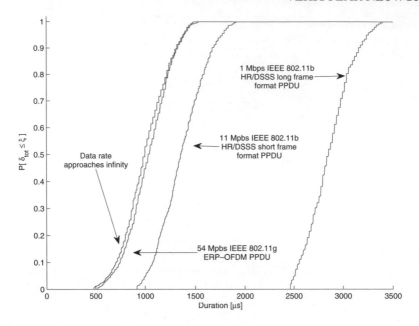

Figure 10.5 Duration of the IEEE 802.11 authentication and association

Table 10.1   Theoretical performance limits of IEEE 802.11 (2007a) handover

	Expected mean of duration ($E[\delta]$)		
	In compliance with IEEE 802.11 (2007a)		
Handover phase	Assuming infinity data rate	1 Mbit/s HR/DSSS long frame format PPDU	IEEE 802.11p (2009)
Handover decision	0 μs	0 μs	0 μs
Network discovery	0 μs	0 μs	0 μs
Link layer (re-)establishment (total)	1441 μs	4887 μs	0 μs
Synchronization	440 μs	2050 μs	0 μs
Authentication & (re-)association	1001 μs	2837 μs	0 μs

BSS procedure' mode. STAs operating therein do not use synchronization procedures as required by the IEEE 802.11 (2007a) baseline standard. For those STAs, authentication and (re-)association are also not required (IEEE 802.11p 2009, Cls. 5.2.10, 11.3, and 11.20). If this amendment to the standard is ratified in its current version, the duration associated with link layer (re-)association also diminishes to zero.

In conclusion, IEEE 802.11 (2007a) in combination with the IEEE 802.11p (2009) amendment will theoretically be able to support fast handover for CBTC and other telemetry vehicular applications. However, this comes at the cost of obtaining all the information necessary to conduct a handover, starting from handover detection and decision and also including network discovery and up to link layer (re-)establishment, by means outside the scope of the standard. There still remains the challenge of designing such methods. The following subsection summarizes related work on fast handover support for IEEE 802.11-based networks, and thereby shows how close current approaches come to the theoretically achievable performance of a zero delay handover. Instead of providing a full literature survey, only best performing practices are summarized.

Related work on IEEE 802.11 fast handover support

Only a few authors have explicitly evaluated the decision phase while focusing on whether zero-delay handover is achievable. A very common approach to detecting the need to conduct a handover is based on assessing the RSSI of the signal received from the current AP. If the signal drops below a certain threshold, a handover is initiated. Though using extremely simplifying assumptions, Zhang and Holtzman (1994) lay the framework for analyzing the smallest possible overlap between adjacent radio cells required to enable a zero-delay handover (see also Zhang and Holtzmann 1996, for an extended version). Their work was brought forward by Emmelmann (2005a,b,c), refining their assumptions and analytical approach to reflect design constraints of existing wireless devices. The analytical framework was also finally validated using channel traces derived from a high speed vehicular environment (Emmelmann et al. 2008). Though not optimized for fast handover, many of today's implementations trigger a handover after n consecutively lost beacons. Hence, the work of Velayos and Karlsson (2004) is noteworthy as it characterizes the probability of missing n beacons due to collisions rather than having left the coverage area of an AP, and thus presents a way to distinguish between mobility and congestion as the causes for this trigger. Based on this work, deciding on a handover after three lost beacons is extremely accurate; the associated delay hence varies between two and three TBTTs.

Since, in practice, network discovery is the most expensive handover phase in terms of causing service   interruption time (a.k.a. handover delay) (Mishra et al. 2003; Velayos and Karlsson 2004), much effort has been invested to reduce its duration. One approach is to provide a priori knowledge of the set of APs that are potential handover candidates. Such information can be used to reduce the number of channels. Neighbor graphs are one approach to this solution. Shin et al. (2004) record handover between APs to create such graphs dynamically. They show the probing time of a *single* channel to be reducible to values in the range 1.8–11 ms. For a typical cell deployment using three non-overlapping frequencies, the discovery phase reduces to an expected mean of 7.3 ms. Another class of approaches moves the network discovery phase before the handover decision. Ongoing communication is only interrupted for such a short timespan that the QoS of upper layer applications is not affected. One example is   the SyncScan algorithm (Ramani and Savage 2005). It continuously tracks nearby APs, following a deterministic approach as it synchronizes short listening periods with the periodic transmission of the beacons of neighboring APs. This requires knowledge of when such beacon transmission occurs, which is realized by a centralized management entity accumulating such knowledge. The authors quantify the associated delay in the order of 15 ms. Singh et al. (2006) as well as Chui and Yue (2006) move away from the centralized to an entirely decentralized approach of SyncScan. They introduce an AP coordination and signaling scheme, synchronizing beacon transmission of all APs in the distribution system to occur within a given, short timeframe. This allows STAs to align their scanning attempts with the potential transmission of beacons, although they scan for beacons regardless if they are within the coverage area of APs transmitting those beacons, which results in unnecessary scanning attempts. Chen et al. (2008) address the latter drawback. Their DeuceScan algorithm combines neighborhood graphs with the SyncScan approach and auxiliary (geographic) position information of APs, achieving an associated service interruption time of up to 3.5 ms. This is only 1450 μs away from the theoretically achievable service interruption associated with actively scanning a single channel (see Figure 10.3). Adding the time to retune the RF transceiver as measured by Ramani and Savage (2005), the total handover delay for DeuceScan ends up as 13.5 ms.

OpportunisticScanning is another novel approach for network discovery (Emmelmann et al. 2009a,b) that does not impose any requirements on the infrastructure to synchronize beacon transmissions. The general idea behind this scheme is that higher layer applications experience a 'seamless, zero-delay' network discovery if the interruption time is so short that the QoS of the application is not affected. Hence, OpportunisticScanning pauses the ongoing communication using the standardized Power Save feature to passively scan another channel for a very short time. As the reception of a beacon is not guaranteed within this short time frame, the probability of finding a beacon within n scan attempts is given. Although the analysis of the scheme for a vehicular environment with mobile users has not yet been conducted, OpportunisticScanning can achieve service interruption times of less than 2.3 ms even for the most robust MCS. In contrast with active scanning, this scheme does not impose any system load on the channel to be scanned, at the cost of requiring the mobile STA to reside within the overlap of adjacent APs for 0.2 to 1.2 seconds (depending on the system load).

Pushing the handover delay further and further towards zero, authors have started to explore non-standard-compliant ways of avoiding the link layer (re-)establishment. For example, Amir et al. (2006) describe an 802.11-based mesh network in which *all* APs operate on the same channel. Hence, neighboring APs receive transmissions from STAs located in the overlap of their coverage area. The authors impose multicast-based signaling over the distribution system to exchange information on which APs can be heard from which other APs as well as information on the received signal strength. Based on signal strength measurements, the AP that is the most likely handover candidate, in addition to the one currently serving the STA, transmits data destined for it. This 'bi-casting' approach implies an overhead, since the capacity of the wireless media has to be allocated twice for the transmission of each packet. Also, packet duplication occurs frequently at the mobile STA. The proposed solution omits the need for network discovery and provides zero-delay handover performance. However, in order to detect mobiles within the overlap of adjacent APs, the authors report a required time of approximately 2–3 seconds, which is only feasible for low or moderate user mobility.

In summary, one can say that neither of the proposed solutions provides a zero-delay handover when considering *all* handover phases commutatively. The only promising, reactive approach eliminating the link layer (re-)establishment has its limits when considered for large-scale infrastructure for vehicular applications, since it is only applicable to single-frequency deployments. Eventually, a handover from one frequency to another is required. However, a combination of certain existing approaches with novel, predictive handover schemes dealing with the 'multiple frequency issue' is a promising approach to enabling zero-delay handover for fast mobile users, for example in a high speed train environment.

10.2.4 Challenges of CBTC for Wi-Fi-based fast handover support

Communications-Based Train Control services require handover latencies well below 10 ms and delay-bound media access. The high velocity of the mobile STA also results in a rather short dwell time within the overlap between adjacent radio cells, since deployment cost usually forbids a high density of APs, which would of course increase the sizes of the overlaps. The same reason also underlies another design constraint: mobiles usually have only a single network interface card (Fettweis 2007a,b).

In view of this, the main challenge to Wi-Fi for fast handover support in a mobile environment is threefold: first, the acceptable handover delay is already in the order of a *single handover phase*, for example the reassociation. Second, the dwell time in the overlap of adjacent cells is too small to trigger and complete the handover process. Third, sooner or later, the mobile will cross the boundary of a (single) frequency domain and have to switch its RF transceiver from one channel to another. Although existing approaches partially address some of these aspects, an integrated solution considering *all* handover phases has so far neither been proposed nor evaluated. This will be done in the following part of this chapter.

10.3   System Concept and Design

Analyzing the technological limits of standard IEEE-compliant devices with respect to achievable handover performance has shown that system performance required for CBTC is achievable under one constraint: none of the information acquired during the handover decision and network discovery phase may result in any signaling at the MAC layer, as the resulting service interruption would violate the QoS constraints of CBTC. Even the signaling required by today's standard to (re-)establish a link layer connection lasts too long and hence must be avoided. Related work has shown that moving this information acquisition away from the MAC is possible, though at the cost of sacrificing a certain amount of system capacity (cf. the bi-casting approach of Amir et al. (2006)). Hence, the system concept and design is driven by five key postulates for achieving best handover behavior:

- The handover 'as seen by the mobile STA' shall occur as seldom as possible; that is, system behavior as experienced by the train shall be the same as if the latter is not moving.

- The mobile is only equipped with a single IEEE 802.11 NIC. (If present, a second card's purpose is merely to increase redundancy and availability of the system. It must not be essential for successful handover support.)

- The system shall support handover between cells operating on different channels.

- The system shall use low cost, commercial off-the-shelf 802.11 equipment on the wireless path. However, all devices operate in a 'closed environment', so modifications of the firmware (which are feasible at negligibly small costs) are permissible.

- System performance shall be stable regardless of the imposed network load.

The following sections describe the resulting system architecture, MAC, and schemes for fast handover support.

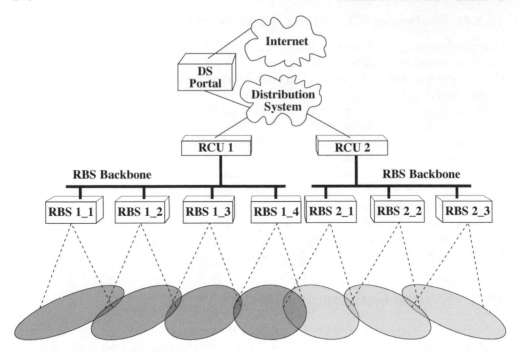

Figure 10.6 System architecture for fast and seamless handover support (Emmelmann et al. (2008) © ACM)

10.3.1 System architecture

The proposed architecture consists of a micro/macro cellular system. Several micro cells operating on the same frequency are grouped into one macro cell such that adjacent macro cells operate on non-interfering channels. A standard IEEE 802.11 NIC is used to span each micro cell. All micro cells are connected to a centralized controller via an Ethernet-based Distribution System (DS). The role of the Radio Control Unit (RCU) within this architecture is to impose a deterministic 'overlay' MAC on top of the CSMA/CA MAC scheme of 802.11. To avoid confusion with traditional deployments of 802.11 equipment, we denote in the following the entity forming the micro cell as Remote Base Station (RBS) and the centralized controller of a macro cell as RCU. Actually, this design paradigm has been adapted from the research on radio-over-fibre-based networks (Kim et al. 2005) providing zero delay handover performance. Figure 10.6 illustrates one example of the overall system architecture.

The RBS is a standard Linux PC with two interfaces: an IEEE 802.11 NIC on the wireless side and a standard Ethernet interface on the (wired) network side. A transparent layer-3 connection directly connects the interface of the 802.11 driver within the RBS to the 'overlay MAC' at the RCU (see Figure 10.7). The modified firmware of the 802.11 NIC directly bridges packets received from the RCU to the wireless interface and vice versa. For the wireless transmission, packets and management information for the 'overlay MAC' are simply encapsulated in 802.11 broadcast frames. This is required to avoid the immediate acknowledgement mandatory for unicast transmissions, and allows a communication without

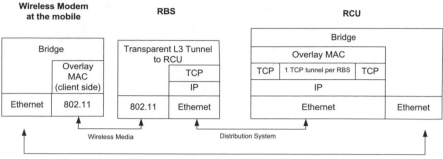

Figure 10.7 Fast handover support: protocol architecture of system components

going through the link layer (re-)establishment procedure outlined above. (Note that in the future a similar 'transparent mode' will be directly available as part of the 802.11p standard.) On the mobile client side, the architecture of the RBS is mirrored except that the contents of the 802.11 broadcast frame is bridged directly to the Ethernet interface. The client-side counterpart of the 'overlay MAC' is thus co-located within the bridging unit.

As a result, this architecture provides a transparent connection from the Ethernet interface of the modem on board the train via the wireless link to the RCU of the macro cell. The user obtains 'plug-and-play components' that internally only use low cost, commercial off-the-shelf components, especially mass market 802.11 NICs for the wireless communication path.

10.3.2 MAC scheme

The 'overlay MAC' at the RCU imposes a deterministic Time Division Multiple Access (TDMA) within its macro cell by regularly scheduling downlink transmission as illustrated in Figure 10.8. The mobile may respond immediately after the reception of a downlink packet. Transmissions are upper bounded by a fixed (transmission slot) length. If the RCU does not have pending downlink traffic for a mobile, it polls the latter within the scheduled transmission slot using an 802.11 Null Data frame. If no uplink traffic is pending, the mobile also sends an 802.11 Null Data Frame acting as a pilot signal. This allows the RBS to assess the current link quality. Also, every n slots, the RCU announces a random medium access period, within which mobiles may employ the 802.11 Distributed Coordination Function (DCF) to signal their presence to the RCU, requesting an initial slot assignment during association. IEEE 802.11 framing and addressing as provided by the commercial off-the-shelf NIC cards is used for the actual transmission.

As the RBS immediately forwards packets received from the RCU, the latter has to compensate for the transmission time between the RBS and the RCU when scheduling transmission slots. This is achieved by the Precision Time Protocol IEEE 1588 (2002), which synchronizes all local clocks in the system to give an estimation of the forwarding delay with an accuracy of a few microseconds (Kannisto et al. 2005).

Another feature of this MAC scheme in combination with the micro-cellular architecture should be noted: even though packets are always only transmitted via exactly one RBS,

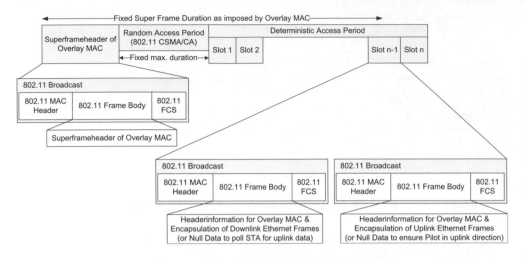

Figure 10.8 Overlay MAC on top of 802.11

they may be received via several RBSs if the mobile is within the overlap of adjacent micro cells. As all RBSs forward the received packet to the 'overlay MAC' of the RCU, the latter can apply packet combining to increase reception probability. Also, the feature of receiving packets via *all* RBSs having connectivity to a mobile is inherently used for seamless, transparent micro-cell handover.

10.3.3   Predictive fast handover

Two kinds of handover occur for the proposed system concept: an intra-macro-cell handover, which can be entirely handled under the control of a single RCU, and an inter-macro-cell handover, which requires interaction between RCUs of several macro cells.

Intra-macro-cell handover

The handover process within a macro cell is transparent to the mobile user and does not require any signaling via the wireless media: as all RBSs within a macro cell operate on the same frequency, uplink traffic may be received by more than one RBS if a mobile is within the overlap of their coverage area. For each received packet, the RBS forwards the received radio signal strength along with the packet to the RCU, which in turn may choose on a packet-by-packet basis the RBS to be used for the next transmission. In general, this approach resembles the one presented in Amir et al. (2006) but does not require the mobile to reside within the overlap of adjacent micro cells for a long time, as the decision for the intra-macro-cell handover is centralized within the RCU. Rather, the overlap of adjacent micro cells may be reduced to a bare minimum, according to Emmelmann (2005a,b,c).

Inter-macro-cell handover

The system requirements regarding the handover delay are already of the order of the channel switch time and the time of the link layer reestablishment (see Section 10.2.4). Accordingly, the predictive fast handover approach avoids all of the traditionally known handover phases involving signaling on the wireless MAC layer.

As the RCU always has knowledge of which RBS to use for the next transmission of packets, it inherently tracks the mobile's movement within its macro cell. It hence detects when a mobile enters the boundary of the macro cell. The boundary is defined as a micro cell that overlaps with a micro cell of another macro cell, as shown for RBS_1_4 and RBS_2_1 in Figure 10.6. Upon a mobile entering the boundary of a macro cell, the latter's RCU signals the neighboring RCU via the DS that a handover might be imminent and that the neighboring RCU should *predictively* start transmitting downlink traffic destined for the mobile. At the same time, the RCU signals the frequency allocated to the neighboring macro cell to the mobile. The mobile may then decide whether to hand over to the neighboring macro cell by observing the current radio signal strength. This decision scheme itself does not add any handover latency (Emmelmann 2005a), given a certain overlap of the cells. Once switched to the new frequency, the mobile immediately receives downlink traffic and may transmit uplink traffic without any prior link re-establishment. Two aspects remain to be solved: how to gain knowledge of the proximity of adjacent micro cells of different macro cells; and how to ensure that traffic for the mobile is received simultaneously by all involved RCUs during the predictive handover phase.

The proximity knowledge may be pre-configured. Alternatively, it may be dynamically learned if a fast handover fails. In the case of a failed inter-macro cell handover, the destination RCU does not predictively serve the mobile, and this forces the mobile to initialize a new link layer connection using the random access period. During this initialization process, the mobile informs the new RCU about the identity of the RCU that was serving it just before the handover. The new RCU can in turn inform the former RCU about the failed handover, giving the identifier of the new RCU and RBS currently being used to serve the mobile. This information exchange establishes a new proximity relation. Other schemes (Chen et al. 2008; Shin et al. 2004) adopting neighbor graphs in combination with movement prediction may also be applied.

A multicast-based approach is employed to assure that mobile-specific traffic is received by all involved RCUs. We therefore assign a unique multicast MAC address to each mobile during its initial association to the system. This MAC address in combination with the mobile's IP address is signaled to the DS portal. The latter in turn manipulates its Address Resolution Protocol (ARP) tables to use the multicast MAC address to forward *any* traffic for the mobile. During the predictive fast handover phase, involved RCUs subscribe to the associated multicast group using the Generic Attribute Registration Protocol (GARP) or Multicast Registration Protocol (GMRP) (IEEE 802.1ak 2007), which is frequently available in commercial off-the-shelf switches and routers. Such multicast-based approaches have been intensively studied by Festag (2003), reused by Amir et al. (2006), and are known not to add any handover latency.

10.4 Implementation

10.4.1 Methodology

The methodology employed for the design, specification and implementation of the fast handover support was driven by the goal of achieving a working implementation while having the highest possible confidence in the correctness of the code. Figure 10.9 depicts

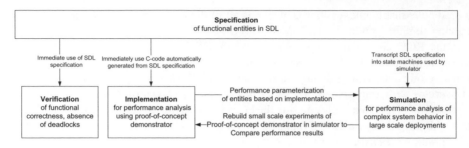

Figure 10.9 Design methodology for fast handover support

the resulting workflow: all functional components added to 'standard functionality' as found on a Linux system were specified using Specification and Description Language (SDL). The specification was then verified for functional correctness, including a deep-search for possible occurrences of deadlock in the protocol. This was automated using Telelogic's commercial toolset TAU (Telelogic TAU 2008). After assuring functional correctness, the same SDL code was used to automatically generate C code that can be executed on a Linux system. Obviously this requires interaction between the automatically generated code of the 'novel' functional entities and the Linux operating system (Langgärtner 2007). Also, functionality provided by the Linux system has to be made available within the SDL code in order to allow automatic testing of the specification for functional correctness and absence of deadlocks. This was achieved by introducing a functional block in SDL that acts as an adaption layer to the Linux environment, as shown in Figure 10.10. Its interface with all the 'novel' functional blocks specified in SDL is invariant. This adaption layer is either replaced by an implementation interacting with external C code providing an interface to a Linux system driver, or replaced by a SDL dummy block emulating the functionality of the Linux environment, depending whether the automatically generated code of *all* components is to be run as native code on Linux or simply to be analyzed within the Telelogic TAU tool. Such a SDL dummy block could be as simple as forwarding SDL signals from one block within the SDL environment to another. Following this approach, another lesson was learned: it was impossible to run automatically generated C code on Linux when the SDL specification made use of its internal timer signals. Hence, only self-defined SDL signals for setting and triggering timer events were used. They again interact with a dummy layer, which is transparently replaced either by external C code or by a realization of timer functionality using the native SDL timer.

The entire code for the proof-of-concept demonstrator discussed in the following section was automatically generated using the methodology outlined above. We acknowledge that, due to the automatic code generation, rather more emphasis was put on guaranteeing functional correctness than on creating optimized code for the target system.

10.4.2   Proof-of-concept demonstrator

The prototype system corresponds to a subset of the architecture illustrated in Figure 10.6. It consists of three micro cells (RBS_1_3, RBS_1_4, and RBS_2_1), which span two macro cells. We use an IEEE 802.11 prototyping card produced by IHP Microelectronics (2008) to build the functionality of each RBS in the prototype. The latter NIC gives us full access to its firmware and comes with a Linux driver that exchanges messages between the card and

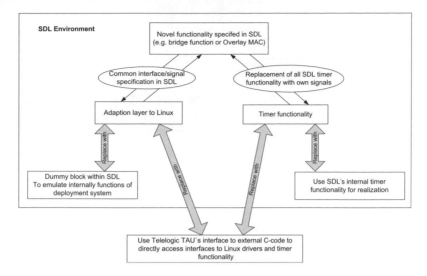

Figure 10.10 Interfacing SDL with external C code and Linux

the host system. On the mobile side, a single STA is added to the system. All the involved wireless cards are connected to a channel emulator that allows the imposition of a stochastic channel model between any transmitter–receiver pair. Also, the channel characteristic can be adjusted according to prerecorded channel traces deriving from a real-world system.

On the backbone side, we connect the proof-of-concept demonstrator via a router (DS portal) to the infrastructure at our research institute. A host acting as a source/destination for UDP packets is placed within the latter network and communicates with the mobile STA. This UDP traffic is later on used to characterize the handover performance.

10.5 Performance Evaluation

10.5.1 Metric design

In order to classify the performance of (end-) systems – including applicable protocols supporting fast handover – the IEEE measures the APTT of the handover process (Bangolae et al. 2005). The standard defines the transition time as the interval between the last successful transmission/reception of a data frame via the originating AP (here the RBS) and the first successful transmission/reception of a data frame via the destination AP, as illustrated in Figure 10.11 (IEEE 802.11.2 2008, Cls. 6.7.2). Hence, the APTT includes all the time required to establish a 'working' link connection for the user/above the MAC.

Obviously, the smallest observable APTT is lower bounded by the time ($t_{switchF}$) required by the mobile to tune its radio transceiver from the frequency the originating AP operates on to that of the destination cell. Additionally, as the successful arrival/transmission of data frames is used to trigger the measurement of the APTT, the precision of this approach depends on the IAT of data frames via the same AP, that is without experiencing a handover. Therefore, we derive in the following the theoretical minimum and maximum of the APTT.

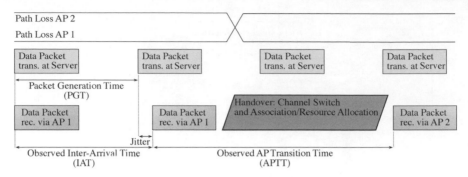

Figure 10.11   Access point transition time performance metric (Emmelmann et al. (2008) © ACM)

Let μ_{IAT} and μ_{APTT} be random variables representing the measured packet-inter-arrival time (i.e. without the occurrence of a handover), and the access point transition time. $E\{\cdot\}$, $Max\{\cdot\}$, and $Min\{\cdot\}$ denote correspondingly the expected mean, maximum, and minimum values.

As the handover – i.e. switching from the old to the new transmission frequency – occurs randomly, a single packet might be lost during the handover if it was transmitted immediately after the mobile started the handover, when it was in the phase of retuning its radio transceiver. Assuming synchronized packet transmissions at the old and the new AP, we derive – as illustrated in Figure 10.12 – the theoretical minimum and maximum of the APTT:

$$\varepsilon := Max\{\mu_{IAT}\} - Min\{\mu_{IAT}\}$$

$$Min\{\mu_{APTT}\} = E\{\mu_{IAT}\} - \varepsilon \tag{10.2}$$

$$Max\{\mu_{APTT}\} = 2 * E\{\mu_{IAT}\} + \varepsilon \tag{10.3}$$

For asynchronous transmissions of packets at the involved APs, (10.2) simplifies to

$$Min\{\mu_{APTT}\} = t_{switchF}, \tag{10.4}$$

since the handover could occur immediately after the reception of a packet via the originating AP and right before the destination AP transmits another one. As these upper and lower limits of the APTT are based on a worst case error propagation, we do expect the measured APTTs to be well within these limits.

In addition to the APTT metric, we define the (mean) handover delay (*HOD*) as

$$HOD := E\{\mu_{APTT}\} - E\{\mu_{IAT}\}, \tag{10.5}$$

which represents how long a packet is delayed after a handover from its (expected mean) time of arrival in the absence of a handover.

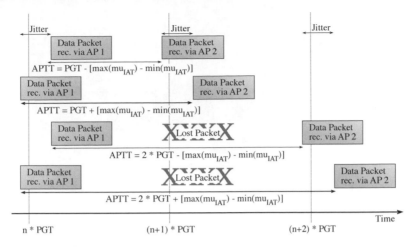

Figure 10.12 Deriving the theoretical minimum and maximum of the APTT (Emmelmann et al. (2008) © ACM)

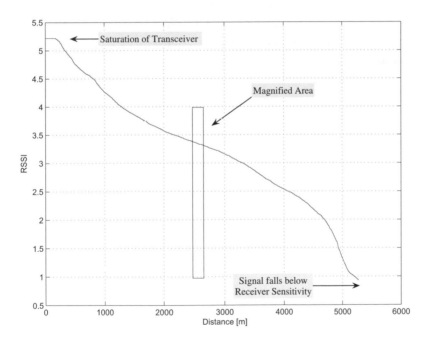

Figure 10.13 Moving average of empirical radio signal strength (Emmelmann et al. (2008) © ACM)

10.5.2   Empirical evaluation

Experimental set-up of the proof-of-concept demonstrator

We emulate the movement of the mobile terminal by changing the attenuation between each transceiver pair connected to the channel emulator as if the mobile has moved from

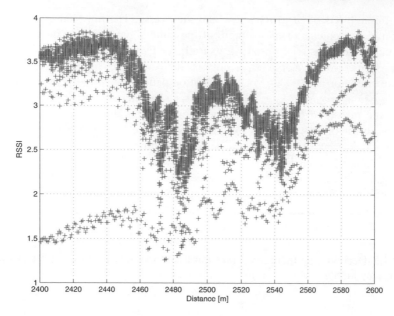

Figure 10.14 Illustration of deep fades in empirical signal strength values (Emmelmann et al. (2008) © ACM)

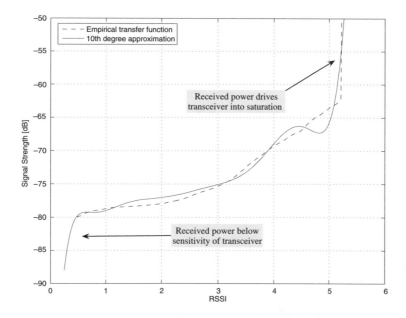

Figure 10.15 RSSI characteristic of transceiver (Emmelmann et al. (2008) © ACM)

RBS 1_3 through RBS 1_4 to RBS 2_1 and vice versa. The imposed attenuation pattern is based on measurements of the strength of the radio signal between the Transrapid high speed train and radio base stations along its trail (see Figure 10.13). The signal experienced is subject to severe deep, short-term fades, as the example in Figure 10.14 illustrates. As the channel emulator used requires an attenuation pattern given in dB as an input, we use the 10th-degree approximation of the transceiver characteristic as shown in Figure 10.15 to convert the measured RSSI values. The chosen degree of the approximation function results in an error of less than 1% over the RSSI range from 0.5 to 1.5 that is relevant for handover decisions. Also, it almost perfectly approximates the power drop-off at the receiver's sensitivity level, hence giving a reasonable compromise between accuracy and the computational complexity of higher degree approximations. The channel emulator uses the approximation function to convert empirical RSSI values and stops forwarding packets to any connected card if the experienced attenuation at the receiver falls below −80 dBm. The mobile initiates a handover once the received signal falls below −77 dBm, hence employing a 3 dBm hysteresis margin. The overlap of adjacent macro cells is calculated according to Emmelmann (2005a,c), whereas the overlap of RBSs belonging to the same macro cell is merely large enough to ensure a signal reception above the −77 dBm handover trigger.

Measurement results

The first measurements quantify the IATs while the mobile resides within one macro cell. Please note that the results for a stationary mobile and a mobile only moving between

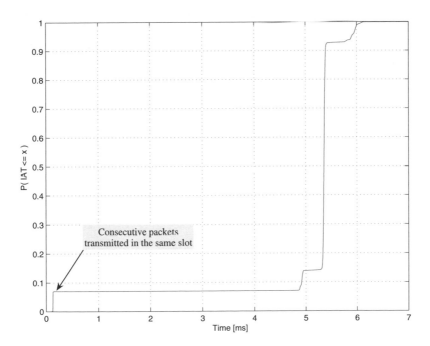

Figure 10.16 Packet inter-arrival times (empirical CDF, without handovers, targeted $PGT = 5$ ms) (Emmelmann et al. (2008) © ACM)

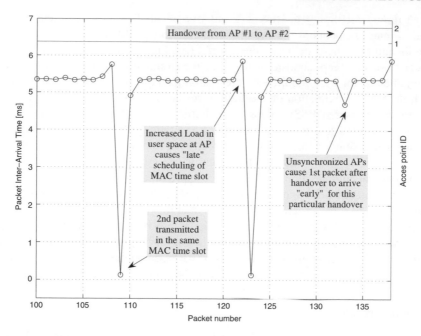

Figure 10.17 Packet inter-arrival times before and after handover (Emmelmann et al. (2008) © ACM)

RBSs of the same macro cell do not differ and are not separately discussed. Even though the expected mean ($E\{\mu_{IAT}\} = 5.01$ ms) does not significantly differ from the 5.00 ms time interval at which the packets are generated at the server, the IAT cumulative distribution function in Figure 10.16 shows two anomalies, namely a median of 5.36 ms and a probability of 8% that consecutive packets arrive within 0.03 ms.   These two phenomena can be explained as follows: since all the protocol components of the proof-of-concept demonstrator – including the bridge connecting the Ethernet interface of the RBS backbone to the wireless interface as well as the bridge between RBS backbone and distribution system at the RCU – run in the user space, we noticed that we reached the capacity limits of our implementation, imposing a forwarding load of 200 packets/s. A further increase of the load resulted in unpredictable packet losses. As a result, our MAC process is regularly interrupted by kernel threads and hence cannot provide a 'real-time scheduling' of packets to be sent to the wireless system at a peaceful five-millisecond rate. As packets arrive faster than the MAC can schedule transmission time slots, we see two packets residing in the transmission queue after every 10th transmitted packet. As the MAC transmission slots per mobile are large enough to hold more than one packet, the MAC forwards the two packets in the queue within the same MAC time slot. Hence, the IAT of these two (data) packets is recorded to be 0.03 ms. Figure 10.17 illustrates this situation occurring for packet numbers 108/109 and 122/123. As a MAC frame (in this specific case holding two data packets) is either correctly received as a whole or entirely discarded, these short IATs cannot be experienced during an inter-macro-cell handover and hence have to be discarded when calculating the theoretical minimum and maximum APTT according to (10.2), (10.3), and (10.5). Table 10.2

Table 10.2 Statistical properties of packet inter-arrival times (without handovers, targeted $PGT = 5$ ms) (© ACM)

	Raw data	Deleted IATs corresponding to packets transmitted within the same MAC time slot
$E\{\mu_{IAT}\}$	5.01 ms	5.37 ms
$Min\{\mu_{IAT}\}$	0.03 ms	4.86 ms
$Max\{\mu_{IAT}\}$	6.27 ms	6.27 ms

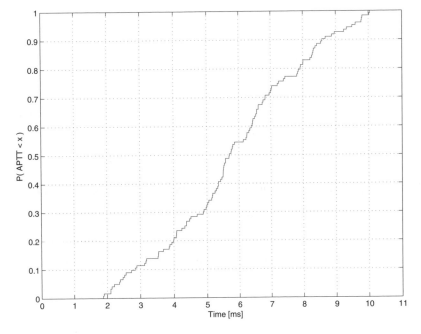

Figure 10.18 Access point transition times (empirical CDF, $E\{IAT_{noHO}\} = 5$ ms) (Emmelmann et al. (2008) © ACM)

summarizes the corresponding statistical properties of the IATs for the recorded raw data and for the case where the IATs belonging to packets received within the same MAC time slot were discarded.

Based on the assessment of the IATs, we derive according to (10.3) and (10.4) the theoretical lower and upper bounds for the APTTs coping with a worst case propagation of errors. As macro cells transmit their MAC time slots asynchronously, we obtain $Max\{\mu_{APTT}\} = 12.15$ ms and $Min\{\mu_{APTT}\} = 1.70$ ms. Here, $t_{switchF}$ is again a measurement-based value accounting for the interval between indicating to the MAC board to switch its operating frequency and the time at which the latter confirms success and is hence ready for a new transmission of user data. Note that this duration also copes with the effects of the driver implementation and of possible process changes within the operating system itself.

Figure 10.18 depicts the empirical CDF of the APTTs. The samples are uniformly

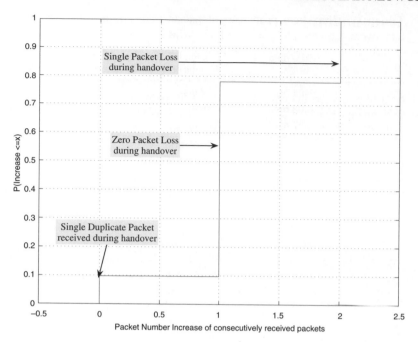

Figure 10.19 Packet loss during handover (Emmelmann et al. (2008) © ACM)

distributed from 1.88 ms up to 10.04 ms, which meets our expectations, since two neigh-boring macro cells are not synchronized regarding the transmission of MAC time slots. The theoretical lower bound is only missed by 0.18 ms, whereas the upper bound accounting for a worst case error propagation was set too conservatively. As the lower bound is determined by the time interval in which the mobile's radio is switched from one frequency to another (including processing overhead due to the firmware and operating system implementations), the mobile cannot receive or transmit packets during this time. Hence, we should experience a packet loss. We observe that the probability of losing exactly one packet is 22%, as shown in Figure 10.19. Notably, this is of the same order as the relation of the frequency switching time to the maximum observed APTT ($Min\{\mu_{APTT}\}/Max\{\mu_{APTT}\} = 19\%$). Also note that in 10% of the cases a single data packet is received twice, once via the originating cell before the handover and once via the designating cell afterwards. As expected, the probability to guarantee packet losses ≤ 1 is 100%.

Finally, we use the empirical results to calculate the handover delay according to (10.5):

$$HOD := E\{\mu_{APTT}\} - E\{\mu_{IAT}\}$$
$$= (5.84 - 5.37) \text{ ms} = 0.47 \text{ ms}$$

Even though the implementation of the proof-of-concept demonstrator did not employ optimized code for improved system performance, this delay is well below the requirements for real-time communication services for CBTC.

10.6   Conclusion

It can be seen from the theoretical limitations imposed by the standard that IEEE 802.11 can very well support fast handover for real-time telemetry applications such as CBTC. Even the last performance constraint of using a handshake between mobile and AP in order to establishing a link layer connection will very likely be removed in the near future by the 802.11p amendment. However, using 802.11 NICs as simple bridges between the DS and the wireless channel comes at a cost: all information required by the handover process has to be obtained without pausing any ongoing communication. Recent work in this field has shown that such information gathering without any noticeable loss of connectivity is possible for all handover phases – even at reasonable cost. The latter only increases if system requirements pile up to nearly 100% availability with highly mobile users facing low AP densities (resulting in low overlaps of the coverage area). The near future will also require advanced schemes enabling fast handover support for system deployments outside the traditional vehicular context. Upcoming WLANs will utilize higher frequency bands (e.g. 60 GHz), which will tremendously shrink cell sizes, resulting in high handover frequencies even for low pedestrian mobility. Since handover will occur very frequently in upcoming systems, research will need to explore cognitive handover schemes that dynamically combine different solutions for fast handover support in such a way that service interruption time will be just short enough to uphold QoS requirements at application level while minimizing associated costs (e.g. allocated resources on the wireless link used for bi-casting towards the mobile). As such a system would intelligently choose the set of handover schemes and algorithms, best handover performance would be achieved when application layer requirements were (just) satisfied at minimal cost and overhead. Such a cognitive, dynamic system has its attractions, since it would allow the deployment of a single system (e.g. a mesh network or roadside-unit deployment) that autonomously adapts its abilities and configuration to different needs. The latter can even change rapidly during the lifetime of a system. For example, a system offering lower handover guaranteed while supporting a larger number of users could reconfigure itself in an emergency situation to an Incident Area Network (IAN) enabling real-time communication for a lower number of military or first-aid workers.

References

Amir, Y., Danilov, C., Hilsdale, M., Musăloiu-Elefteri, R. and Rivera, N. (2006) Fast handoff for seamless wireless mesh networks. *MobiSys '06: Proceedings of the 4th International Conference on Mobile Systems, Applications and Services*, pp. 83–95. ACM, New York, NY, USA.

Bangolae, S., Wright, C., Trecker, C., Emmelmann, M. and Mlinarsky, F. (2005) Test methodology proposal for measuring fast bss/bss transition time. doc. 11-05/537, IEEE 802.11 TGt Wireless Performance Prediction Task Group, Vancouver, Canada. Substantive Standard Draft Text. Accepted into the IEEE P802.11.2 Draft Recommended Practice.

Carvalho, M. and Garcia-Luna-Aceves, J. (2003) Delay analysis of IEEE 802.11 in single-hop networks. *Proceedings of the 11th IEEE International Conference on Network Protocols*, pp. 146–155.

Chen, Y.S., Chuang, M.C. and Chen, C.K. (2008) Deucescan: Deuce-based fast handoff scheme in IEEE 802.11 wireless networks. *Vehicular Technology, IEEE Transactions on* **57**(2), 1126–1141.

Chittester, C. and Haines, Y. (2004) Risks of Terrorism to Information Technology and to Critical Interdependent Infrastructure. *Journal of Homeland Security and Emergency Management* **1**(4), Berkley Electronic Press.

Chui, S.K. and Yue, O.C. (2006) An access point coordination system for improved VOIP/WLAN handover performance. *Vehicular Technology Conference, 2006. VTC 2006-Spring. IEEE 63rd* **1**, 501–505.

Dutta, A., Das, S., Chiba, T., Yokota, H., Idoue, A., Wong, K.D. and Schulzrinne, H. (2006) Comparative analysis of network layer and application layer IP mobility protocols for IPv6 networks. *Wireless Personal Multimedia Communications (WPMC)*, pp. 6–10.

Emmelmann, M. (2005a) Influence of velocity on the handover delay associated with a radio-signal-measurement-based handover decision. *Vehicular Technology Conference (VTC) Fall 2005, Proc. of IEEE 62nd*, vol. 4, pp. 2282–2286 IEEE, Dallas, TX, USA. Extended version.

Emmelmann, M. (2005b) Influence of velocity on the handover delay associated with a radio-signal-measurement-based handover decision. Technical Report TKN-05-003, TU Berlin, Berlin, Germany.

Emmelmann, M. (2005c) Velocity effects on RSM-based handover decision. doc. 11-05/233, IEEE 802.11 TGt Wireless Performance Prediction, Plenary Session, Atlanta, GA, USA.

Emmelmann, M., Langgärtner, T. and Sonnemann, M. (2008) System design and implementation of seamless handover support enabling real-time telemetry applications for highly mobile users. *Proc. of the 6th ACM International Symposium on Mobility Management and Wireless Access (MobiWac 2008)*, pp. 1–8, Association for Computing Machinery (ACM).

Emmelmann, M., Rathke, B. and Wolisz, A. (2007a) *Mobile WiMAX: Toward Broadband Wireless Metropolitan Area Networks*. Auerbach Publications, CRC Press chapter Mobility Support for Wireless PAN, LAN, and MAN. ISBN: 0849326249.

Emmelmann, M., Wiethoelter, S., Koepsel, A., Kappler, C. and Wolisz, A. (2006) Moving towards seamless mobility: State of the art and emerging aspects in standardization bodies. *WPMC 2006*, San Diego, CA, USA. Invited Paper.

Emmelmann, M., Wiethoelter, S., Koepsel, A., Kappler, C. and Wolisz, A. (2007b) Moving towards seamless mobility – state of the art and emerging aspects in standardization bodies. *Springer's International Journal on Wireless Personal Communication – Special Issue on Seamless Handover in Next Generation Wireless/Mobile Networks*.

Emmelmann, M., Wiethölter, S. and Lim, H.T. (2009a) Influence of network load on the performance of opportunistic scanning. *IEEE Conference on Local Computer Networks (LCN)*, Zurich, Switzerland.

Emmelmann, M., Wiethölter, S. and Lim, H.T. (2009b) Opportunistic scanning: Interruption-free network topology discovery for wireless mesh networks. *International Symposium on a World of Wireless, Mobile and Multimedia Networks (IEEE WoWMoM)*, Kos, Greece.

Festag, A. (2003) *Mobility Support in IP Cellular Networks – A Multicast-Based Approach*. PhD thesis, Technical University Berlin.

Fettweis (Ed.), G. (2007a) Wireless gigabit with advanced multimedia support – system concept evaluation part 4: Medium access layer. Project report, Dresden University of Technology.

Fettweis (Ed.), G. (2007b) Wireless gigabit with advanced multimedia support – system concept final version part 1: Overview. Project report, Dresden University of Technology.

IEEE 1588 (2002) IEEE 1588 – IEEE standard for a precision clock synchronization protocol for networked measurement and control systems.

IEEE 802.11 (2007a) IEEE Standard for Information technology – Telecommunications and information exchange between systems – Local and metropolitan area networks – Specific requirements – Part 11: Wireless LAN Medium Access Control MAC and Physical Layer PHY Specifications. IEEE Std 802.11-2007 (Revision of IEEE Std 802.11-1999).

IEEE 802.11 (2007b) Wireless LAN Medium Access Control (MAC) and Physical Layer (PHY) Specifications.

IEEE 802.11.2 (2008) IEEE 802.11.2 – Recommended practice for the evaluation of 802.11 wireless performance.

IEEE 802.11e (2005) *Std 802.11 Information Technology Telecommunications And Information Exchange Between Systems, Local and Metropolitan Area Networks, Specific Requirements, Part 11: Wireless LAN Medium Access Control (MAC) and Physical Layer (PHY) Specifications*. IEEE Press chapter Amendment 8: Medium access control (MAC) quality of service (QoS) enhancements.

IEEE 802.11k-2008 (2009) IEEE 802.11k-2008 – radio resource measurement, draft amendment to standard for information technology – telecommunications and information exchange between systems – LAN/MAN specific requirements – Part 11: Wireless Medium Access Control (MAC) and physical layer (PHY) specifications.

IEEE 802.11p (2009) Wireless Access in Vehicular Environment, Draft Amendment to Standard for Information Technology – Telecommunications and Information Exchange Between Systems – LAN/MAN Specific Requirements – Part 11: Wireless Medium Access Control (MAC) and physical layer (PHY) specifications.

IEEE 802.1ak (2007) IEEE standard for local and metropolitan area networks – Virtual bridged local area networks – Amendment 7: Multiple registration protocol.

IHP Microelectronics (2008) Homepage of the institute for innovations for high performance microelectronics. http://www.ihp-microelectronics.com/.

Kannisto, J., Vanhatupa, T., Hannikainen, M. and Hamalainen, T. (2005) Software and hardware prototypes of the IEEE 1588 precision time protocol on wireless LAN. *The 14th IEEE Workshop on Local and Metropolitan Area Networks, LANMAN 2005*.

Kim, H.B., Emmelmann, M., Rathke, B. and Wolisz, A. (2005) A radio over fiber network architecture for road vehicle communication systems. *Proc. of IEEE Vehicular Technology Conference (VTC 2005 Spring)*, vol. 5, pp. 2920–2924, Vol. 5, Stockholm, Sweden.

Langgärtner, T.V.F. (2007) *Design of a software development environment for the IHP MAC prototyping development board*. Master's thesis, Technical University Berlin.

Lewis, T. (2006) *Critical infrastructure protection in homeland security: defending a networked nation*. John Wiley & Sons, Ltd.

Miller, L.E. (2009) Wireless Technologies and the SAFECOM SoR for Public Safety Communications. http://www.antd.nist.gov/wctg/manet/docs/WirelessAndSoR060206.pdf.

Mishra, A., Shin, M. and Arbaugh, W. (2003) An empirical analysis of the IEEE 802.11 MAC layer handoff process. *SIGCOMM Comput. Commun. Rev.* **33**(2), 93–102.

Mitola, J.I. (2000) SDR architecture refinement for JTRs. *Proc. 21st Century Military Communications MILCOM 2000*, vol. 1, pp. 214–218.

Nijkamp, P. (1995) From missing networks to interoperable networks : The need for European cooperation in the railway sector. *Transport Policy* **2**(3), 159–167.

Ramani, I. and Savage, S. (2005) Syncscan: practical fast handoff for 802.11 infrastructure networks. *INFOCOM 2005. 24th Annual Joint Conference of the IEEE Computer and Communications Societies. Proceedings IEEE* **1**, 675–684, vol. 1.

RSAC (1999) Report of the Railroad Safety Advisory Committee to the Federal Railroad Administrator, Implementation of Positive Train Control Systems. Federal Railroad Administrator, US Department of Transportation, Washington, DC.

Shin, M., Mishra, A. and Arbaugh, W.A. (2004) Improving the latency of 802.11 hand-offs using neighbor graphs. *MobiSys '04: Proceedings of the 2nd International Conference on Mobile Systems, Applications, and Services*, pp. 70–83. ACM, New York, NY, USA.

Singh, G., Atwal, A.P.S. and Sohi, B.S. (2006) An efficient neighbor information signaling method for handoff assistance in 802.11 wireless. *Mobility '06: Proceedings of the 3rd International Conference on Mobile Technology, Applications & Systems*, p. 14. ACM, New York, NY, USA.

Telelogic TAU (2008) Rational Tau: A standards-based, model-driven development solution for complex systems. www.telelogic.com/Tau.

Velayos, H. and Karlsson, G. (2004) Techniques to reduce the IEEE 802.11b handoff time. *IEEE International Conference on Communications* **7**, 3844–3848, Vol. 7.

Virtual Automation Networks Consortium (2006) Real time for embedded automation systems
    including status and analysis and closed loop real time control. Deliverable D04.1-1, EC Information
    Society Technology.

Zhang, N. and Holtzman, J. (1994) Analysis of handoff algorithms using both absolute and relative
    measurements. *Vehicular Technology Conference, 1994 IEEE 44th* **1**, 82–86, vol. 1.

Zhang, N. and Holtzmann, J. (1996) Analysis of handoff algorithms using both absolute and relative
    measurements. *IEEE Transactions on Vehicular Technology* **45**(11), 174–179.

11

New Technological Paradigms[1]

Bernd Bochow

Fraunhofer Institute for Open Communication Systems

Convergence of vehicular networks and general multi-purpose wireless networks can be assumed on a mid-term timescale: vehicular networks interoperate with general wireless networks or integrate with general wireless networks providing vehicular-specific network services. In this process the evolution of vehicular networks and general wireless networks will be linked more tightly than today, and it can be expected that some degree of cross-fertilization will occur. For example, new developments in wireless communications such as IMT-Advanced, upcoming Dynamic Spectrum Access (DSA) methods and TV whitespace access will find their way into vehicular networking, first as hybrid or overlay solutions and then in a convergence process. Conversely, dedicated applications and services can be expected to evolve from current vehicular networks: for example, location-based services that benefit from vehicular mobility, unifying personal and vehicular mobility from the perspective of the application.

This chapter addresses upcoming technological challenges in vehicular networks that are likely to demand more than an evolutionary improvement. Topics discussed here are mainly related to the evolution, integration and convergence of vehicular communications involving, for example, cognitive methods. After discussing the key aspects of vehicular network evolution, an overview of future challenges in the field is given. Because of the wide range of research topics contributing to future vehicular networks, challenges are roughly outlined and are not addressed too deeply, so as to keep within reasonable bounds. Instead, several references that underline the relevance of topics discussed in this chapter are given. Next, the motivation for applying new paradigms to address these challenges is presented, by focusing on specific upcoming topics in Vehicular Communication (VC). The chapter closes with an outlook on the role of vehicular networks in the future Internet.

---

[1]The views expressed are those of the author, and do not necessarily represent those of the Fraunhofer Institute for Open Communication Systems.

Vehicular Networking   Edited by Marc Emmelmann, Bernd Bochow, C. Christopher Kellum
© 2010 John Wiley & Sons, Ltd

11.1   Evolution and Convergence of Vehicular Networks

The evolution of wireless networks in general is driven by user demands for enhanced or new applications, requesting even more bandwidth, connectivity and availability of the network. Evolution of wireless networks thus is determined mainly by the balance of interests and capacities of user, network operator, service provider, device manufacturer and regulatory authority (regarding the availability of Radio Frequency (RF) spectrum).

For vehicular networks the situation is slightly different, since the main purpose of the network is in the scope of operations of the hosting transportation system. The reliable and dependable operation of the network here often coincides with stakeholders' responsibilities for safety of life and goods, clearly putting the emphasis on reliability rather than on technical novelty. Thus, stakeholders drive network development from their specific scope, perspective and needs, mostly considering the communication system as an integral part of the transportation system.

Consequently, the network, service and transportation system operator are often the same, taking also the roles of network user and authority if necessary. The communication system is thus specifically optimized to a given purpose (i.e. is defined in its features and capacities by the transportation system operator), resulting in dedicated and potentially divergent application-specific network developments for the various types of transportation systems. Limited interoperability with general purpose wireless networks is a likely outcome of this approach. For example, the lack of General Packet Radio Service (GPRS) support in the initial GSM-Rail (GSM-R) specification now limits parallel use of GPRS and GSM-R, in that GSM-R advanced features are disabled while terminals are in GPRS mode. Advancing GSM-R has been considered recently, since the coexistence of GPRS with GSM-R is needed to implement IP-based services in European Train Control System (ETCS) level 2 (Aitken et al. 2003; RailXperts 2009; UIC 2008).

Wherever feasible, VC systems now are increasingly opened up to support the sharing of resources with the vehicular network for passenger and third party communication demands, enabling, for example, on-board communications to connect to the Internet. This allows an improvement in the efficiency of network utilization and the sharing of infrastructure costs, but also aims to create additional business opportunities beyond the inherent socio-economic benefits of an Intelligent Transportation System (ITS) (cf. e.g. EUNET/SASI Consortium 2001; Jraiw 2005).

In contrast with general multi-purpose wireless networks that also support vehicular communications as a specialized service, vehicular networks provide dedicated security and dependability features by design. Opening up these networks to passenger communications further weakens the limited protection provided by utilizing a proprietary communication system and might even disclose unexpected vulnerabilities in the basic design. Hence, it is of major interest to the vehicular communication network operator that passenger communication support must not introduce uncontrollable security threats, and that it takes place non-intrusively. Passengers are considered as secondary users of a vehicular network, which clearly prioritizes its mission-critical purpose by, for example, strictly isolating safety-relevant from passenger communication functions. In addition, connectivity and enhanced interoperability with external networks in general adds a risk of also opening the vehicular network to malicious use or to attacks by adversaries on on-board or infrastructure communications, which could take place, for example, via the Internet.

On the other hand, supporting passenger communication potentially speeds up further improvement of the vehicular network by more closely connecting the evolution of vehicular networks with the technology improvements of general wireless networks driven by passengers' communication demands. For example, vehicular communication can immediately benefit from evolving end-to-end security enhancements developed as a countermeasure for arising security threats observed in general wireless networks. Upcoming DSA methods (see Akyildiz et al. 2006) potentially extending Vehicle-to-Infrastructure (V2I) communications wireless access into TV whitespaces (cf. the digital dividend, e.g. Ofcom 2009) can significantly increase resilience and dependability of vehicular networks but will be deployed in general purpose wireless communications first. IMT-Advanced (i.e. 3G beyond LTE, see 3G Americas 2009; UMTS Forum 2009) is already considering support of both high-speed mobility and TV whitespace operation in its current road map.

Enhancing wireless access methods in general wireless networks as a means of efficiently handling the ever increasing number of simultaneous wireless users (e.g. by context-based control of emitted RF power, or by utilizing agile directional antennas) can help to solve scalability and interference issues in dense wireless node population situations. For example, network robustness and dependability on the routing layer as well as for applications in high penetration urban Vehicle-to-Vehicle/Infrastructure (V2X) scenarios can be improved by means of spatial adaptation of radio coverage (see e.g. Eichler and Schroth 2007; Ho et al. 2008; Kosch et al. 2006 for a discussion of scalability focusing on the information dissemination problem in Vehicular Ad hoc Networks, or VANETs).

A convergence of vehicular networks and general purpose wireless networks thus can be expected to increase robustness and scalability of vehicular communications and potentially will also enable a convergence of applications. Future location-based applications here may benefit from the physical mobility of the network regarding the collection, distribution and presentation of information along the path of a vehicle. Car-to-Car (C2C) networks here are among the first that create a practical benefit from geographical addressing and geocast message transport as well as from keeping information distribution local, based on the relevance of information to a given situation of a vehicle (c.f. Rybicki et al. 2007; Yang and Recker 2005).

Due to increasing demands caused by upcoming applications, by a growing number of wireless users, by increasing bandwidth requirements and by the need to use scarce spectrum resources more efficiently, wireless communication is currently undergoing a change of paradigm. To achieve efficiency, flexibility, autonomicity and evolvability (in terms of power consumption and reduced radiated power, agility in spectrum access and protocol support, network self-configuration, self-management and self-learning), for example by applying cognitive methods, is becoming reasonably practical, especially for ad hoc topologies. It can be assumed that this will also be valid for VC, thus defining the challenges of next generation vehicular networks.

11.2 Future Challenges

11.2.1 Handling network growth

This issue is to some degree specific for commercially deployed public V2X communication (e.g. C2C VANETs) because a large number of wireless nodes here competes within a limited geographical area in sharing strictly limited (licensed) RF spectrum. Since the number of nodes must be seen in relation to the geographical area occupied as well as the radio coverage and communication bandwidth allocated to each node, even a small number of nodes in a given area might be able to saturate the allotted spectrum. Hence, this issue is closely related to resource management efficiency and thus may apply for any unmanaged, decentralized or ad hoc topology. Resource management can address the issue of scarce spectrum (see Section 11.2.2) but cannot solve all related issues.

Future vehicular networks will be faced with a huge number of communicating nodes. For example, VANETs will evolve in terms of numbers of communicating vehicles and roadside units, easily accumulating in dense urban scenarios to several hundred wireless nodes within the theoretical range of their short-range wireless communications (see e.g. Hu et al. 2009, 2008 for a capacity estimation of urban roads, to obtain a rough idea of the number of nodes involved in the scenarios to be considered here).

The following challenges arise specifically for dense scenarios:

- handling RF Line-of-Sight (LoS) obstruction due to other vehicles in the close vicinity;

- providing increased accuracy of positioning and time synchronization;

- enabling context-dependent optimization of message Round Trip Time (RTT).

11.2.2   Managing resources in ad hoc scenarios

Even more than for general purpose wireless networks, efficient resource management of vehicular networks is crucial for network operations. The demand to allocate communication resources within a given deadline – i.e. messages needing to be delivered by the network within a restricted time frame – characterizes industrial networks rather than present-day wireless networks. For cellular or infrastructure-based topologies this can be achieved by resource reservation or pre-allocation. Additionally, the network topology is quite well known, and the number of simultaneously active wireless users can be controlled and, eventually, restricted to some degree. Well understood algorithms exist here (cf. Stanczak et al. 2006).

For decentralized or ad hoc topologies, the number of concurrent users is usually unpredictable and resource management is distributed, depending on the detail of network state information locally available. Research on this area is at a rather early stage and algorithms considered until now are highly iterative (i.e. are based on genetic algorithms, particle swarm optimizations, or game theory – see for example Fang and Bensaou 2004; Kennedy et al. 1995; Xue et al. 2003). Most of these algorithms depend on observing the current resource utilization by other nodes in the geographical (e.g. for radio resources) and topological (e.g. for network resources) vicinity.

In wireless networks resource management efficiency is basically a matter of knowledge required to identify available resources in the temporal, spatial and spectral dimensions, as well as of the capacity to coordinate and collaborate between resource users. Additionally, knowledge needs to be available on time, which is challenging for highly dynamic ad hoc topologies. In general, the timeliness of knowledge is an issue for both centralized and decentralized topologies, but in centralized topologies effects are mitigated by a-priori knowledge or by increasing predictability (e.g. using already known time frames, as with TDMA and/or TDD schemes). In consequence, efficient resource management in dynamic topologies can be achieved only within a very limited geographical area (set by communication latency of the knowledge required) and for a time frame where topology and resource availability can be considered quasi-stationary.

Thus specific challenges arise for distributed resource management in dynamic topologies:

- gaining and distributing knowledge on topology and resource availability in temporal, spatial and spectral dimensions;

- efficient wireless node collaboration and cooperation in resource utilization.

11.2.3 Enabling interworking, integration and convergence

Vehicular networks benefit from a diversity of available wireless access methods, optimized for dedicated needs. For example, short-range communication allows for low-delay Vehicle-to-Vehicle (V2V) communication in ad hoc multi-hop configurations, given a sufficiently high density of vehicles situated in the vicinity. But short-range communication does not provide sufficient reliability for low-density, large-hop-count, high-speed or wide-range scenarios such as V2X in rural scenarios. Hybrid wireless access combining both short-range VANET and cellular V2X (see Aguado et al. 2008; Ashida and Frantti 2008; Costa et al. 2008) can relax the situation to a certain degree but potentially introduces additional communication delay, system complexity and infrastructure cost.

At the present time interworking is achieved based on the common use of IP protocols. Integration and convergence is challenging since it comprises the convergence of both infrastructure (including infrastructure services) and wireless access. While the latter can be achieved by utilizing multiple wireless interfaces or reconfigurable RF interfaces, switching between different wireless links in a way that is transparent to the application may be more demanding for resource-constrained and mission-critical VC.

Thus, integration and convergence is first of all an issue of efficiently managing multiple access methods in parallel to satisfy the demands and requirements set by relevance and constraints of information, vehicle and application context, and connectivity and dependability of communication links. In ad hoc multi-hop configurations a decision might be needed for each node to select a suitable communication path for the next hop autonomously, based on local policies (e.g. on spectrum use) and context (e.g on resource utilization by neighboring nodes) if required (see e.g. Bochow and Bechler 2005; Kutzner et al. 2003; Wu et al. 2008).

Thus specific challenges arise in enabling interworking, integration and convergence with vehicular networks:

- concurrently handling multi-channel and multi-interface access techniques;

- enabling context-based decision making on how to utilize multiple access techniques;

- providing transparency to applications regardless of communication path selected.

11.2.4 Providing integrated on-board and vicinity communications

Vehicles can be considered as mobile networks, providing wireless access to on-board passengers but also to wireless users situated not on board but in the vicinity of a vehicle. The provision of wireless access to in-car networks using vehicles as mobile Internet routers has been among the first applications of vehicular networks (Cisco Systems, Inc. 2009). Additionally, remote maintenance as a 'classic' V2X application may involve both remote and nearby clients for different purposes.

Although in-vehicle communication is often considered in the context of passenger information and entertainment applications, or in providing Internet access to passengers in public transportation, it is also crucial for certain safety-relevant applications such as eCall (European Commission 2009) or in Incident Area Network (IAN) applications (Delaney 2005; Refaei et al. 2008). The vehicle here is automatically providing context information (e.g. position and neighboring vehicles) and communication services without relying on passengers' support. Opening up a vehicle to vicinity communications allows a greater area to be served but adds security and coexistence issues as well as the demand to support additional wireless standards.

Access control and authorization functions to in-vehicle services and to the vehicular network are thus mandatory. Dependable management of vicinity communications seems necessary, since attackers may utilize the vicinity communications capacity of a vehicle to compromise the privacy-preserving features of a vehicular network (e.g. by tracking vehicles without participating in the network). This leads to the assumption that vicinity communications should be an 'on-demand' rather than an 'always-on' feature of a vehicle.

In contrast with wireless on-board communications, vicinity communication enlarges the geographical area covered, the number of users potentially communicating with the vehicle and the number of wireless standards to support. Precautions must be taken not to compromise the performance and dependability of the V2X wireless link and of other vital functions of the vehicle. It is mainly a platform issue to ensure non-intrusive operation, coexistence on the wireless medium and on the protocol level.

Thus specific challenges arise in providing integrated on-board and vicinity communications:

- increased flexibility in terms of standards and protocols supported;

- self-management or self-organization of vicinity communications;

- integration of coexistence and security features to avoid interfering with V2X communications.

11.3   New Paradigms

The main shift in paradigm that has been introduced into vehicular networks by VANETs is in the change from a centralized and well-managed infrastructure-based network towards open, self-organizing and flexible ad hoc configurations that may or may not rely on – or collaborate with – fixed infrastructures. Along with this paradigm shift a number of new research topics have been directed towards vehicular networks, addressing, for example, self-organization, collaboration, learning and cognition. This might be seen as another paradigm shift, since, from the networking perspective, vehicles are no longer merely wireless network terminals but are now learning, deciding and acting autonomously. In this they collect and share knowledge, acting as a wireless mobile sensor network as well.

Vehicular network evolution and growth will further increase the demand for collaborative schemes in distributed resource management and information acquisition and dissemination. Observation of neighboring vehicles, learning from their behavior and sharing information obtained are prerequisites for efficient collaborative interference mitigation, relevance estimation and prioritization of information to forward. The following discussion thus takes up some of the challenges identified in the previous section, and completes the discussion with respect to this paradigm shift.

11.3.1   RF line of sight obstruction due to other vehicles in close vicinity

Despite the short distance between receiver and transmitter in dense scenarios, LoS conditions are usually not matched easily since there are always vehicles nearby moving in and out of the Fresnel zone. Fading conditions are comparable to a bad multi-path fading situation and need attention at the RF receiver and antenna sections of the communication platform. Transmitter per-packet power control relieves the interference situation, but decreases the chance for indirect reception if the LoS is blocked. Proposals for optimizing the communication range by minimizing interference (e.g. Li et al. 2004) based on this approach do not yet consider inherent packet loss for short-distance links and assume that all nodes can receive the information as long as they are situated within the communication range. LoS obstruction is a physical channel property and can potentially be mitigated either by making use of smart antennas, by having antenna diversity or spatial diversity on the communication platform or by means of collaboration on the protocol level, detecting and notifying hidden nodes, for example, by utilizing multi-hop forwarding even on short distances.

11.3.2   Increased demand for accuracy of positioning and time synchronization

A significant number of collaborative applications (e.g. cooperative Collision Avoidance Systems, or CASs for active safety in VANETs) rely on the knowledge about a vehicle's position and the time interval between event and reaction (potentially applying to different vehicles). Accuracy required for the proper operation of these applications (e.g. for event ordering, relevance estimations and subsequent information distribution) depends on both vehicle speed and spatial distance between vehicles. Assuming for example a C2C VANET, Shladover and Tan (2006) report requirements of 0.5 m for positioning accuracy and 10 Hz for the position update rate. Given the need to utilize commercial-grade equipment and current (or soon available) Global Navigation Satellite Systems (GNSS), this demands the use of assisted GNSS such as Differential GPS (DGPS), or the application of communication-based collaborative positioning methods. Both approaches have their specific drawbacks (i.e. visibility of satellites for GNSS and bandwidth limitations for message-based collaborative positioning) and may need combining in dense scenarios to satisfy accuracy requirements (see e.g. Efatmaneshnik et al. (2009) for a discussion on channel capacity requirements for collaborative positioning).

11.3.3   Optimization of message round trip time

This is an issue closely related to the problem of increasing efficiency of medium access, based on the context of vehicle and application and the relevance of information to forward (see e.g. Delot et al. (2008) for an approach to estimating the relevance of information in VANET information distribution). In dense scenarios this is significant, since the number of vehicles in competition on the available channel bandwidth will cause fading, collisions and/or congestion, resulting either in an increased number of packet drops or message repetitions (regardless of whether they are associated with the communication system or the application).

For example, the figure given above for the minimum update rate of position information in collaborative C2C applications is valid only if the transmission time of a message is significantly shorter. That is, information forwarded from an issuing application might be outdated before it can reach the receiver, which might not be tolerated by some applications. Prioritization of messages on the wireless link is mandatory in this case (e.g. Torrent-Moreno et al. 2004), but will likely fail if neighboring vehicles do not collaborate in the scheme. Optimization of message RTT thus demands a context-based allocation and reservation of bandwidth resources to minimize access delay and transmission latency with respect to the relevance of information communicated.

11.3.4   Gaining and distributing knowledge on topology and resource availability in temporal, spatial and spectral dimensions

This comprises a node-local sensing and sounding strategy and a communication strategy enabling the exchange of context information acquired. Sensing (focused on observation) and sounding (focused on probing) includes radio channel estimation and radio scene analysis, as well as link layer and above traffic metering aspects. RF channel estimation for highly mobile environments such as VANETs (and hence channel equalizer design and Doppler compensation for the RF interface) is a complex issue on its own, given that channel parameters may quickly change within a time frame comparable to that used for message transmission. In the context of Multiple-Input Multiple-Output (MIMO) for high mobility applications such as high-speed train to wayside communication, the topic is addressed recently by Klenner et al. (2009); Peiker et al. (2009). Radio scene analysis (see Haykin 2005) extends the topic towards observing the complete radio neighborhood and can be directed to focus on RF interference mitigation (e.g. by means of beam- and null-steering of smart antennas) as well as towards the observation and estimation of resource utilization by neighboring nodes.

Back pressure algorithms used in opportunistic and congestion-based routing schemes for ad hoc multi-hop schemes (e.g Neely and Urgaonkar 2009; Scheuermann et al. 2007) can be seen as a way of probing the network state and adapting route and packet rate accordingly. Detailed traffic analysis and user modeling on the protocol layer (i.e. OSI link layer and above) can increase predictability of resource utilization for observing nodes (see e.g. Riihijarvi et al. 2009). Collaboration between nodes here is necessary since traffic analysis is also considered an attack scenario. Future protocols might be resilient to malicious traffic analysis (see e.g. Deng et al. 2005), potentially also hampering traffic analysis as a means to optimize resource utilization.

Distribution of context information is part of communication-based collaboration and is a cross-layer issue involving intra-layer exchange of sensing information as well as cross-layer optimization of context communication and context aggregation. This aims to reduce protocol overhead as much as possible, given that the amount of context information necessary to describe the network state is tremendous.

11.3.5   Efficient collaboration and cooperation in resource utilization

Communication-based collaboration is seen as an enabler to significantly increase efficiency of wireless communications in general, but is a complex issue for high mobility scenarios due to rapid changes in the network topology and in the radio scene. Applications are foreseen in mitigating collisions in wireless channel access (e.g. by collaborative transmission), opportunistic use of the channel and collaborative channel estimation. The latter is related to macro-MIMO and is relevant for ad hoc multi-hop configurations (see e.g. Shin and Yun 2008; Tila et al. 2003). Node collaboration here allows the selection of suitable neighboring vehicles for relaying to implement macro-MIMO.

11.4   Outlook: the Role of Vehicular Networks in the Future Internet

Vehicular networks evolved from their very beginnings in tethered low rate data and voice communications for command and control purposes to multi-purpose mobile networks. In mass-market applications such as public V2X communications large scale deployment is still not yet around the corner, but is foreseeable within a timescale of a few years ahead.

Vehicular networks are now emerging from a role as coexisting dedicated-purpose networks, towards becoming a part of the future Internet. More information about current research activities and a concise overview of potential future Internet architectures can be found in Paul et al. (2009). Hourcade et al. (2009), for example, emphasizing the Content Distribution Network (CDN) perspective of the future Internet, consider no special role for C2C networks and assume seamless integration on the application and service level. To realize this vision the research area of Disruption- and Delay Tolerant Networks (DTNs) as discussed in Paul et al. (2009) is of high relevance for the design of future vehicular networks and for evolving VANETs, respectively.

In the service-oriented perspective of the future Internet (see Chandrashekar and Zhang 2006) vehicular networks can take multiple roles. They may be considered as multiple service-oriented overlay networks simultaneously providing machine-to-machine communication services for improving safety and energy efficiency, wireless Internet access for in-vehicle users and applications, and access to sensor data, for example, from road-traffic, environmental or RF spectrum sensors. In the latter role vehicular networks also relate in some aspects to the Internet of things (e.g. seen as a dynamic collection of networked smart objects, cf. Orefice et al. 2010; Sekkas et al. 2010). Section 8.4.2 of this book provides a brief discussion of the role of Wireless Sensor Networks (WSNs) in the context of vehicular networks.

Isolation between these roles on the level of services and applications as well as on the network and network infrastructure level is becoming more and more crucial as VC evolves. That is, resilience and dependability of evolving vehicular networks is directly affected by sharing platforms, wireless links and infrastructure for cost reduction. This is even more valid if vehicular networks themselves collaborate with or go into symbiosis with other external networks (see Brownfield and Davis 2005; De Poorter et al. 2008).

Sophisticated methods are at hand to detect the malicious use of the wireless link and to ensure reliability, security and privacy on the network side. See, for example, Section 5.7.1 of this book for a brief discussion on detecting forged GNSS signals based on information previously obtained from the wireless receiver, and Section 5.5 for a discussion of active adversary attacks on the wireless link, for example utilizing relay attacks.

It should be noted here that despite adversary attacks posing a significant challenge to VC (as can be seen from the discussion in Section 6.5.1), the most important threat to the integrity of the wireless communication system is still in undetected defective equipment.

Nevertheless, the wireless access to the infrastructure and the wireless link itself will remain weak elements in dependable VC.

Recent developments evolving from the military domain use of Software Defined Radio (SDR) and Cognitive Radio (CR) (Fette 2006; JPEO JTRS 2009; Mitola 2000) have opened the way towards achieving increased robustness and resilience by providing a kind of redundancy on the wireless link supporting spectrum handover through DSA. Introduced as an energy- and spectrum-efficient method to increase communication bandwidth, coexistence level and radio coverage on the wireless access, DSA (see Akyildiz et al. 2006) also enables reaction to a malicious interferer by simply evading the affected frequency bands. Additionally, collaborative RF sensing (cf. Noguet et al. 2009) allows the interferer to be located and network-level countermeasures to be taken if network integrity is affected.

The ongoing discussion within the future Internet research community considers the current IP, UDP and TCP protocols more and more as a bottleneck, and not sufficiently flexible to allow the plethora of existing and upcoming applications to utilize the capacity of the underlying communication link efficiently. The evolution of VANETs so far has clearly shown some of these shortcomings, especially in the context of highly dynamic C2C communications (Chandran et al. 2001; Holland and Vaidya 2002; Lim et al. 2003). As with proposals discussed by the future Internet community, vehicular networks will be among the first that implement optimized protocols on the OSI network and transport layer that coexist with IP, UDP and TCP. Sections 3.8 and 8.4 of this book discuss the coexistence of IP-based protocols in VC and the Wireless Access in Vehicular Environments (WAVE) protocol stack from different perspectives.

These concepts have not been developed with seamless integration in mind, but rather aim for isolation between safety-relevant and user communications. Future applications, for example demanding the monitoring of safety-relevant V2X data traffic in the vicinity of a vehicle, may be difficult to realize if this isolation cannot be weakened in a well-controlled manner. Chapters 1 and 2 of this book focus on applications and use case scenarios, providing a concise overview and classification of current and potentially upcoming applications for vehicular networks in different areas. The methodology applied for classification of applications presented in Chapter 1 of the book can be extremely helpful when aiming to define the requirements for application-driven integration of vehicular networks into the future Internet.

The evolution of vehicular networks fostered a new perspective on communication-based collaboration between vehicles. Research activities such as Kognimobil (2009) started efforts to utilize V2X communication for collaboratively sensing a vehicle's vicinity, distributing knowledge acquired in this process and acting autonomously based on this knowledge. The concept of this long-term research activity is very demanding, and the approach goes far beyond cognitive networking in its scope.

Although the suitability of autonomicity and cognition for countering the ever increasing complexity of networks is not undisputed (Stavrakakis and Panagakis 2006), it is commonly agreed that these will be key enablers for the future Internet and will also be necessary components of future vehicular networks. Research and development here is still at an early stage, and the potentials are not yet fully explored. For now it seems to be clear that the various self-x features, ranging from self-organization through self-healing to self-management and self-learning, just to name a few, will be distributed over the network as well as over the whole protocol stack (a cross-layer/cross-network approach).

The coordination of self-x features in such a vehicular network will be challenging to avoid conflicting optimizations that potentially can drive – in a very practical sense – a network down. Collaboration and cooperation between cognitive functions local to network nodes (i.e. vertically distributed) and between network nodes (i.e. horizontally distributed) becomes crucial. Since collaboration is usually accompanied by significant control and management protocol overhead, cognitive vehicular networks may thus suffer from scarce resources (for example, communication bandwidth), from transmission delay or from communication losses more severely than is observed within other domains of the wired or wireless future Internet.

In conclusion, the discussion above intends to raise awareness that vehicular networks unquestionably will be an inherent part of the future Internet, despite the fact that a large number of technological challenges are still unresolved or even not yet approached. Vehicular networks will thus for a long time remain dedicated rather than general purpose networks. This book may be able to provide a broader view of the topic and potentially assist in recognizing connected areas of technology and application, as well as potentials and synergies from the various facets of vehicular communications discussed, and beyond.

References

3G Americas (2009) The Mobile Broadband Evolution: 3GPP Release 8 and Beyond. http://3gamericas. org/documents/3GPP_Rel-8_Beyond_02_12_09.pdf.

Aguado, M., Matias, J., Jacob, E. and Berbineau, M. (2008) The WIMAX ASN Network in the V2I scenario. *IEEE VTC 2008-Fall*.

Aitken, J., Lehrbaum, M. and Owen, G. (2003) GSM-R, Advanced, Approved, Available and Applicable. http://www.jja.com.au/index.php/technical-papers-/21-gsm-r-advanced-approved-available-and-applicable.

Akyildiz, I., Lee, W., Vuran, M. and Mohanty, S. (2006) NeXt generation/dynamic spectrum access/cognitive radio wireless networks: a survey. *Computer Networks* **50**(13), 2127–2159.

Ashida, M. and Frantti, T. (2008) *System Architecture for C2C Communications Based on Mobile WiMAX*, pp. 558–567.

Bochow, B. and Bechler, M. (2005) Internet Integration. In *Inter-Vehicle-Communications Based on Ad Hoc Networking Principles – The FleetNet Project* (ed. Franz, W., Hartenstein, H. and Mauve, M.) Universitätsverlag Karlsruhe, pp. 175–211. ISBN 3-937300-88-0.

Brownfield, M.I. and Davis, N.J. (2005) Symbiotic highway sensor network. *Proc. VTC-2005-Fall Vehicular Technology Conference 2005 IEEE 62nd*, vol. 4, pp. 2701–2705.

Chandran, K., Raghunathan, S., Venkatesan, S. and Prakash, R. (2001) A feedback-based scheme for improving TCP performance in ad hoc wireless networks. *IEEE Personal Communications* **8**(1), 34–39.

Chandrashekar, J. and Zhang, Z. (2006) Towards a Service Oriented Internet. *IEICE Transactions on Communications* **89**(9), 2292–2299.

Cisco Systems, Inc. (2009) Cisco Mobile Network Solutions for Public Safety. http://www.cisco.com/en/US/prod/collateral/routers/ps272/prod_white_paper0900aecd806220af_ps6591_Products_White_Paper.html.

Costa, A., Pedreiras, P., Fonseca, J., Matos, J., Proenca, H., Gomes, A. and Gomes, J. (2008) Evaluating WiMax for vehicular Communication Applications. *IEEE Conference on Emerging Technologies and Factory Automation*, pp. 1185–1188.

De Poorter, E., Latré, B., Moerman, I. and Demeester, P. (2008) Symbiotic networks: Towards a new level of cooperation between wireless networks. *Wireless Personal Communications* **45**(4), 479–495.

Delaney, W.J. (2005) (wo/2005/119972) mobile temporary incident area network for local communications interoperability.

Delot, T., Cenerario, N. and Ilarri, S. (2008) Estimating the relevance of information in inter-vehicle ad hoc networks. *IEEE Int. Conf. on Mobile Data Management–Workshops*.

Deng, J., Han, R. and Mishra, S. (2005) Countermeasures against traffic analysis attacks in wireless sensor networks. *Proceedings of the First International Conference on Security and Privacy for Emerging Areas in Communications Networks*, pp. 113–126 Citeseer.

Efatmaneshnik, M., Balaei, A. and Dempster, A. (2009) A Channel Capacity Perspective on Cooperative Positioning Algorithms for VANET. *ION GNSS 2009*.

Eichler, S. and Schroth, C. (2007) A Multi-Layer Approach for Improving the Scalability of Vehicular Ad-Hoc Networks. *Kommunikation in Verteilten Systemen – KiVS 2007*.

EUNET/SASI Consortium (2001) Eunet socio-economic and spatial impacts of transport. Final Project Report ST-96-SC037, European Union.

European Commission (2009) eCall – saving lives through in-vehicle communication technology. http://ec.europa.eu/information_society/doc/factsheets/049_eCall_august 09_en.pdf.

Fang, Z. and Bensaou, B. (2004) Fair bandwidth sharing algorithms based on game theory frameworks for wireless ad hoc networks. *IEEE INFOCOM*, vol. 2, pp. 1284–1295 Citeseer.

Fette, B. (2006) *Cognitive Radio Technology*. Newnes. (Communications Engineering Series).

Haykin, S. (2005) Cognitive radio: brain-empowered wireless communications. *IEEE Journal on Selected Areas in Communications* **23**(2), 201–220.

Ho, Y.H., Ho, A.H. and Hua, K.A. (2008) *Routing Protocols for Inter-vehicular Networks: A Comparative Study in High-mobility and Large Obstacles Environments*, vol. 31. Elsevier. Mobility Protocols for ITS/VANET.

Holland, G. and Vaidya, N. (2002) Analysis of TCP performance over mobile ad hoc networks. *Wireless Networks* **8**(2), 275–288.

Hourcade, J.C., Saracco, R., Neuvo, Y., Wahlster, W., Posch, R. and Sharpe, M. (2009) Future internet 2020 – visions of an industry expert group. Technical report, European Commission, DG Information Society and Media Directorate for Converged Networks and Service.

Hu, M., Jiang, R., Wang, R. and Wu, Q. (2009) Urban traffic simulated from the dual representation: Flow, crisis and congestion. *Physics Letters A* **373**(23–24), 2007–2011.

Hu, M.B., Jiang, R., Wu, Y.H., Wang, W.X. and Wu, Q.S. (2008) Urban traffic from the perspective of dual graph. *The European Physical Journal B - Condensed Matter and Complex Systems* **63**(1), 127–133.

JPEO JTRS (2009) Joint tactical radio system website. http://jpeojtrs.mil/.

Jraiw, K. (2005) Socioeconomic impact of sustainable road transport system – case study. Technical report, Transport Division (EATC), East Asia Department, Asian Development Bank (ADB), Philippines.

Kennedy, J., Eberhart, R. *et al.* (1995) Particle swarm optimization. *Proceedings of IEEE International Conference on Neural Networks*, vol. 4, pp. 1942–1948, Piscataway, NJ: IEEE.

Klenner, P., Reichardt, L., Kammeyer, K. and Zwick, T. (2009) MIMO–OFDM with Doppler Compensating Antennas in Rapidly Fading Channels. *Multi-Carrier Systems and Solutions 2009: Proceedings from the 7th International Workshop on Multi-Carrier Spread Spectrum, May 2009, Herrsching, Germany*, p. 69, Springer.

Kognimobil (2009) Transregional collaborative research center 28 – cognitive automobiles. http://www.kognimobil.org/index.php.

Kosch, T., Adler, C.J., Eichler, S., Schroth, C. and Strassberger, M. (2006) The scalability problem of vehicular ad hoc networks and how to solve it. *IEEE Wirel Commun* **13**(5), 22–28.

Kutzner, K., Tchouto, J., Bechler, M., Wolf, L., Bochow, B. and Luckenbach, T. (2003) Connecting vehicle scatternets by Internet-connected gateways. *MMC'2003*.

Li, X., Nguyen, T. and Martin, R. (2004) An analytic model predicting the optimal range for maximizing 1-hop broadcast coverage in dense wireless networks. *Lecture Notes in Computer Science*, pp. 172–182.

Lim, H., Xu, K. and Gerla, M. (2003) TCP performance over multipath routing in mobile ad hoc networks. *Proc. of IEEE ICC, Anchorage, Alaska.*

Mitola, J.I. (2000) SDR architecture refinement for JTRS. *Proc. 21st Century Military Communications MILCOM 2000*, vol. 1, pp. 214–218.

Neely, M. and Urgaonkar, R. (2009) Optimal backpressure routing for wireless networks with multi-receiver diversity. *Ad Hoc Networks* **7**(5), 862–881.

Noguet, D., Demessie, Y.A., Biard, L., Bouzegzi, A., Debbah, M., Haghighi, K., Jallon, P., Laugeois, M., Marques, P., Murroni, M., Palicot, J., Sun, C., Thilakawardana, S. and Yamaguchi, A. (2009) Sensing techniques for cognitive radio – state of the art and trends. White paper, IEEE SCC41 – P1900.6 Working Group.

Ofcom (2009) Digital dividend: cognitive access – Statement on licence-exempting cognitive devices using interleaved spectrum. http://www.ofcom.org.uk/consult/condocs/cognitive/statement/statement.pdf.

Orefice, P., Paura, L. and Scarpiello, A. (2010) Inter-vehicle communication qos management for disaster recovery. *20th Tyrrhenian International Workshop on Digital Communications.*

Paul, S., Pan, J. and Jain, R. (2009) Architectures for the future networks and the next generation internet: A survey. Wustl technical report, wucse-2009-69.

Peiker, E., Teich, W. and Lindner, J. (2009) Windowing in the Receiver for OFDM Systems in High-Mobility Scenarios. *Multi-Carrier Systems and Solutions 2009: Proceedings from the 7th International Workshop on Multi-Carrier Spread Spectrum, May 2009, Herrsching, Germany*, p. 57, Springer.

RailXperts (2009) ETCS Technology. http://www.etcs.eu/en/zielsetzungen.htm.

Refaei, M.T., Souryal, M.R. and Moayeri, N. (2008) Interference avoidance in rapidly deployed wireless ad hoc incident area networks. *Proc. INFOCOM Computer Communications Workshops IEEE Conference on*, pp. 1–6.

Riihijarvi, J., Wellens, M. and Mahonen, P. (2009) Measuring complexity and predictability in networks with multiscale entropy analysis. *Proc. INFOCOM 2009. The 28th Conference on Computer Communications. IEEE*, pp. 1107–1115.

Rybicki, J., Scheuermann, B., Kiess, W., Lochert, C., Fallahi, P. and Mauve, M. (2007) Challenge: peers on wheels – a road to new traffic information systems. *MobiCom '07: Proceedings of the 13th Annual ACM International Conference on Mobile Computing and Networking*, pp. 215–221. ACM, New York, NY, USA.

Scheuermann, B., Transier, M., Lochert, C., Mauve, M. and Effelsberg, W. (2007) Backpressure multicast congestion control in mobile ad-hoc networks. *CoNEXT*, p. 23.

Sekkas, O., Piguet, D., Anagnostopoulos, C., Kotsakos, D., Alyfantis, G., Kassapoglou-Faist, C. and Hadjiethymiades, S. (2010) Probabilistic Information Dissemination for MANETs: the IPAC Approach. *20th Tyrrhenian International Workshop on Digital Communications.*

Shin, O. and Yun, S. (2008) Apparatus and method for operating relay link in relay broadband wireless communication system.

Shladover, S.E. and Tan, S. (2006) Analysis of Vehicle Positioning Accuracy Requirements for Communication-Based Cooperative Collision Warning. *Journal of Intelligent Transportation Systems* **10**(3), 131–140.

Stanczak, S., Wiczanowski, M. and Boche, H. (2006) Resource Allocation in Wireless Networks-Theory and Algorithms. *Lecture Notes in Computer Science.*

Stavrakakis, I. and Panagakis, A. (2006) Panel 1 Report: Autonomicity Versus Complexity. *Autonomic Communication, Second International IFIP Workshop, WAC 2005, Athens, Greece, October 2–5, 2005, Revised Selected Papers*, pp. 286–292, Lecture Notes in Computer Science. Springer, Berlin/Heidelberg.

Tila, F., Shepherd, P. and Pennock, S. (2003) Theoretic capacity evaluation of indoor micro-and macro-MIMO systems at 5 GHz using site specific ray tracing. *Electronics Letters* **39**, 471.

Torrent-Moreno, M., Jiang, D. and Hartenstein, H. (2004) Broadcast reception rates and effects of priority access in 802.11-based vehicular ad-hoc network. *Proceedings of the 1st ACM International Workshop on Vehicular Ad Hoc Networks (VANET 2004)*, Philadelphia, PA, USA.

UIC (2008) GPRS for ETCS test cases. http://uic.asso.fr/IMG/zip/GPRSforETCS-3.zip.

UMTS Forum (2009) Mobile Broadband Evolution: the roadmap from HSPA to LTE. http://www.umts-forum.org/component/option,com_docman/task,doc_download/gid,2089/Itemid,12/. UMTS Forum White Paper.

Wu, M., tao Yang, L., yi Li, C. and Jiang, H. 2008 Capacity, collision and interference of VANET with IEEE 802.11 MAC. *International Workshop on Intelligent Networks and Intelligent Systems* **0**, 251–254.

Xue, Y., Li, B. and Nahrstedt, K. (2003) Price-based resource allocation in wireless ad hoc networks. *Lecture Notes in Computer Science* pp. 79–96.

Yang, X. and Recker, W. (2005) Simulation studies of information propagation in a self-organizing distributed traffic information system. *Transportation Research Part C: Emerging Technologies* **13**(5–6), 370–390.

Further Reading

Socio-economic aspects of vehicular networking

Bekiaris, E. and Nakanishi, Y. (2004) *Economic impacts of intelligent transportation systems: innovations and case studies*. JAI Press.

Commission of the European Communities (2006) Communication from the Commission to the Council, the European Parliament, the European Economic And Social Committee and the Committee of the Regions on the Intelligent Car Initiative "Raising Awareness of ICT for Smarter, Safer and Cleaner Vehicles".
http://eur-lex.europa.eu/LexUriServ/LexUriServ.do?uri=COM:2006:0059:FIN:EN:PDF.

eIMPACT (2008) Assessing the impacts of intelligent vehicle safety systems. http://www.eimpact.info/.

Jraiw, K. (2005) Socioeconomic impact of sustainable road transport system – case study. Technical report, Transport Division (EATC), East Asia Department, Asian Development Bank (ADB), Philippines.

The EUNET/SASI Consortium (2001) EUNET socio-economic and spatial impacts of transport. Final Project Report ST-96-SC037, European Union.

Wenger, J. (2005) Business Models for Vehicle Infrastructure Integration (VII). The Fully Networked Car. *A Workshop on ICT in Vehicles. International Telecommunications Union, Geneva*.

Vehicular networking applications and requirements

Atlas Elektronik GmbH Press Release (2009) ATLAS announce a quantum leap in maritime communications technology. http://www.maritime-index.com/pressdet.php?id=2837.

Barth, M., Xue, L., Chen, Y. and Todd, M. (2002) A hybrid communication architecture for intelligent shared vehicle systems. *Intelligent Vehicle Symposium, 2002. IEEE*, vol. 2, pp. 557–563, vol. 2.

Desourdis, R., Smith, D.R. and d. Speighs, W. (2001) *Emerging Public Safety Wireless Communication Systems*. Artech House Inc.

Garrun, D. (2009) New Hybrid Technology Boosts High-Speed Naval Networks.
http://www.naval-technology.com/news/news50752.html.

Hartenstein, H. and Laberteaux, K. (2008) A tutorial survey on vehicular ad hoc networks. *Communications Magazine, IEEE* **46**(6), 164–171.

Hartenstein, H. and Laberteaux, K. (2010) *VANET: Vehicular Applications and Inter-Networking Technologies*. Intelligent Transportation Series. John Wiley & Sons, Ltd.

Homeland Security, U.S. Dept. of (2009) SAFECOM Homepage.
http://www.safecomprogram.gov/SAFECOM.

Nijkamp, P. (1995) From missing networks to interoperable networks: The need for European cooperation in the railway sector. *Transport Policy* **2**(3), 159–167.

Stacey, D. (2008) *Aeronautical Radio Communication Systems and Networks*. John Wiley & Sons, Ltd. ISBN: 978-0-470-01859-0.

U.S. Department of Justice (2001) A guide for applying information technology in law enforcement. Technical report, U.S. Dept. of Justice.

Zhou, M.T., Harada, H., Kong, P.Y., Ang, C.W., Ge, Y. and Pathmasuntharam, J.S. (2009) Multi-channel transmission with efficient delivery of routing information in maritime WiMAX mesh networks. *IWCMC '09: Proceedings of the 2009 International Conference on Wireless Communications and Mobile Computing*, pp. 426–430. ACM, New York, NY, USA.

Attributes and limitations of today's vehicular networking

Ho, Y.H., Ho, A.H. and Hua, K.A. (2008) *Routing protocols for inter-vehicular networks: A comparative study in high-mobility and large obstacles environments*, vol. 31. Elsevier. Mobility Protocols for ITS/VANET.

Li, F. and Wang, Y. (2007) Routing in vehicular ad hoc networks: A survey. *Vehicular Technology Magazine, IEEE* **2**(2), 12–22.

Olariu, S. and Weigle, M. (2009) *Vehicular Networks from Theory to Practice*. Chapman & Hall.

Interoperability, standardization, and cooperation aspects of vehicular networks

Chen, J.C. and Zhang, T. (2004) *IP-Based Next-Generation Wireless Networks: Systems, Architectures, and Protocols*. John Wiley & Sons, Ltd. ISBN: 978-0-471-23526-2.

IEEE Std 1474.1 (1999) IEEE Standard for Communications-Based Train Control (CBTC) performance and functional requirements.

IEEE Std 1474.3 (2008) IEEE Recommended Practice for Communications-Based Train Control (CBTC) System Design and Functional Allocations.

ITU-R (2006) *Land Mobile Handbook (including Wireless Access) – Volume 4: Intelligent Transport Systems – Edition 2006*. ITU-R.

Lin, X., Lu, R., Zhang, C., Zhu, H., Ho, P.H. and Shen, X. (2008) Security in vehicular ad hoc networks. *Communications Magazine, IEEE* **46**(4), 88–95.

Sinha, N., Bera, R. and Mitra, M. (2009) Hybrid Technology Providing Concurrent Vehicular Safety and Communication. *Progress In Electromagnetics Research C* **6**, 53–65.

Implementation aspects of vehicular networks

Andrews, S. and Cops, M. (2009) Vehicle infrastructure integration proof of concept technical description – vehicle : final report. Technical report, United States. Dept. of Transportation. Research and Innovative Technology Administration. FHWA-JPO-09-043.

Harmon, T., Marca, J., Martini, P. and Klefstad, R. (2008) Design, implementation and test of a wireless peer-to-peer network for roadway incident exchange. *International Journal of Vehicle Information and Communication Systems* **1**(3), 288–305.

Kandarpa, R. (2009) Vehicle Infrastructure Integration proof-of-concept technical description – infrastructure: final report. Technical report, United States. Dept. of Transportation. Research and Innovative Technology Administration. FHWA-JPO-09-048.

Lewis, T. (2006) *Critical infrastructure protection in homeland security: defending a networked nation*. John Wiley & Sons, Ltd.

Miller, R. and Huang, Q. (2002) An adaptive peer-to-peer collision warning system. *IEEE Vehicular Technology Conference*, vol. 1, pp. 317–321, Citeseer.

VII Consortium (2009) Vehicle Infrastructure Integration – Proof of Concept – Results and Findings – Summary – Vehicle. http://ntl.bts.gov/lib/31000/31100/31135/14477_files/14477.pdf.

Xu, Y., li Yang, Q. and hua You, Z. (2009) Design of CAN/GPRS gateway for vehicle remote communications. *Proc. International Conference on Measuring Technology and Mechatronics Automation ICMTMA '09*, vol. 1, pp. 131–133.

Acronyms and Abbreviations

3G	Third Generation
3GPP	Third Generation Partnership Project
4CIF	$4 \times$ CIF (*four times the size of a CIF image*)
8-PSK	8 level PSK (*modulation with 3 bits per symbol*)
AAR	Association of American Railroads
ACC	Adaptive Cruise Control
ACELA	Amtrak's Express Train Service
ACK	Acknowledge
ACSES	Advanced Civil Speed Enforcement System
ADA	Advanced Driver Assistance
AF	Audio Frequency
Amtrak	National Passenger Rail Corporation
AODV	Ad hoc On Demand Distance Vector Routing
AP	Access Point
API	Applications Programming Interface
APSC TELEMOV	Advisory Panel for Standards Cooperation on Telecommunications related to Motor Vehicles
APTT	Access Point Transition Time
AR	Access Router
ARIB	Association of Radio Industries and Businesses
ARP	Address Resolution Protocol
ARQ	Automatic Repeat Request
ASL	Application Sub Layer
ASN	Access Service Network
ASTC	Automatic Self-Time-Correcting
ASTM	American Society for Testing and Materials
ATC	Automatic Train Control
ATP	Automatic Train Protection
ATS	Automatic Train Stop
BA	Binding Acknowledgment
BID	Binding Identifier
BNSF	Burlington Northern Santa Fe
BoF	Birds of a Feather

Vehicular Networking   Edited by Marc Emmelmann, Bernd Bochow, C. Christopher Kellum
© 2010 John Wiley & Sons, Ltd

BPA	Baseline Pseudonymous Authentication
BPSK	Binary Phase-Shift Keying
BS	Base Station
BSM	Basic Safety Message
BSS	Basic Service Set
BU	Binding Update
C2C	Car-to-Car
C2C-CC	CAR-2-CAR Communication Consortium
CA	Certification Authority
CAB Radio	In-train GSM-R Mobile Station
CACC	Cooperative Adaptive Cruise Control
CAD	Computer Aided Dispatch
CALM	Communications Access for Land Mobiles
CAN	Controller Area Network
CAS	Collision Avoidance System
CBTC	Communications-Based Train Control
CBTM	Communications-Based Train Management
CC	Cruise Control
CCH	Control Channel
CCTV	Closed Circuit TV
CCW	Cooperative Collision Warning
CDF	Cumulative Distribution Function
CDMA	Code Division Multiple Access
CDN	Content Distribution Network
CEN	European Committee for Standardization
CENELEC	European Committee for Electrotechnical Standardization
CEPT	European Conference of Postal and Telecommunications Administrations
CFR	Code of Federal Regulation
CM	Configuration Management
CMDD	Content, Map or Database Download
CN	Correspondent Node
CO	Colorado
CoA	Care-of Address
COMeSafety	Communications for eSafety
COOPERS	Cooperative Systems for Intelligent Road Safety
CR	Correspondent Router (*in IP mobility*)
CR	Cognitive Radio (*in advanced wireless communications*)
CRC	Cyclic Redundancy Code
CRL	Certificate Revocation List
CRN	Congested Road Notification
CSI	Channel State Information
CSIBS	Channel-Switching-Induced Broadcast Synchronization
CSMA	Carrier Sense Multiple Access
CSMA/CA	Carrier Sense Multiple Access Collision Avoidance
CSXT	CSX Transportation

CTC	Centralized Traffic Control
CVIS	Cooperative Vehicle–Infrastructure Systems
CVW	Cooperative Violation Warning
CW	Continuous Wave
DHAAD	Dynamic Home Agent Address Discovery
DAC	Discretionary Access Control
DC	District of Columbia
DCF	Distributed Coordination Function
DCH	Dedicated Channel
DCS	Data Communication System
DCS	Digital Control Systems (*in railway traffic control*)
DDOS	Distributed Denial of Service
DGPS	Differential GPS
DHCP	Dynamic Host Configuration Protocol
DIC	Driver Information Center
DiffServ	Differentiated Services
DIFS	Distributed Interframe Space
DL	Downlink
DMCC	DSRC Multi-Channel Coordination
DOD	United States Department of Defense
DOS	Denial of Service
DOT	United States Department of Transportation
DS	Distribution System
DSA	Dynamic Spectrum Access
DSP	Digital Signal Processor
DSL	Digital Subscriber Line
DSR	Dynamic Source Routing
DSRC	Dedicated Short Range Communications
DSSS	Direct Sequence Spread Spectrum
DTC	Direct Traffic Control
DTN	Disruption- and Delay Tolerant Network
DTV	Digital Television
EAN	Extended Area Network
EC-DSA	Elliptic Curve Digital Signature Algorithm
EDGE	Enhanced Data Rates for GSM Evolution
EEBL	Emergency Electronic Brake Light
EFC	Electronic Fee Collection
EIRENE	European Integrated Railway Radio Enhanced Network
EIRP	Equivalent Isotropic Radiated Power
EMS	Emergency Medical Service
ERM	Electro-Magnetic Compatibility and Radio Spectrum Matters
ERTMS	European Rail Traffic Management System
ETC	Electronic Toll Collection
ETCS	European Train Control System
ETMS	Electronic Train Management System
ETSI	European Telecommunications Standards Institute

EUL	Enhanced Up Link
FC	Foreigner Certificate
FCC	Federal Communications Commission
FDD	Frequency-Division Duplexing
FIPS	Federal Information Processing Standards
FPGA	Field-Programmable Gate Array
FR	Fixed Router *in a moving network*
FRA	Federal Railroad Administration
FTP	File Transfer Protocol
GAO	Government Accountability Office/Government Accounting Office
GARP	Generic Attribute Registration Protocol
GCOR	General Code of Operating Rules
GI	Guard Interval
GMRP	Multicast Registration Protocol
GMSK	Gaussian Minimum Shift Keying
GNSS	Global Navigation Satellite Systems
GPRS	General Packet Radio Service
GPS	Global Positioning System
GPSR	Greedy Perimeter Stateless Routing
GS	Group Signature
GSM	Global System for Mobile Communication
GSM-R	GSM-Rail
HA	Home Agent
HAZMAT	Hazardous Materials
HDLC	High-Level Data Link Control
HGCW	Highway Grade Crossing Warning System
HMI	Human–Machine Interface
HMIPv6	Hierarchical Mobile IPv6
HoA	Home Address
HPA	Hybrid Pseudonymous Authentication
HR/DSSS	High Rate Direct Sequence Spread Spectrum
HSDPA	High Speed Downlink Packet Access
HSM	Hardware Security Module
HSUPA	High Speed Uplink Packet Access
HTTP	Hyper Text Transfer Protocol
IAGO	Informatisation et Automatisation par Guide d'Onde (Waveguide transmission system for computer and automation applications)
IAN	Incident Area Network
IAT	Inter-Arrival Time
IATF	Information Assurance Technical Framework
IATFF	Information Assurance Technical Framework Forum
ICT	Information and Communications Technology
IEC	International Electrotechnical Commission
IEEE	Institute of Electrical and Electronics Engineers
IETF	Internet Engineering Task Force

IGMP	Internet Group Management Protocol
IL	Illinois
IMT	International Mobile Telecommunications
IP	Internet Protocol
IPsec	Internet Protocol Security
IPv4	Internet Protocol Version 4
IPv6	Internet Protocol Version 6
IRSA	Integrated Full-Range Speed Assistant
ISM	Industrial, Scientific, and Medical
ISO	International Organization for Standardization (Organisation internationale de normalisation)
ISP	Internet Service Provider
ITCS	Incremental Train Control System
ITS	Intelligent Transportation System
ITS-RS	Intelligent Transportation System Radio Service
ITU	International Telecommunication Union
ITU-R	International Telecommunication Union – Radiocommunication Sector
ITU-T	International Telecommunication Union – Telecommunication Standardization Sector
JAN	Jurisdictional Area Network
KVB	Controle de Vitesse par Balise (Transponder/beacon (balise) based speed control system used on French trains)
LCD	Liquid Crystal Display
LED	Light Emitting Diode
LFN	Local Fixed Node
LMR	Land Mobile Radio
LoS	Line-of-Sight
LSAP	lower layer Service Access Point
LTE	Long-Term Evolution
MA	Massachusetts
MAC	Medium Access Control
MAC	Mandatory Access Control (*in railway traffic control*)
MANEMO	MANET and NEMO
MANET	Mobile Ad hoc Network
MARS	Multi-Agent Real-time Simulator
MCG	Mobile Communication Gateway
MCS	Modulation and Coding Scheme
MEA	Monitor–Evaluate–Act
METRA	North East Illinois Commuter Railroad
MEXT	Mobility Extensions
MI	Michigan
MIB	Management Information Base
MIMO	Multiple-Input Multiple-Output
MIO	Most Important Object
MITRAC	Train computer product family made by Bombardier

MLD	Multicast Listener Discovery
MLME	MAC sublayer management entity
MN	Mobile Node
MNN	Moving Network Node
MNP	Mobile Network Prefix
MODCOMM	MODURBAN Communication System
MODURBAN	Modular Urban Guided Rail Systems
MP-89	Matériel roulant sur pneumatique (Commuter train rolling stock on tires designed in 1989, used on lines 1 and 14 in Paris)
MPEG	Moving Picture Experts Group
MR	Mobile Router
MS	Mobile Station
MSC	Message Sequence Chart
MVB	Multifunction Vehicle Bus
NACK	Negative Acknowledge
NAJPTC	North American Joint Positive Train Control System
NAT	Network Address Translation
NEC	North East Corridor
NEMO	Network Mobility
NIC	Network Interface Card
NIST	United States National Institute of Standards and Technology
NORAC	Northeast Operating Rules Advisory Committee
NOW	Network On Wheels
NS	Norfolk Southern
NSA	United States National Security Agency
NSTAC	National Security Telecommunications Advisory Committee
NTSB	National Transportation Safety Board
OBDII	On-Board Diagnostic systems
OBU	On-Board Unit
OCRS	Ohio Central Railroad System
OEM	Original Equipment Manufacturer
OFDM	Orthogonal Frequency Division Multiplexing
OFDMA	Orthogonal Frequency Division Multiple Access
OH	Ohio
OS	Operating System
OSI	Open Systems Interconnection
OTC	Optimized Train Control
PA	Public Address
PA-BF	Pilote Automatique Basse Fréquence (Automatic Pilot based on Low Frequency communication)
PACK	Piggybacked implicit ACK
PAN	Parking Availability Notification
PATH	Port Authority of New York and New Jersey
PCN	Post Crash Notification
PD	Process Data (*in communication protocols*)
PDA	Personal Digital Assistant

PD	Proportional-Differential (*in ACC*)
PDCCH	Physical Downlink Control Channel
PER	Packet Error Rate
PET	Privacy Enhancing Technologies
PHY	Physical
PIM-SM	Protocol Independent Multicast–Sparse Mode
PIS	Passenger Information System
PLCP	Physical Layer Convergence Protocol (*see also: Physical Layer Convergence Procedure*)
PNP	Pseudonym Provider
PPDU	Physical Layer Protocol Data Unit
PSDU	Physical Layer Service Data Unit
PSK	Phase Shift Keying
PSL	Parking Spot Locator
PTB	Post Transmission Backoff
PTC	Positive Train Control
QAM	Quadrature Amplitude Modulation
QoS	Quality of Service
QPSK	Quadrature Phase-Shift Keying
RATP	Régie Autonome des Transports Parisiens (Paris Area Transport Authority)
RBAC	Role-based Access Control
RBS	Remote Base Station
RCU	Radio Control Unit
RER	Le Réseau Express Régional (Paris Area Rapid Transit System)
RF	Radio Frequency
RFID	Radio Frequency Identification
RFN	Road Feature Notification
RHCN	Road Hazard Condition Notification
RMS	Root Mean Square
RL	Revocation List
RO	Route Optimization
ROI	Region-of-Interest
ROLL	Routing Over Low Power and Lossy Networks
RRC	Regional Radiocommunication Conference
RSIA 2008	Rail Safety Improvement Act of 2008
RSSI	Receive Signal Strength Indication
RSU	Road-Side Unit
RT	Railway Telecommunication
RTCP	Real-Time Control Protocol
RTS/CTS	Request to Send/Clear to Send
RTT	Round Trip Time
RTVR	Real-time Video Relay
RVP/D	Remote Vehicle Personalization/Diagnostics
SA	Service Announcement
SACEM	Système d'aide à la conduite, à l'exploitation et à la maintenance

	(Automated assistance system for train driving, exploitation and maintenance used on RER line A in Paris)
SAE	Society of Automotive Engineers
SAET	Système Automatisé d'Exploitation des Trains (Automated Train Exploitation System used on Paris metro line 14)
SAFECOM	US Homeland Security Office for Interoperability and Compatibility (OIC) Communications Program
SAFESPOT	Cooperative Vehicles and Road Infrastructure for Road Safety
SAM	Scalable Adaptive Modulation
SAP	Service Access Point
SC	South Carolina
SCADA	Supervisory Control and Data Acquisition
SCH	Service Channel
SDL	Specification and Description Language
SDMA	Spatial Division Multiple Access
SDO	Standards Development Organization
SDR	Software Defined Radio
SDU	Service Data Unit
SeVeCom	Secure Vehicular Communication
SOFDMA	Scalable Orthogonal Frequency Division Multiplex Access
SoR	Statement of Requirements
SRD	Seat Reservation Display
STA	Terminal Station
STB	Surface Transportation Board
STRACNET	Strategic Rail Corridor Network
SUMMITS	Sustainable Mobility Methodologies for Intelligent Transport Systems
SVA	Stopped or Slow Vehicle Advisor
SZ	Service Zone
TBTT	Target Beacon Transmission Time
TC	Technical Committee *in Standardization*
TC	Trusted Component *in Security*
TCN	Train Communication Network
TCP	Transmission Control Protocol
TCS	Traffic Control System
TDD	Time-Division Duplexing
TDMA	Time Division Multiple Access
TFT	Thin Film Transistor liquid crystal display
TG	Task Group
TGV	Train à grande vitesse (French high-speed train)
TIA	Telecommunications Industry Association
TICS	Transport Information and Control System
TIH	Toxic by Inhalation
TNO	Netherlands Organization for Applied Scientific Research (Nederlandse Organisatie voor Toegepast Natuurwetenschappelijk Onderzoek)
TOLL	Free Flow Tolling

TORNAD	Token Ring Network Alsthom Device
TP	Traffic Probe
TS	time slot
TS	Train Sentinel (*in railway traffic control*)
TSF	Timing Synchronization Function
TTC	Transportation Technology Center
TTC	Time-To-Collision (*in ACC*)
TTL	Time-to-Live
TWC	Track Warrant Control
TX	Texas
UAV	Unmanned Aerial Vehicle
UDP	User Datagram Protocol
UHF	Ultra High Frequencies
UIC	Union Internationale des Chemins de Fer
UL	Uplink
UMTS	Universal Mobile Telecommunications System
UP	Union Pacific
USB	Universal Serial Bus
V2I	Vehicle-to-Infrastructure
V2V	Vehicle-to-Vehicle
V2X	Vehicle-to-Vehicle/Infrastructure
VANET	Vehicular Ad hoc Network
VANEMO	VANET and NEMO
VC	Vehicular Communication
VeHIL	Vehicle Hardware In the Loop
VHF	Very High Frequencies
VII	Vehicle Infrastructure Integration
VoIP	Voice over IP
VSC	Vehicular Safety Communication
VTMS	Vital Train Management System
WAVE	Wireless Access in Vehicular Environments
WCDMA	Wideband Code Division Multiple Access
WG	Working Group
WiMAX	Worldwide Interoperability for Microwave Access
WLAN	Wireless Local Area Network
WMAN	Wireless Metropolitan Area Network
WRC	World Radiocommunication Conference
WSA	WAVE Service Advertisement
WSM	WAVE Short Messages
WSN	Wireless Sensor Network
WSMP	WAVE Short Message Protocol
WTB	Wire Train Bus
WY	Wyoming

Subject Index